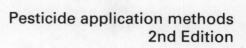

Pesticide application methods
2nd Edition

Pesticide application methods
2nd Edition
G A Matthews

Longman Scientific & Technical

Copublished in the United States with
John Wiley & Sons, Inc., New York

Longman Scientific & Technical,
Longman Group UK Ltd,
Longman House, Burnt Mill, Harlow,
Essex CM20 2JE, England
and Associated Companies throughout the world

Copublished in the United States with
John Wiley & Sons, Inc., 605 Third Avenue, New York, NY10158

First published 1979
Second Edition 1992

British Library Cataloguing in Publication Data

A catalogue record for this book is
available from the British Library

Library of Congress Cataloging-in-Publication Data
Matthews, G. A.
    Pesticide application methods / G.A. Matthews – 2nd ed.
      p. cm.
    Includes bibliographical references and index.
    ISBN 0-582-40905 5
    ISBN 0-470-21818-5 (USA only)
    1. Pesticides – Application.   2. Spraying and dusting in
agriculture.   3. Spraying equipment.   I. Title.
SB953.M37   1992
632′.94 – dc20

91-29746
CIP

Set in 10/12 Linotron 202 Times Roman at 9

Produced by Longman Singapore Publishers (Pte) Ltd.
Printed in Singapore

Contents

Preface to second edition

The general public has become more vocal in its criticisms of pesticide use, yet continues to demand higher quality and cheap agricultural produce. Despite the interest in 'organic' farming, pesticide usage continues to be a vital factor in reducing crop losses, but the need for more restrained pesticide use has led to the International Code of Conduct on the Distribution and Use of Pesticides, published by the Food and Agriculture Organization of the United Nations. In consequence Product Stewardship has received more attention from the international agrochemical companies.

In the United Kingdom greater emphasis has been given to the design of engineering controls to minimize operator contamination, to the selection of spray nozzles to reduce spray drift and to operator training to improve the standard of application. Newly registered pesticides are generally less hazardous and are marketed and packaged in formulations which are safer to use. However, even with significant improvement in the use of existing technology, research and development continues to be needed to provide more efficient systems to transfer less pesticide precisely when and where it is needed.

Funding for application research is difficult to attract while an effective range of pesticides can be applied against diverse crop pests, an interdisciplinary approach is required and novel systems have not always been adopted commercially. Equipment manufacturers have fewer resources than the large agrochemical companies to devote to R and D, and farmers' investment in their machinery means that changes are far less frequent than choice of chemicals.

Nevertheless, over the last twelve years since the first edition of this book was published, some significant changes in application technology have occurred. Some of the new equipment is primarily for specialized markets, while adoption of new features on sprayers has been accepted more generally to meet the requirements of legislation.

Research has improved our understanding of the principles of pesticide application, with much effort directed at the use of electrostatic charging of sprays. A new chapter has been added to reflect this, although commercial use of charged sprays is still very limited. Developments in droplet sizing and changes in equipment design are reflected in the changes to other chapters.

Acknowledgements

As with the previous edition, I have been assisted by discussions with many of those involved in pesticides and their application. Much information has also been gained from visits to many countries to see how pesticides are being applied to crops and advise on integrated pest management in a range of agricultural environments.

I thank Mr Evan Thornhill for his continued support at IPARC and Dr Roy Bateman for droplet size data. In updating the text of aerial application I am grateful to Mr Tim Sander of Micronair (Aerial) Ltd for his comments and contribution on navigational aids. Mr Tom Bals very kindly reviewed the chapter on Controlled Droplet Application. I am most grateful to Mrs Janet Biddlecombe for typing the text of this edition. I owe special thanks to my wife, Moira, for her encouragement and support.

Note The author has endeavoured to ascertain the accuracy of the statements in this book. However, facilities for determining such accuracy with absolute certainty in relation to every particular statement have not necessarily been available. The reader should therefore check local recommendations before implementing in practice any particular pesticide technique or method described herein.

We are grateful to the following for permission to reproduce copyright material:
Academic Press Inc. and the author, Dr I. J. Graham-Bryce for Fig. 2.17 (Hartley & Graham-Bryce, 1980); Academic Press Inc. (London) Ltd. for Fig. 2.9 (Fryer, 1977); Academic Press Inc. (London) Ltd. and the respective authors for Figs 2.19 (Courshee, 1967), 5.9 (Dombrowski, 1961) & 7.24 (Frost, 1990); Air Lloyd (Germany) for Fig. 13.2; E. Allman & Co. Ltd. for Figs 6.5, 7.1, 7.2 & 16.5; *Annals of Applied Biology* and the author, Dr C. A. Hart for Fig. 2.14 (Hart & Young, 1987); Blackwell Scientific Publications Ltd. for Fig. 13.19 (Quantick, 1985a); BP Co. Ltd. for Figs 8.6, 8.8 & 8.9; British Crop Protection Council for Figs 1.8a (Greaves & Marshall, 1987), 2.1a (Doble *et al.*, 1985), 6.1a & b (BCPC), 6.7 (BCPC), 9.7 & 9.8 (Law, 1990), 13.4a & c (Spillman, 1977), 17.2 (Smith, 1984), Tables 3.6 (Barlow & Hadaway, 1974), 12.2 (Walker, 1976), 12.5 & 12.6 (Wheatley, 1976);

Butterworth-Heinemann Ltd. and the author, Prof. J. J. Spillman for Fig. 13.11 (Spillman, 1982) © 1982 Butterworth-Heinemann Ltd.; Cambridge University Press and the author, Prof. G. K. Batchelor for Fig. 4.4 (Batchelor, 1967); J. W. Chafer Ltd. for Fig. 7.17; Ciba-Geigy Agrochemicals for Fig. 3.1; Cleanacres Ltd. for Fig. 5.8; Delavan Ltd. for Figs 7.4a, 7.5c & 7.7a; the author, Prof. N. Dombrowski for Fig. 5.1a; Elsevier Applied Science Publishers Ltd. for Fig. 4.17 (Uk, 1977) & Table 2.3 (Holloway, 1970); Elsevier Sequoia S.A. for Fig. 5.20a (Dombrowski & Lloyd, 1974); Elsevier Science Publishers B.V. and the respective authors for Figs 4.10 (Johnstone *et al.*, 1974b) & 13.20 (Bache, 1975); Entomological Society of America for Fig. 4.18 (Himel, 1969a); Evers & Wall Ltd. for Fig. 10.14a; Fluid Technology (Australia) for Fig. 5.6; Food & Agriculture Organization of the United Nations for Figs 2.10, 13.3, 13.4b, 13.5a, 13.6, 13.18, 13.23 & 13.26 (FAO, 1974), Graticules Ltd. for Fig. 4.19; Hardi UK Ltd. for Fig. 7.3a; Horstine Farmery Ltd. for Figs 12.9 & 12.10; H.D. Hudson Manufacturing Co. (USA) for Fig. 6.9; ICI Agrochemicals for Figs 3.2, 4.11, 4.12, 9.5a, 11.1 & 12.1a; International Institute for Applied Systems Analysis for Fig. 1.2 (Comins, 1977c); International Rice Research Institute for Table 12.1 (Kiritani, 1974); the editor, *Journal of the American Mosquito Control Association* for Table 2.2 (Mount, 1970); Lurmark Ltd. for Fig. 5.3c; Meteorological Office for Table 17.1 (Spackmann & Barrie, 1982); Microgen Corp. (USA) for Figs 10.15 & 10.17; Micronair (Aerial) Ltd. for Figs 13.1, 13.12, 13.14 & 13.15; Micron Sprayers Ltd. for Figs 5.20b & 8.19; Ministry of Agriculture, Fisheries & Food for Figs 16.1a–d © Crown Copyright; Motab gmbh for Fig. 11.2; Natural Resources Institute for Fig. 12.8a (Sutherland, 1980) & Table 8.1 (Johnstone & Johnstone, 1976); Pergamon Press PLC for Fig. 10.13 (Morgan, 1981) Copyright 1981 Pergamon Press PLC; The Royal Aeronautical Society for Fig. 4.3 (Spillman, 1984); The Royal Society and the author, Dr I. J. Graham-Bryce for Table 2.1 (Graham-Bryce, 1977); Schering Agrochemicals Ltd. for Fig. 7.19; Simplex Manufacturing Co. (USA) for Figs 13.8 & 13.9; Society of Chemical Industry for Fig. 2.15 (Ford & Salt, 1987), Spraying Systems Co. (USA) for Figs 5.3a, 5.4, 5.5a & b, 5.12a–c, 7.8, 7.9a, 7.11a & b & 7.12a–g; Taylor & Francis Ltd. and the author, A. Lavers for Fig. 8.5 (Cowell & Lavers, 1988); Taylor & Francis and the respective authors for Tables 4.6 (Matthews, 1975a), 10.1 (Clayphon, 1971) & 15.1 (Clayphon & Matthews, 1973); Tifa (CI) Ltd. for Figs 11.4 & 11.6; Turbair Ltd. for Figs 8.15 & 8.16; 'Vibratak' for Fig. 10.9; Wellcome Environmental Health for Fig. 10.16; World Health Organization for Figs 6.10 (WHO/R. Da Silva) & 13.17 (WHO/R. Witlin); the editor, *Zimbabwe Agricultural Journal* (Government of the Republic of Zimbabwe) for Fig. 1.4 (Duncombe, 1973).

Conversion tables

	A	B	A→B	B→A
Weight	oz	g	×28.35	×0.0353
	lb	kg	×0.454	×2.205
	cwt	kg	×50.8	×0.0197
	ton (long)	kg	×1016	×0.000984
	ton (short)	ton (long)	×0.893	×1.12
Surface area	in^2	cm^2	×6.45	×0.155
	ft^2	m^2	×0.093	×10.764
	yd^2	m^2	×0.836	×1.196
	yd^2	acres	×0.000207	×4840
	acres	hectares	×0.405	×2.471
Length	μm	mm	×0.001	×1000
	in	cm	×2.54	×0.394
	ft	m	×0.305	×3.281
	yd	m	×0.914	×1.094
	mile	km	×1.609	×0.621
Velocity	ft/s	m/s	×0.305	×3.281
	ft/min	m/s	×0.00508	×197.0
	mile/h	km/h	×1.609	×0.621
	mile/h	ft/min	×88.0	×0.0113
	knot	ft/s	×1.689	×0.59
	m/s	km/h	×3.61	×0.277
	cm/s	km/h	×0.036	×27.78
Quantities/area	lb/acre	kg/ha	×1.12	×0.894
	lb/acre	mg/ft^2	×10.4	×0.09615
	kg/ha	mg/m^2	×100	×0.01
	mg/ft^2	mg/m^2	×10.794	×0.093
	oz/yd^2	cwt/acre	×2.7	×0.37
	gal (Imp.)/acre	litres/ha	×11.23	×0.089
	gal (USA)/acre	litres/ha	×9.346	×0.107
	fl. oz (Imp.)/acre	ml/ha	×70.05	×0.0143
	fl. oz (USA)/acre	ml/ha	×73.14	×0.0137
	oz/acre	g/ha	×70.05	×0.0143
	oz/acre	kg/ha	×0.07	×14.27
Dilutions	fl. oz/100 gal (Imp.)	ml/100 litres	×6.25	×0.16
	pint/100 gal (Imp.)	ml/100 litres	×125	×0.008
	oz/gal (Imp.)	g/litre	×6.24	×0.16
	oz/gal (USA)	g/litre	×7.49	×0.134
	lb/100 gal (Imp.)	kg/100 litres	×0.0998	×10.02
Density of water	gal (Imp.)	lb	×10	×0.1
	gal (USA)	lb	×8.32	×0.12
	lb	ft^3	×0.016	×62.37
	litre	kg	×1	×1
	ml	g	×1	×1
	lb/gal (Imp.)	g/ml	×0.0997	×10.03
	lb/gal (USA)	g/ml	×0.1198	×8.34
	lb/ft^3	kg/m^3	×16.1	×0.0624

Volume	in³	ft³	×0.000579	×1728
	ft³	yd³	×0.037	×27
	yd³	m³	×0.764	×1.308
	fl. oz (Imp.)	ml	×28.35	×0.0352
	fl. oz (USA)	ml	×29.6	×0.0338
	gal (Imp.)	gal (USA)	×1.20	×0.833
	gal (Imp.)	litre	×4.55	×0.22
	gal (USA)	litre	×3.785	×0.264
	cm³	m³	×10^{-6}	×10^6
	cm³	μm³	×10^{12}	×10^{-12}
Pressure	lb/in²	kg/cm²	×0.0703	×14.22
	lb/in²	bar	×0.0689	×14.504
	bar	kPa	×100	×0.01
	lb/in²	kPa	×6.89	×0.145
	kN/m²	kPa	×1	×1
	N/m²	kPa	×0.001	×1000
	lb/m²	atm	×0.068	×14.696
Power	hp	kW	×0.7457	×1.341
Temperature	°C	°F	$\frac{9}{5}$ °C+32	$\frac{5}{9}$ (°F−32)

Pesticide calculations

1. To determine the quality (X) required to apply the recommended amount of active ingredient per hectare (A) with a formulation containing B percentage active ingredient.

$$\frac{A \times 100}{B} = X.$$

Example: Apply 0.25 kg ai/ha of 5 per cent carbofuran granules

$$\therefore \frac{0.25 \times 100}{5} = 5 \text{ kg granules/ha}$$

2. To determine the quantity of active ingredient (Y) required to mix with a known quantity of diluent (Q) to obtain a given concentration of spray

$$Q \times \frac{\text{per cent concentration required}}{\text{per cent concentration of active ingredient}} = Y.$$

(a) *Example*: Mix 100 litres of 0.5 per cent ai, using a 50 per cent wettable powder

$$100 \times \frac{0.5}{50} = 1 \text{ kg of wettable powder}$$

(b) *Example*: Mix 2 litres of 5 per cent ai using a 75 per cent wettable powder

$$2000 \times \frac{5}{75} = 133 \text{ g of wettable powder.}$$

Units abbreviations and symbols

Units

A	ampere
atm	atmospheric pressure
bar	baropmetric pressure
cm	centimetre
dB	decibel
fl. oz	fluid ounce*
g	gramme
g	acceleration due to gravity (9.8 m/s^{-2})
gal	gallon*
h	hour
ha	hectare
hp	horsepower
kg	kilogram
km	kilometre
kN	kilonewton
kPa	kilopascal
kW	kilowatt
l	litre
m	metre
mg	milligramme
ml	millilitre
mm	millimetre
μm	micrometre
N	newton
μP	micro-poise
P	poise
psi	pounds per square inch
pt	pint
pto	power take-off (tractor)
s	second
V	volt

*Volume measurements may be in Imperial or American units as indicated by (Imp.) or (USA).

Abbreviations and symbols

A	area
a	average distance between airstrip or water supply to fields
ac	alternating current
adv	average droplet volume
ai	active ingredient
AN	Antanov aircraft
BPMC	fenobucarb
C	average distance between fields
CDA	controlled droplet application
D	diameter of centrifugal energy nozzle or opening of nozzle
d	droplet diameter
DCD	disposable container dispenser
'D' cell	a standard size dry battery
dc	direct current
DUE	deposit per unit emission
ec	emulsifiable concentrate
EDX	energy dispersive X-ray
EPA	Environmental Protection Agency (USA)
F	average size of field
FAO	Food and Agriculture Organization of the United Nations
FN	flow number
FP	fluorescent particle
GIFAP	Groupement International des Associations Nationales de Fabricants de Produits Agrochimiques. (International Group of National Associations of Manufacturers of Agrochemical Products)
H	height
HAN	heavy aromatic naphtha
HCN	hydrogen cyanide
HLB	hydrophile-lipophile balance
HP	high power battery
HV	high volume
Hz	hertz
ID	internal diameter
IGR	Insect growth regulator
IPM	Integrated pest management
IRM	Insecticide resistance management
ISA	International Standard atmosphere
K, k	constant
kV	kilovolt
L	length
LAI	leaf area index
LD$_{50}$	median lethal dose
LV	low volume
MCPA	4–chloro–o–tolyloxyacetic acid
MRL	Maximum residue level
MV	medium volume
N, n	number of droplets
nmd	number median diameter
NPV	nuclear polyhedrosis virus
OES	occupational exposure standards
P	particle parameter

PIC	prior informed consent	U, u	wind speed
PMS	particle measuring system	ULV	ultra low volume
PRV	pressure-regulating valve	UR	unsulfonated residue
PTFE	polytetrafluoreothylene	UV	ultra-violet light
PVC	polyvinyl chloride	V	velocity
Q	application rate (litre/ha)	V_f	velocity of sprayer while ferrying
q	application rate (litre/m²)	V_s	velocity of sprayer while spraying
Q_a	volume of air	vad	volume average diameter
Q_f	quantity of spray per load	VLV	very low volume
q_n	throughput of nozzle	vmd	volume median diameter
Q_t	volume applied per minute	VRU	variable restrictor unit
rev	revolution	W	width
rpm	revolutions per minute	w	angular velocity
S	swath	WHO	World Health Organization
s	distance droplet travels	wp	wettable powder
SP	single power battery	γ	surface tension
SMV	spray management values	n	viscosity of air
SR	stability ratio	P_a	density of air
T	temperature	P_d	density of droplet
T_r	time per loading and turning	$>$	is less than
T_w	turn time at end of row	$<$	is greater than

Multidisciplinary nature of pesticide application.

1

The role of chemical control

Introduction

At the present time the general public is becoming increasingly concerned about the use of pesticides (Fig. 1.1) to the extent that some consumers buy *organically* grown food 'free of pesticides'. Calculations show that despite that, our diet contains 10 500 times more natural pesticide than manufactured pesticide (Ames 1983; Berry 1990). In practice the population is routinely exposed to many substances with a carcinogenic potential greater than that of pesticides (Graham-Bryce 1989). 'Organic' food is more expensive to produce and may require cultivars that are naturally resistant to pests due to their chemical composition. In fact one potato variety resistant to insect pests had to be withdrawn because of its solanine content (Jadhav et al 1981). To obtain an economic return on their agricultural production, despite the competition of weeds, pathogens and insect pests, farmers will continue to need to apply pesticides which can be sold only after rigorous testing including the detection of any residues that may persist after its initial

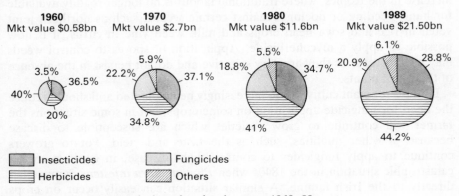

Fig. 1.1 The agrochemical market development 1960–89

application. This change in public opinion has led to a greater awareness of our environment, and has promoted the need to conserve natural enemies and use cultural control of 'pests'.

Recent advances in plant breeding and genetic engineering to improve resistance of crop plants to insect pests and diseases will undoubtedly reduce the need to rely on the conventional pesticides, and incorporating resistance in crops to herbicides may make chemical control of weeds easier than at present. Nevertheless, there will still be a need to apply 'pesticides' as a relatively easy and quick method of regulating an upsurge in a pest population. Such 'pesticides' will not be the highly persistent, broad-spectrum organochlorines or similar products from the 1940s and 1950s, which prompted Rachel Carson to predict a '*Silent Spring*', but will be increasingly pesticides which are more readily degraded and be more selective in their mode of action, so that they present less risk to the environment.

Crop losses remain severe in many areas of the world, particularly in the developing countries, and where irrigation allows pests to survive throughout the year. Weed competition in the initial stages of crop growth can be so severe that plants remain stunted and final yields are a mere fraction of their true potential. As one example, the competition by the weed *Chenopodium album* in lettuce reduced the weight of lettuce (11 plants/m²) by 90 per cent when there were 38 weeds/m² and the number of marketable lettuce by 58 per cent with only 2 weeds/m² (Lacey 1985). Virtual weed-free conditions are now possible using the range of herbicides available. Ploughing of some well-structured soils is no longer necessary every year as seed can be direct drilled after applying a broad action herbicide, that is inactivated on contact with the soil. Herbicide use has increased most where labour costs are high, there is a peak demand for labour, or where mechanical hoeing will cause damage to the young crop. In conjunction with other agronomic practices such as tie-ridging and planting along contours, herbicide use can reduce soil erosion by minimizing soil disturbance. In West Africa improved row weeding either by hand hoeing or by application of a herbicide increased yields by up to 35 per cent (Carson 1987). Thus there is the expectation that herbicide usage will increase in the tropics, where traditional labour is no longer readily available for hand-weeding or hoeing. Against certain weeds, such as northern joint vetch in rice and soya bean crops and milk weed vine in citrus, it is now possible to apply a mycoherbicide. Application of spores to control weeds may well increase as they are very selective and do not persist in the absence of their host plants.

Disease-resistant cultivars are increasingly being planted and should reduce the need for fungicide application on some crops, but in some situations the farmer will continue to grow varieties which are susceptible to disease because of other qualities, such as the taste and yield. Potato growers continue to apply fungicides to control blight disease, in contrast to the catastrophic situation in the 1840s when *Phytophthora infestans* damage led directly to the Irish famine. A similar situation can easily occur on crops which have a relatively narrow source of germplasm, especially if a new pathogenic race of the disease occurs. Nevertheless the trend is to minimize fungicide applications by monitoring the climatic conditions which favour the build-up of the disease so that spray applications can be timed more accurately.

Insect pests continue to cause losses of yield and quality either directly by their feeding, or indirectly as vectors of disease. Whereas these losses can be extremely serious and can result in total loss of a crop in some fields, for example the effect of an invasion of locusts or armyworms, the extent of damage is usually far less due to the intervention of natural enemies. The difficulty for the farmer is knowing when a pest population has reached a level at which economic damage will occur so that preventive action can be taken. The farmer therefore needs chemical control when quick action is needed. Over one-third of insecticides used in agriculture have been applied to cotton crops as production in most countries can be drastically reduced by insect damage, but as on other crops, the trend is to minimize insecticide use by adopting integrated pest management to utilize different control tactics in a harmonious manner and avoid as far as possible undesirable effects on the environment (Matthews 1984; Flint and van den Bosch 1981; van Emden 1989)

Pesticides continue to play an important role in controlling vector-borne diseases, such as malaria, onchocerciasis and trypanosomiasis. Large-scale control programmes have drastically reduced the incidence of disease, but subsequent maintenance of low levels of the vector also requires other control tactics to avoid excessive use of pesticides and selection of resistant populations.

R. F. Smith (1970) pointed out that despite intensive research into alternative methods of controlling pests, pathogens and weeds, pesticides remain our most powerful tool in pest management. This remains true today, but they need to be regarded as a valuable resource and used more wisely if we are going to reduce the amount of chemical applied and the number of applications (Southwood 1977). This approach is urgently needed to decrease selection pressure for resistance, prolong the useful life of each pesticide and reduce environmental contamination and residues in food. The latest pesticides to be developed are generally far more selective and less toxic to humans if they are to be accepted by the registration authorities. They are also much more active (ie only a few grams per hectare are required), so greater adoption of improved methods of applying them will be required if they are to be applied more efficiently.

Integrated control

Prior to the development of synthetic organic pesticides, scientists had developed techniques of exploiting biological control, such as the introduction of *Rodolia cardinalis* into California from Australia to control the cottony cushion scale on citrus. Farmers had also developed techniques of cultivation, crop rotation and the use of closed seasons to minimize the effects of pests. Some of these techniques tended to be forgotten when it was relatively easy to apply a chemical, despite early warnings that over-reliance on chemical control could have undesirable side effects (W.E. Ripper, 1944; 1956). Today pesticides are accepted as a component of an *integrated control* approach, but there is often conflict between what is theoretically the optimum solution to a pest problem and the practicalities confronting the person who has to make the decision in the field. In some situations there may be overuse of pesticides, mainly as an insurance against loss of revenue, while in many areas of the world more pesticide may be needed to increase crop yields and reduce the incidence of disease. Ideally, integrated control is based on a complete understanding of the population dynamics of the pest so that appropriate control strategies can be developed to exploit biological and cultural controls and minimize pesticide use. In practice this is seldom (if ever) achieved, as population changes will be influenced by so many different factors, but practical decisions can be made on the basis of current knowledge. Van Emden (1972) pointed out that plant resistance, if only partial, can be an important restraint on damage compatible with the use of chemicals. He showed that on plants with slight resistance to the cabbage aphid *Brevicoryne brassicae* half the recommended dose of a selective aphicide, pirimicarb, was adequate for control because the natural enemies were unaffected and checked the population of aphids which survived. When a non-persistent chemical is applied to reduce the level of a pest population, a natural enemy can be released later to cope with lower pest population. Thus resmethrin or the fungus *Verticillium* can be used to reduce the whitefly *Trialeurodes vaporariorum* population prior to introduction of the parasitoid *Encarsia formosa* which is ineffective early in the season due to low light intensities and other factors (Parr et al 1976; Hussey and Scopes 1985).

The effect of a persistent, broad-spectrum chemical has been repeatedly demonstrated by the upsurge of pest populations when the balance between the pests and their natural enemies is destroyed. For example, cottony cushion scale increased following inappropriate application of DDT on citrus in California (De Bach 1974). Bagworms increased on oil palms (Wood 1971) and cocoa (Conway 1972) following the use of dieldrin. Consequently the less persistent trichlorphon was used against young bagworms as they emerged, thus allowing sufficient natural enemies to survive and attack the remaining pests. More recently there have been reports of upsurges of whiteflies and aphids on cotton following applications of pyrethroids alone or as mixtures with organophosphate insecticides.

Integrated control can be used by individual farmers; thus a farmer may use a resistant cultivar, monitor the pest population and apply pesticides if the pest reaches economic significance, and subsequently destroy crop residues harbouring pests in the off-season. For example, cotton farmers in Central Africa grow a pubescent jassid resistant cultivar (Parnell et al 1949), uproot or destroy their cotton plants after harvest and bury crop residues by ploughing, and then time insecticide applications according to crop monitoring data (Tunstall and Matthews 1961; Matthews and Tunstall 1968; Matthews 1989a). The success of uprooting since the 1930's (Peat and McKinstry 1938) has undoubtedly depended on the full co-operation of all the farmers, as was shown by the outbreak of pink bollworm *Pectinophora gossypiella*, when some farmers failed to uproot their cotton by the prescribed date (Matthews et al 1965). Recommendations on the choice of insecticides and quantity applied have been updated regularly.

The selection of control techniques and their subsequent regulation throughout a given area, or ecosystem, irrespective of county or national boundaries is regarded as pest management. Whereas many people consider *pest management* and integrated control, or indeed pest control, as synonymous, a distinction is here made between the use of integrated control by individuals and co-operative use of the same strategy by all concerned. Thus a 'pest management' system attempts to regulate all aspects of the different control techniques employed. It may give particular emphasis to one particular technique, especially where success is achieved without conscious integration of methods (Way 1972). As far as chemical control in a pest management programme is concerned, the choice of chemical, or a group of chemicals with the same mode of action, or alternation of chemical groups, is determined for the whole area and advice is given on when chemicals should be applied. This is particularly important in adopting strategies to minimize the risk of selecting resistant pest populations as discussed in the next section. Pest management, like integrated control, must also be a dynamic system requiring continual adjustment as information on the pest complex in an area and new pesticides increases. This can be facilitated by maintaining computer databases and 'expert' systems to provide up-to-date information to farmers and their advisers.

Pest management is essential with the need to increase agricultural production per unit area as the area available for cultivation is limited. The necessity for pest management to involve non-pesticide methods, such as the use of a closed season, the application of pheromones and biological control, is stressed as more pests become resistant to the most widely available chemicals, especially since users tend to be blinkered by short-term needs and apply the cheaper products rather than consider the long-term consequences.

Resistance to pesticides

Application of a pesticide preferentially removes susceptible individuals of a pest population and inevitably results in an increase in the proportion of those resistant to a particular chemical (Fig. 1.2). This selection for resistance to a particular chemical or chemical group may occur if it is applied at frequent and regular intervals to a given pest population, and this restricts

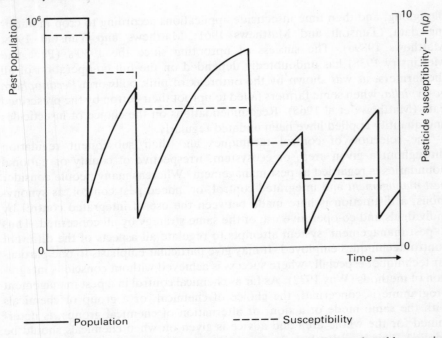

Fig. 1.2 Changes in pest population and pesticide susceptibility produced by a series of pesticide treatments (after Comins 1977c)

the number of chemicals subsequently available to control the pest. Initially resistance to insecticides and fungicides was reported more frequently but there has been an increase in the number of weed species that are now resistant to herbicides. Often resistance occurs over a limited area and not throughout the range or distribution of the pest species.

Resistance develops rapidly if most of the pest population is exposed to the specific pesticide, if the pest can multiply quickly or if there is limited immigration of unexposed individuals. The latter is especially liable to occur with intensive and continuous cropping, as in glasshouses and on irrigated crops in the tropics. Resistance of red spider mites to acaricides can develop so quickly that the useful 'life' of a new acaricide may be as short as four seasons, even in orchards (Kirby 1973).

Selection pressure for resistance is reduced if part of the pest population is on alternative host plants or other crops which are not treated with the chemical. Thus resistance to insecticides in the cotton bollworm *Helicoverpa armigera* has not been a problem in Africa, where only a proportion of farmers treat their cotton and where alternative host plants including maize, sorghum, tobacco and tomatoes are not at present treated with insecticides. In contrast *H. armigera* rapidly developed resistance to the pyrethroids in parts of Australia, where intensive control was practised (Gunning et al 1984).

In extreme cases farmers have had to abandon growing certain crops because of their inability to control the pests. An example of this was the cessation of cotton growing in the Ord valley in Australia (Basinski and Wood 1987). This has led to the introduction of Insecticide Resistance Management schemes (IRM) (Sawicki and Denholm 1987), where there is the infrastructure to supervise and enforce the strategy. There is considerable controversy over which strategy to adopt to conserve susceptible populations. Strategies that have been adopted have been based largely on pragmatic assumptions.

The user has been tempted to increase either the dosage, or frequency of application, or both, when resistance is suspected, but this merely aggravates the situation. One answer has been to use an alternative chemical with a different mode of action, but cross resistance may occur, even with chemicals of dissimilar actions. This has happened with red spider mite in various parts of the world, with *Helicoverpa* in Australia and also with peach potato aphid *Myzus persicae*, ticks and other pests.

Laboratory experiments have indicated that resistance in a population to two chemicals develops over about the same period, irrespective of whether first one chemical is used; then, when resistance has developed, a change is made to the second chemical, or they are used alternately (Brown 1971). When the genes for resistance are recessive, Comins (1977b) considers that it may be advantageous to spread out the selection for each resistance gene by using both chemicals on each generation. In Central Africa, carbaryl or DDT, and subsequently pyrethroids instead of DDT, were applied on cotton according to which bollworm was present and the size of their population in the crop, and this may have been an important contributory factor in maintaining susceptibility since the original recommendation was adopted thirty years ago (Tunstall and Matthews 1961). In Australia a similar resistance management strategy has been introduced, but after resistance was detected. The use of pyrethroids is restricted to a relatively short period each year for all crops (Forrester and Cahill 1987) (Fig. 1.3). Each year the level of resistance rises rapidly as soon as pyrethroid sprays are applied.

Red spider mite is a particularly serious pest, as its natural enemies are susceptible to a wide range of pesticides and the period for one generation can be less than one week. The mite has developed resistance to a number of acaricides very rapidly, and an attempt has been made in some countries to regulate their use so that farmers in one area can use only one particular acaricidal chemical group at any one time. In Zimbabwe, there are now three zones in each of which only one group of acaricides may be used for two seasons, before the groups are rotated to a different area (Fig. 1.4) (Duncombe 1973). This may reduce the risk of multiple resistance over a short period as there is a four year gap when populations within that area are not exposed to any further selection pressure for one of the chemical groups. By then, susceptibility to the first group of chemicals may have been sufficiently re-established to enable it to exert effective control, particularly if the numbers of mites can be reduced by controlling other pests with more

7

(a)

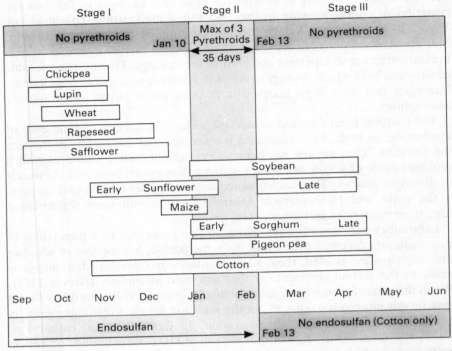

Fig. 1.3 (a) Restriction of pyrethroids to one part of the season (b) the level of resistance in *Helicoverpa armigera* in Australia (data from Neil Forrester)

selective pesticides which have less effect on the natural enemies of mites. The success of a rotation scheme really depends on the importance of the resistance genes in the absence of selection pressure. In a perennial crop situation, apple orchards, the adoption of integrated control has slowed the development of resistance in mites (Hoyt 1970).

Another approach has been to advocate the use of certain mixtures (Curtis 1985), but substantial delays in selection of resistance are achieved only if the components of the mixture are equally effective against the target species, have similar levels of persistence and the pest is initially as susceptible to each component. In practice the use of cocktails can lead to multiple resistance. Once selection for a particular mechanism of resistance has taken place, resistance often remains, even when selection pressure is removed. Tests may indicate that a pest has again become susceptible, but resistance usually develops quickly as soon as the selection pressure is reapplied. Clearly whatever strategy is adopted, careful monitoring of levels of resistance at regional agricultural centres is needed, so that an appropriate strategy can be determined for that locality under field conditions.

(b)

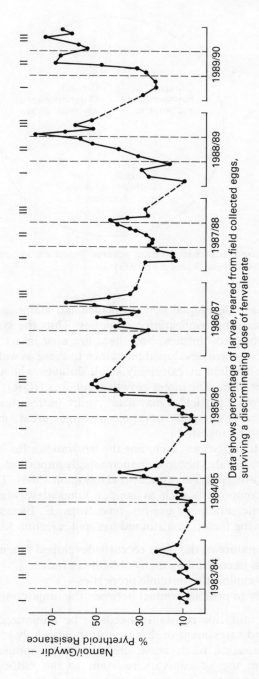

Data shows percentage of larvae, reared from field collected eggs, surviving a discriminating dose of fenvalerate

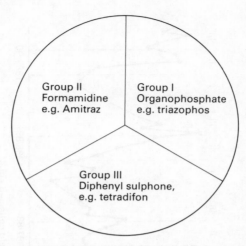

Fig. 1.4 Idealized acaricide rotation scheme based on system used in Zimbabwe (updated from Duncombe 1973)

Chemists have developed new pesticides and maintained some degree of control with existing application techniques. Thus the synthetic pyrethroids replaced the organochlorines, and there are now insect growth regulators, but some species have developed resistance to these as well. Some of the new pesticides are applied at extremely small dosages which necessitates more accurate application. The insect growth regulators (IGR), as well as the more selective microbial pesticides may also require more attention to application to ensure there is adequate coverage of target areas, especially when the pesticide has to be ingested.

Resistance to fungicides, including the benzimidazoles, dicarboximides and phenylamides, has also become an increasingly important problem, for which few solutions are available at present (Hollomon 1986). The exception is the non-specific compounds such as copper fungicides, which have remained effective despite prolonged use on some crops. E. Evans (1971) considered that the following factors contributed to rapid selection for resistant strains:

1 the specific nature of the most recently developed fungicides
2 their success in controlling many obligate parasites
3 the need to exploit the systemic properties
4 the tendency to prolong contact between the fungus and fungicide.

Selection for fungicide resistance needs to be minimized by better disease forecasting and assessment of disease thresholds likely to cause crop loss or quality deterioration to decrease the number of applications needed and integrating the use of cultivars resistant to the pathogen (Brent 1987). Industrial co-operation on the development of strategies to optimize the effectiveness of fungicides has now been possible on an international scale (Ruscoe 1987).

Changes in the weed species often follow frequent use of a herbicide in one particular area, as the species tolerant to the chemical can grow without competition. This has resulted in the need for different and often more expensive herbicides or combinations of herbicides. Resistance to a particular herbicide may show more slowly compared to insecticides or fungicides, as the generations of weeds overlap due to dormant seeds and there are fewer generations each year, but development of resistance to triazines is already evident. It occurs where the same herbicide has been used in the same fields over many years, thus in some areas excessive amounts of atrazine now fail to control fat-hen (*Chenopodium album*) or chickweed (*Stellaria media*). In 1981 there were 29 species with resistance to s-triazine herbicides (Le Baron and Gressel 1982), but there are now many more resistant species and reports of resistance to other herbicides, such as paraquat, trifluralin and diclofop-methyl (Gressel 1987).

The cost of introducing any new pesticide has now escalated to meet the registration requirements, especially the toxicological data and environmental impact studies, so the number of agrochemical companies with the resources needed to do the development work has declined. Thus to optimize the life of any new pesticide, its judicious use in pest management programmes is essential.

Timing of spray application

Pesticides are frequently applied as a prophylactic or on a fixed calendar schedule irrespective of the occurrence or level of the pest infestation. Such a policy is favoured, since forward planning is so much easier; the user knows exactly which chemical to purchase and when delivery is required so that stock is not carried over to the next year. However, fewer applications may be needed if they are timed more accurately and this will reduce selection pressure for resistance. A routine pest assessment is required, preferably aided by a pest forecast of the probable level of infestation to avoid fixed schedules.

Forecasting

Accurate forecasting of pest incidence will depend on the collection of a vast amount of appropriate biological data and integrating them with meteorological data. Very few long-term forecasts, such as that established for estimating the probability of an infestation of *Aphis fabae* in field beans (Way et al 1981) have been possible (Fig. 1.5). In east Africa pheromone traps have been used instead of light traps to forecast outbreaks of *Spodoptera exempta*, as the pest can occur in large numbers very suddenly. Pheromone traps are being used to detect the presence of pests in a number of crops, for example

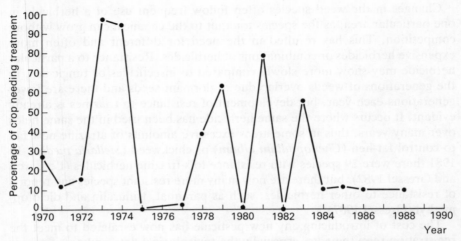

Fig. 1.5 Percentage of bean crop in United Kingdom requiring spray treatment against *Aphis fabae* (after Way et al., 1977; updated to 1988 by M. E. Cammell) No data for years 1976 and 1987

to determine when to spray against pea moth, by alerting the grower to the presence of the pest in their crops 10–13 days before control may be necessary (Wall and Greenway 1981). Dissemination of information from a central laboratory such as the forecasting of aphids from Rothamsted based on data from a grid of 23 suction traps distributed throughout the UK depends on widespread farmer ownership of computer and viewdata systems to receive the information (Lewis 1987).

Forecasting of certain crop diseases has been advocated; thus two days with a temperature not less than 10°C and a relative humidity above 75 per cent form the basis of a potato blight forecasting system (Beaumont 1947). An accurate knowledge of disease epidemiology is needed for reliable forecasts to be made (Fry 1987), but there is increasing awareness of the use of mini-meteorological stations to measure the conditions in crops sensitive to certain pathogens.

Economic thresholds

The effects of different numbers of pests on crop yield need to be measured so that the decision when to spray can be based on estimates of the pest population. The population density at which control measures should be applied to prevent an increasing pest population from reaching the economic injury level is the **economic threshold**. Economic injury level is the lowest population density that will cause economic damage (Stern 1966; Onstad 1987; Pedigo et al 1986). Such an approach requires considerable skill in interpreting data. Not only does the value of a crop vary due to the yield level in the locality, but also the economic injury level is inversely related to the market price of the crop and directly to the cost of control (Headley 1972). Sometimes it is difficult to use economic injury levels under commercial conditions (Poston et al 1983).

Low economic thresholds are used in grain storage and in cases where high-quality produce free from pest attack or mould is required, for example pre-packaged goods in supermarkets. If the crop has a high value, more money for pest control can naturally be justified in order to protect it. In vector control an initial infestation, however small, must be controlled to prevent or at least reduce the spread of disease.

Sprays have been applied in relation to data obtained by monitoring a number of crops including cotton (Fig. 1.6) (Boyer et al 1962; Matthews and Tunstall 1968; Benedict et al 1989). Further use of the system has been limited by lack of data on the amount of damage done by particular pest populations at various stages of crop development. Relatively simple techniques of monitoring pest populations and/or damage are needed if spraying according to an economic threshold is to be more widely used (Poston et al 1983).

The time needed to sample a crop can be reduced by using a sequential sampling system. Extensive preliminary investigations are needed to develop a method in which sampling can stop as soon as adequate data have been obtained to give a clear indication whether or not to spray (Southwood 1966; Wilson et al 1989). Sequential sampling has been used for a number of pests, including the coffee shield bug *Antestiopsis* (Rennison 1962) and bollworms (Ingram and Green 1972). Based on data for sequential sampling a pegboard was devised so individual farmers could, with some training, scout for bollworm eggs on their own cotton (Fig. 1.7) (Beeden 1972). The trend is also to sample natural enemies where possible and adjust the threshold accordingly. Apart from counting in the crop, various trapping techniques have been used for sampling populations. The use of pheromone traps as a means of timing sprays may be particularly important, owing to their selectivity and effectiveness when pest densities are low (Shorey 1973), but pheromone traps may indicate only the likelihood of an infestation and scouting within the crop may still be necessary when moth activity has been noted.

One method of sampling insect populations is to spray non-persistent pyrethroids to knock down the insects from trees and bushes. This method is particularly useful for low-density pests such as the coffee shield bug.

The time of sampling and the stage of the life cycle sampled are most important if pesticides are to be applied to prevent economic damage. Detection of eggs is most important to avoid delay in treatment. Some larvae, like bollworms, are very difficult to find until they reach the third or fourth instar, while others such as codling moth or stem borers remain feeding inside the plant, fruit or stems and can be counted only by dissecting the plant. A pesticide application should be applied early, at the start of an infestation of first-instar larvae, otherwise less control is achieved. A larger

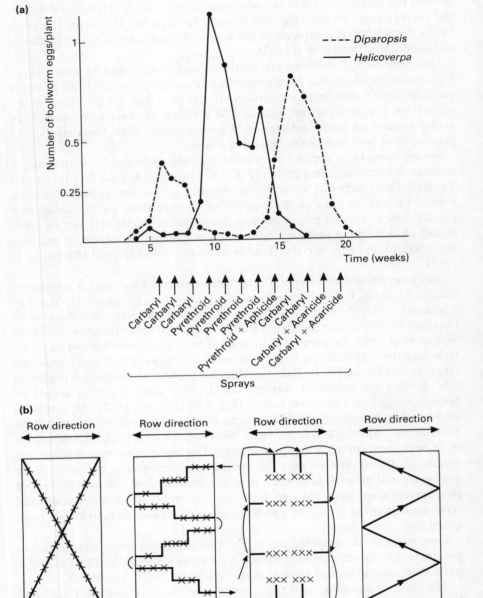

Fig. 1.6 (a) Timing of spray treatments on cotton based on crop monitoring of bollworm eggs in Central Africa (updated from Matthews and Tunstall 1968) (b) Sampling schemes for eggs in a cotton crop

Fig. 1.7 Pegboard for small-scale farmer to record insect pests of cotton

dose is required to kill later instars, even if the amount of active ingredient per unit weight of the insect remains constant. Changes in behaviour or morphology of the later instars of an insect may necessitate an increase in the dosage (Gast 1959). Chemical control of the first-instar larvae may be enhanced by any natural resistance to the pest if applications are timed according to counts of eggs or adults.

R.F. Smith and van den Bosch (1967) have argued that economic thresholds should be based on damage rather than pest numbers, for example the dead hearts of rice caused by stem borers. A pest may be present in large numbers yet have little effect on yield, especially if there is sufficient time for the plant to compensate. On the other hand, very low populations of certain pests can do considerable damage, especially if a disease is transmitted, for example the virus disease, rosette of groundnuts. Other factors such as moisture stress, temperature, lack of nutrients or a combination of factors may limit yield, irrespective of pest damage.

Application sites and placement

Scouting for a pest may show up particular foci of infestations in a crop and these can often be treated separately to reduce spread of the pest and avoid the cost of applying pesticides to the whole area. Pink bollworm (*Pectinophora gossypiella*) infestations may be found on the edges of fields near villages where cotton stalks are stored for fuel. Many wind-borne insects collect on the lee side of hedges (Lewis, 1965) or other topographical features. Red spider mite outbreaks are sometimes close to isolated trees within a field, partly because of the effect of air movements but also because pesticide deposits are often lower in such areas, especially after aerial application. These sections of a field marked out by scouts can be successfully spot-treated with acaricides applied with a knapsack sprayer (Burton 1972).

A spray directed at the blooms and apical foliage of chrysanthemums has been shown to give control of leaf miners without affecting the parasitoids and their hosts on the lower part of the plant and undersides of leaves, where they were protected by the dense canopy of foliage (Scopes and Biggerstaff 1973).

Control of thrips and the tomato leaf miner has been achieved by spraying a mixture of polybutenes plus deltamethrin on the glasshouse floor, thus avoiding treatment of the foliage (Helyer 1985).

Leaving one or two swaths untreated around the field margins has not only assisted game and wildlife conservation but also provided a refuge for natural enemies of pests (Oliver-Bellasis and Southerton 1986). A sterile strip can be created around the field margins by spraying a herbicide or clean cultivation (Fig. 1.8)

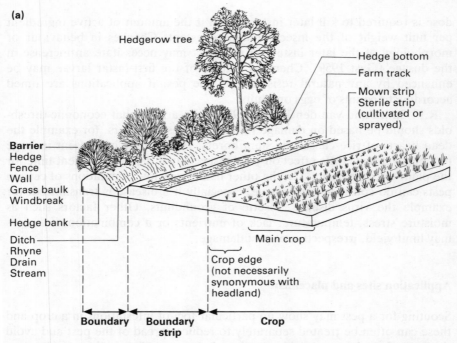

Fig. 1.8 (a) Principal components of arable field margin (from Greaves and Marshall 1987) (b) Application of herbicide at edge of field to create sterile strip

Contact effects can reduce natural enemies of the pest immediately after insecticide application unless the dose applied is only sufficient to kill the pest without affecting the predator. Applications of discrete droplets with areas of untreated foliage reduce the risk of eliminating natural enemies. Systemic insecticides are less likely to affect natural enemies once they have penetrated plant tissues. Systemics placed in the soil as granules or seed treatment are taken up through the roots, and generally kill sucking pests so they are less likely to have a direct adverse effect on their predators.

When foliar sprays are needed, the effect of a non-persistent insecticide on predators and parasites can be reduced in plantations of perennial crops such as citrus and coffee by applications to different sections on separate occasions. Natural enemies that survive in the untreated sections are given sufficient time to disperse into the treated area that is left unsprayed, when the remainder of the untreated trees are sprayed.

No doubt there is scope for further development of selective application, for example by using electrostatically charged sprays to enhance natural enemy action even with annual crops, especially in relation to multiple-cropping systems.

The importance of restricting pesticides as far as possible to the actual target is fundamental to good pest management and is considered in more detail in Chapter 2.

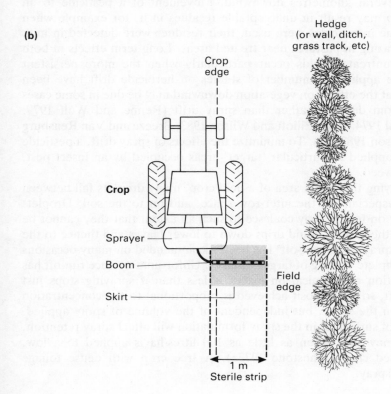

(b)

Hedge (or wall, ditch, grass track, etc)

Crop edge

Crop

Sprayer

Boom

Skirt

Field edge

1 m Sterile strip

2

Targets

There has been considerable criticism of the use of pesticides due to their movement from treated areas by 'drift' at the time of application and their subsequent movement in the atmosphere or in the soil. Great concern is naturally caused when the effects of drift are noticed. Certain insecticides kill bees rapidly, while herbicides such as 2,4-D can damage susceptible crops sometimes several kilometres downwind. Movement of a pesticide to an adjacent crop may result in unacceptable residues in it, for example when organochlorine insecticides were used, their residues were detected in milk from cows grazing in pastures near treated areas. Long-term effects in both treated and untreated fields occur particularly when the more persistent chemicals are applied. A number of studies on herbicide drift have been carried out but the effects on vegetation downwind may be due in some cases to vapour from deposits rather than spray drift (Renne and Wolf 1979; Maybank et al 1974, 1978; Elliott and Wilson 1983; Breeze and Van Rensburg 1988; Thompson 1983a, b). To minimize the effects of spray drift, a pesticide needs to be applied to particular 'target' areas occupied by an insect pest, pathogen or weed.

When spraying the whole area of a field crop, many droplets fall between the foliage, especially in the inter-row space, and on to the soil. Droplets which impact on foliage may coalesce to such an extent that they cannot be retained and the surplus liquid drips down to lower leaves and thence to the soil. Indeed spraying to 'run-off' has been recommended on many occasions to ensure complete wetting of target surfaces. Unfortunately once run-off has started retention of chemical on leaves is less than if spraying stops just before run-off, so the deposit achieved is proportional to the concentration of pesticide in the spray, but independent of the volume of spray applied. The amount of surfactant in the spray formulation will affect spray retention, but run-off may start when as little as 100 litres/ha is applied to a low, sparsely leaved crop (Johnstone 1973a). A tree crop with dense foliage retains more spray.

As much as one-third of the spray applied to a crop may be lost to the soil at the time of application. Himel (1974) has referred to this loss as 'endodrift' in contrast to the losses due to 'exodrift' outside the treated area. Endodrift may be an important contributor to ground water contamination where it occurs.

Pesticide collected on the target may be washed off later by rain or in some cases by overhead irrigation. Some estimates have suggested that up to 80 per cent of the total pesticide applied to plants may eventually reach the soil (Courshee 1960). Pesticide contamination of the soil in this manner has caused major changes in the population of non-target organisms. Earthworm numbers have been reduced by over 60 per cent following application of benomyl in high volume sprays (2 250 litres/ha) (Cooke et al 1974). Dosages are often increased to maintain control of a pest and allow for losses due to drift. For example, if $3 \times 10^{-2} \mu g$ of an insecticide is needed to kill an insect only 30 mg is needed for a population of 1 million if all of it reaches the target, yet over 3 000 times this amount is usually applied for effective control in the field (Brown 1951). Graham-Bryce (1977) has drawn attention to the inefficiency of pesticide application (Table 2.1).

Table 2.1 Utilization of crop protection chemicals (from Graham-Bryce 1977)

Pesticide	Method	Pest/crop	Efficiency of utilization (%)
dimethoate	foliar spray	aphids on field beans	0.03
lindane	foliar spray	capsids on cocoa	0.02
ethirimol	seed treatment	barley (for mildew control)	2.2
disulfoton	soil incorporation	wheat (for aphid control)	2.9
lindane/dieldrin	aerial spraying of swarms	locusts	6.0
paraquat	spray	grass	up to 30

The volume of liquid in which a pesticide has to be applied has been usually left to the user's discretion; some manufacturers have recommended a range as wide as 200–1 000 litres/ha. Many farmers did not change the nozzles or the volume of application once they had set up their sprayer, irrespective of which type of pesticide they were applying. Little or no attention was given to droplet size, so over the years a wide range of different nozzles has been used. Most produce a wide spectrum of droplet sizes, the smallest being subject to exodrift, and the largest being the main component of endodrift. However, with the introduction of new legislation in the UK, efforts have been made to define 'spray quality' produced by different nozzles, based on assessment of their spray spectra (Fig. 2.1) (Doble et al 1985). Some manufacturers have recommended specific nozzles using a code

Targets

(a)

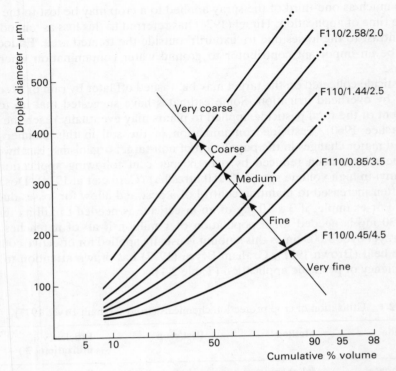

Fig. 2.1 (a) Reference chart for fan spray nozzles in relation to spray quality (from Doble et al 1985) (b) An assessment of spray deposition down wind with different spray qualities; figures on graph lines indicate level of deposit (data supplied by Parkin)

that defines the spray type, angle and output at a given pressure (see p. 104).

Ideally there is an optimum droplet size (Himel 1969b) or spectrum which gives the most effective coverage of the target with minimum contamination of the environment. Greater attention to droplet size is now essential with the trend to using smaller volumes of spray. This trend was initiated mainly by the increasing cost of personnel, which was often wasted in carting large volumes of water to the area to be sprayed. In the tropics, the use of low volumes of spray liquid has been particularly important since the scarcity of water has been a major deterrent to farmers spraying to control their pests and weeds.

Clearly, if pesticides are to be used more efficiently, the actual target needs to be defined in terms of both time and space and the proportion of emitted pesticide that reaches the target and is in a form available to the pest must be greatly increased (Courshee 1960). Hislop (1987) referring to the complexity of the problem (Fig. 2.2) defined our objective as the placement on targets of just sufficient of a selected active ingredient to achieve a desired biological

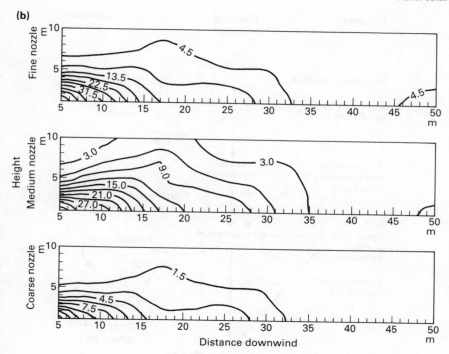

result with safety and economy. Definition of the target requires a knowledge of the biology of the pest, in order to determine at which stage it is most vulnerable to pesticides. Unfortunately, only a small proportion of a pest population may be at the most susceptible stage at any given time. Insects have several distinct stages during their life cycle, for example adults, eggs and nymphs or distinct larvae and pupae. Similarly, with weeds, foliage may be affected by herbicide while seeds can remain unaffected, enabling weeds to recolonize an area. Difficulties such as this in defining the target led to the use of persistent chemicals, but this has increased the risk of selecting pest populations resistant to a particular pesticide. If users are to apply less persistent and more selective pesticides, more attention is needed to define the target and when an application is needed. Where different stages of a pest may be controlled chemically, it is important that different pesticides are used to reduce selection for resistance. Thus against mosquito vectors of malaria, different pesticides are needed as larvicides and adulticides.

Insect control

The concept of crop protection has aimed at reduction in the population of the development stage of the pest directly responsible for damage within

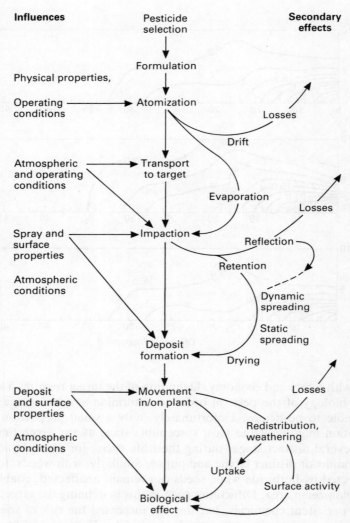

Fig. 2.2 Processes involved in pesticide transfer and deposition (adapted from Young 1986)

individual fields. Crop protection is most efficient when the pesticide is applied economically on a scale dictated by the area occupied by the pest and the urgency with which the pest population has to be controlled (Joyce 1977). Control has been directed principally at the larval stage of many insect pests. This policy has been highly successful when treatments have been early to reduce the amount of larval feeding. If treatment is too late, not only is a higher dose required to kill larger larvae, but also much of the damage may have already been done. Unfortunately, treatments directed at the larval stage may have little or no effect on the eggs, pupae and adults and repeat treatments are often necessary as more larvae develop. Similarly control of adults by spraying may result in 100 per cent mortality within a crop, but subsequently adults can be readily found because immature stage have developed within the crop and other adults may have migrated from elsewhere. Treatment may then have to be repeated. This was shown very clearly with the difficulties of controlling whiteflies (*Bemisia tabaci*) on cotton, where the nymphs on the undersurfaces of leaves were less exposed to insecticides.

In a pest-management system, biological information must extend beyond a simple description of the life cycle to include information on the ecology of the pest. In particular, insect control requires an understanding of the movement of pests, between different host plants and within ecological areas. For a given pest species the target may vary according to

1 the control strategy being adopted
2 the type of pesticide being used
3 the habitat of the pest
4 the behaviour of the pest.

These factors are interrelated, but some examples of insect pests illustrate how particular targets can be defined.

Control strategy

The principal objective in the control of locusts and other grasshoppers is to prevent their immigration from breeding sites, but in many situations this has not always proved possible so protection of farmers' crops is also essential. In each case, the target is the vegetation on which young wingless hoppers typically feed. Droplets with a volume median diameter (vmd) of 70–90 μm are blown by the wind almost horizontally (Fig. 2.3) and are selectively collected by the vertical surfaces of the sparse desert vegetation (Courshee 1959; Symmons et al 1989). Only a small proportion of the spray is deposited on the ground. Courshee (1959) measured the efficiency of application on the biological target in relation to the amount emitted, and referred to the deposit per unit emission (DUE).

This technique was originally developed using dieldrin, which was effective with as little as 0.3 litres per hectare of a 5–10 per cent formulation (Fig. 2.3b). Only slight contamination of the vegetation was needed as hoppers consumed daily their own weight of vegetation and in this way accumulated a toxic dose. This insecticide is no longer recommended, but the same principle has been used with alternative insecticides such as fenitrothion, bendiocarb and the pyrethroids, using up to 3 litres/ha. With the newer less persistent insecticides, greater care is needed to achieve better distribution of a sufficient dosage to ensure hoppers are killed before they move into an untreated area.

As discussed in Chapter 4, droplets with an optimum diameter of 75 μm have a volume of 221 picolitres, so the toxic dosage needs to be conveyed in the mean number of such droplets likely to impact on the locusts. Calculations of this type are needed to determine the volume and concentration of spray required.

An exhaust nozzle sprayer was recommended (see pp. 240–42), but its use requires good vehicle maintenance, so alternative systems that do not rely on using the vehicle exhaust have been developed (Fig. 2.3c) (see pp 205–6). Aircraft have also been used to treat areas against hoppers (Rainey 1974) (Fig. 2.4).

Where locusts have not been controlled in the hopper stage, a different strategy is required if flying swarms of adults pose a threat to farming areas. An aerial curtain is created so that insects flying through it collect a lethal dose of a contact insecticide (Sawyer 1950). The aerial curtain technique known as the 'Porton method' was first used successfully against *Schistocerca gregaria* in Kenya in 1945 (Gunn et al 1948). Efficiency of application, expressed as the fraction of locusts killed compared with the theoretical number that could have been killed from a given volume of insecticide, has

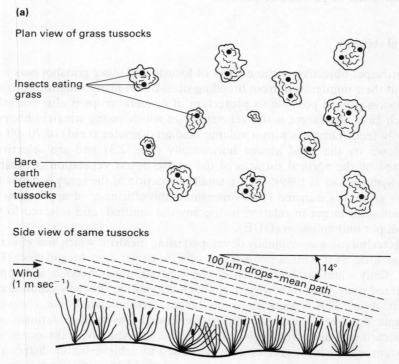

(a)

Plan view of grass tussocks

Insects eating grass

Bare earth between tussocks

Side view of same tussocks

Wind (1 m sec⁻¹)

100 μm drops–mean path

14°

Fig. 2.3 (a) Downwind movement of droplets (b) Track spacing used for locust hopper control with vehicle mounted sprayer (c) Spinning disc sprayer ('Ulvamast') being used to spray locusts

(b)

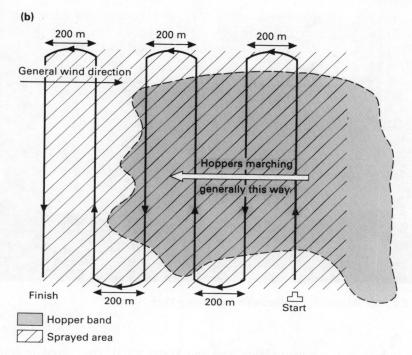

(c)

Fig. 2.4 Aerial spraying of locust swarm (photo Dick Brown)

been as high as 6 per cent (MacCuaig and Watts 1963). One problem is the mobility of the swarms so aerial operations have to be relocated (Bennett 1976), thus the most economical method of halting an upsurge of locusts continues to be the spraying of hopper bands. They are much less mobile than swarms and the infested area is relatively stationary for weeks at a time. An area of 11 000 km² infested with hopper bands could be treated with 35 000 litres of insecticide, whereas over 200 000 litres were needed to destroy about two-thirds of a swarm of *S. gregaria* covering 600 km². Unfortunately during 1988–9 much of the spraying required higher dosages aimed at settled swarms of adult locusts (Skaf et al 1990). It is essential therefore that as much as possible is known about the movement of locusts in relation to potential breeding areas from which locusts can endanger crops. FAO has established a remote sensing facility to process satellite data to assist those monitoring locust populations and those organizing control operations (Hielkama 1990).

Johnstone (1991) used a computer model to assess the selection of an optimum droplet size for locust control in relation to wind speed and emission height.

Type of pesticide

Selection of an application technique may be affected by the mode of action of the pesticide. An insecticide may be effective

1 by contact
2 as a stomach poison
3 by having fumigant effects.

The chemical may be moved systemically within a plant or be translocated across leaves to the site of action.

Some chemicals, such as mevinphos, break down very rapidly in the environment while others remain biologically active over a long period. Naturally, this difference is important when controlling a pest.

In the control of mosquitoes persistent and non-persistent insecticides require different application techniques. One of the principal methods used to control domiciliary mosquitoes and interrupt malaria transmission has been the application of a persistent insecticide as a residual deposit on walls inside houses. Hand-operated compression sprayers are principally used but pyrethroids have been applied experimentally using electrostatically charged sprays (Chadd and Matthews 1988). Pyrethroids have been also applied to bed-nets (Rozendaal 1989; Anon 1989).

Alternatively less persistent insecticides have been applied as space sprays. Droplets over 50 μm fall to the ground too quickly so aerosol droplets are needed. The optimum droplet size for collection on mosquitoes is about 5–15 μm (Mount 1970) (Table 2.2). Some chemicals such as the natural pyrethrins have an irritant effect which disturbs insects and causes them to fly. This is an advantage as flying insects collect more droplets than those at rest (Kennedy et al 1948).

Tables 2.2 Percentage mortality of caged female *Aedes taeniorhynchus* with ULV non-thermal aerosol of technical malathion 92 metres downwind (based on Mount 1970)

Droplet diameter	Dosage * (kg ai/ha)		
	0.005	0.01	0.02
6– 8.0 μm	38	100	100
8–11.0 μm	38	100	100
11–14.0 μm	38	98	100
13–23.0 μm	18	52	84

Note: * based on 184 m swath

The best time to apply aerosol sprays in the open is in the evening, especially when inversion conditions (see pp. 80–1) exist and wind velocity is low so that the spray cloud is not dispersed too rapidly. An insecticide of low mammalian toxicity is obviously needed when applications are to be close to human dwellings.

The contrast between residual deposits and space sprays is also provided in other situations such as in the control of pests of stored products. Repetitive aerosol sprays or fogging of a non-persistent chemical such as 0.4

per cent pyrethrin plus 2 per cent piperonyl butoxide at 50 ml/100 m³ have been used, but in some situations a residual spray may be preferred.

Systemic pesticides

These chemicals are redistributed in the plant by upward movement, so that if the pesticide is applied to the soil and it is absorbed by the roots, control of a pest on the upper leaves is possible. The reverse is not generally true although there are some herbicides such as glyphosate that are translocated from foliage down to the rhizomes.

Many systemic pesticides are applied as sprays but the most toxic chemicals are usually formulated as granules for soil application. Granule application is sometimes prophylactically, that is irrespective of whether the pest is present in sufficient numbers to cause damage, as some farmers find it difficult to monitor crops and treat the whole area sufficiently quickly if pest populations suddenly increase. Some pests are vectors of virus diseases, so crops prone to these diseases, such as sugar beet are very likely to be treated with granules, especially as treatment of the undersurface of the basal leaves with sprays is extremely difficult.

In IPM programmes, instead of application of conventional pesticides, there is an increasingly important role for pheromones which can be used in combination with insecticides as a 'lure and kill' strategy. An aerosol type spray may be used with a pheromone to stimulate the pest to fly into the spray, but more frequently a localized residual deposit is applied to attract the pest to the insecticide. Such a deposit may be applied as a paste-like formulation to selected leaves at random within the field or sprayed as a micro-encapsulated formulation along sections of a field. Targets may be traps with larger dispensers of pheromone to persist over a long period. Distribution of sufficient pheromone without any insecticide is also important as a mass confusion technique which can preserve natural enemies.

Increased use of microbial pesticides is expected. In addition to products such as *Bacillus thuringiensis*, and some baculoviruses, certain weeds can be controlled by mycoherbicides. Other research is in progress on a number of fungi and other biological agents such as entomopathogenic nematodes which can be applied to specific targets. The particulate properties of these products, and in many cases the need to transfer them more accurately to specific targets, will require improved application technology. The dosage required will need to be carefully related to the droplet size and volume that can be deposited where it is effective. Thus if 10^3 spores need to be transferred to an insect in a spray with 100 μm droplets, (ie each with a volume of 524 picolitres) at least two droplets of a spray containing 10^9 spores per millilitre of spray must be deposited on an insect. Entwistle (1986) has discussed the requirements for loading spray droplets with Nuclear Polyhedrosis Virus (NPV) inclusion bodies.

Pest habitat

Tsetse flies (*Glossina* spp.), vectors of pathogenic trypanosomes, are of great economic importance. Different species of tsetse flies live in riverine, forest and savannah areas, in each of which control with insecticides is directed at the adult flies. Tsetse flies are unusual in their breeding habit in that the female does not lay eggs but gives birth at intervals, depending on temperatures, to a single third-instar larva. The larva, which at birth is heavier than the female fly, burrows down to usually 1–3 cm below the soil surface. The larval skin hardens to form the puparium in which the tsetse becomes the fourth instar, pupa and eventually a pharate adult which emerges into the open air. Population densities can be low over extensive areas, but control measures can be designed to exploit the more restrictive range of flies during a dry season.

Tsetse flies seek the shade of woodland and areas where the humidity is higher than in the surrounding vegetation. By discriminative treatment of these restricted localities and within these, spraying a residual insecticide selectively only on the sites such as the underside of horizontal branches, where tsetse flies rest (Fig. 2.5) (Davies 1967) the amount of insecticide needed can be significantly minimized. Sprays were applied with a knapsack compression sprayer fitted with a narrow cone (60°) nozzle so that the surface was completely wetted without run-off. Only one spray was needed in an area unless the treated area was reinvaded, so effects on birds and other non-target organisms was minimal (Koeman et al 1971). This contrasts with the repeated application at intervals of a few days on certain agricultural crops. Some trials have investigated the use of electrostatically charged ultra low volume sprays to treat resting sites to reduce the logistic problems associated with large-scale use of compression sprayers.

Discriminative and selective spraying from helicopters was also attempted in Nigeria where studies had shown that tsetse flies moved to the foliage higher in the trees at night (Scholz et al 1976). The technique was useful for rapid treatment along cattle routes and treating riverine species, but it was more expensive and less selective than ground treatments, with adverse effects on non-target organisms more noticeable in the immediate post-spray period. An alternative to spraying natural resting sites is to treat screens and traps made from black cloth that attracts the tsetse flies. Just touching the surface of cloth treated with pyrethroid insecticides such as deltamethrin can kill the flies, but care is needed to ensure the traps or screens are not vandalized so villagers need to be trained to appreciate their importance.

Large-scale control of tsetse flies in savannah areas has been primarily by sequential treatments of aerially applied aerosol sprays (Fig. 2.6) (Allsopp 1984; Johnstone and Cooper 1986; Clarke 1990). Early studies by Hadaway and Barlow (1965) had determined the need to use 30–40 μm droplets, and subsequent studies have confirmed the importance of droplet size (E.G. Harris et al 1990; Johnstone et al 1987, 1988, 1989a, 1989b, 1990). Recent operations have used Micronair AU4000 rotary atomizers (Allsopp 1990)

Targets

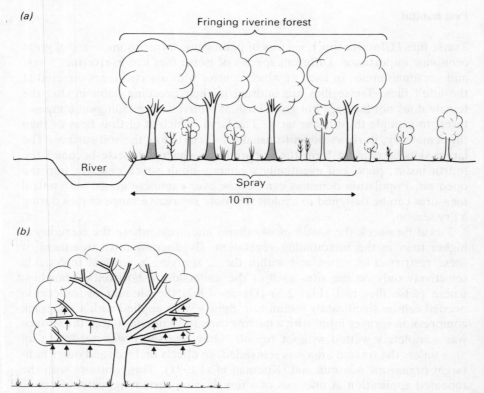

Fig. 2.5 Sites for selective spraying against *Glossina morsitans* in the Sudan savannah zone (*Note:* Against *G. tachinodes* spraying can be confined to area 5 m from each bank of the watercourse) (b) Sites for selective spraying of the undersides of tree branches during the dry season in the northern Guinea savannah zone (from Davies 1967)

(see Chapter 13). Five to six treatments have been needed, the interval between which was determined by temperature conditions and detailed knowledge of the biology of the tsetse flies.

However, these aerially treated areas need to be protected from reinvasion by ground spraying the perimeter with a residual insecticide or by the use of screens (Fig. 2.7), where appropriate.

Behaviour of the pest

As mentioned earlier, chemical control of lepidopteran pests of economic crops has been directed at the larval stages (caterpillars) which cause serious defoliation or damage to the stems or fruit. Larvae penetrate the plant's

30

Fig. 2.6 Aerial spraying of tsetse flies at night under inversion conditions

Fig. 2.7 Treating a screen with insecticide to control tsetse flies

tissues and so the period of their exposure to insecticides on the surface of the plant is often short. One example is the red bollworm *Diparopsis*, a major pest of cotton in Africa. This insect overwinters from one season to the next as a diapause pupa, and subsequent moth emergence takes place over many weeks. Eggs are laid on all parts of the cotton plants although on maturer plants more are found on the bracts. The first-instar larva chews its way out of the eggshell in the early morning when the humidity is high and then actively searches for a bud or boll, inside which it feeds. When the egg is on a bract of a glabrous variety, the period from hatching until the larva starts to chew into a bud or boll, may be as short as 9 min (Tunstall 1962). The young larva does not ingest the first few bites and later it may form a cell with frass and silk threads between the bract and a boll to protect it during the first moult. The larva may come out of the bud at night and re-enter the same bud or transfer to a fresh boll, if necessary. The fully grown *Diparopsis* larva drops to the ground to pupate inside an earthen cocoon, usually in the top 10 cm of the soil. This insect is vulnerable to insecticides only during the period immediately after egg hatch when the first-instar larva is walking along leaves, petioles, stems and bracts in search of a bud or boll (Fig. 2.8)

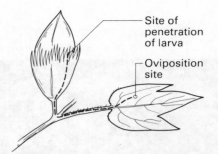

Site of
penetration
of larva

Oviposition
site

Fig. 2.8 Route of bollworm larva on cotton plant before eating its way inside boll

(Matthews 1966). Therefore, the principal target for a spray deposit is the bracts which protect the buds and bolls, but the period of contact may be so limited that a lethal dose may not be picked up. The aim therefore is to distribute insecticide as evenly as possible within the plant canopy so that a lethal dose is picked up from the stems and petioles. The presence of hairs on these surfaces in jassid-resistant varieties, delays the movement of larvae and so lengthens the period of contact with the insecticide.

When cotton plants are examined from above, generally only the upper surfaces of leaves are visible. This foliage presents such a barrier that only a small proportion of spray applied downwards over the top of the plants penetrates the canopy to reach stems and petioles. Ideally nozzles need to direct spray upwards and between branches (see pp. 138–9) (Tunstall et al 1961) so that some underleaf coverage is obtained and a percentage of the deposit is less likely to be washed off by rain. Even under dry conditions, deposits on the undersides of leaves are expected to be more persistent since they are protected from ultra-violet (UV) light. However, effective control has been obtained by applying droplets in the range 50–100 μm vmd as these are carried by air movements and can carry around leaves. Nevertheless, a high proportion of the insecticide is still deposited on the outer part of the canopy.

In contrast, the American bollworm *Helicoverpa* (formerly *Heliothis*) *armigera* prefers the upper part of cotton plants, so larvae are more easily controlled with insecticides because they move frequently to fresh buds. Joyce (1977) has controlled *Helicoverpa* in the Sudan using smaller droplets (40 μm). Chemical control of *Helicoverpa* on maize is much more difficult as larvae burrow down into the tip of the cob.

The presence of several pests and the importance of natural enemies and the growing of insect-resistant varieties complicate the selection of an application technique for rice farmers. The brown planthopper *Nilaparvata lugens* on rice is particularly difficult to control as both adults and nymphs live mostly at the base of plants where it is shaded and moist. Ideally no insecticide should be used, but when chemical control is needed, the proportion of spray collected where the hoppers are is small when the plant is sprayed at the heading stage, since the foliage of adjacent plants effectively covers the whole area. Nozzles should be directed sideways between rice plants especially when using an insect growth regulator such as buprofezin directed at the nymphs. In Japan there is extensive use of microgranules as these are collected in the leaf axils (see pp. 268–9).

Determination of the most appropriate target for deposition of a pesticide requires careful examination of the behaviour of the pest in the field throughout its life cycle. Observations should not be confined to daytime since many insect pests are more active at night. Jassid nymphs on cotton are readily controlled at night, whereas in sunlight the nymphs are confined to the lower surfaces. In practice, a very irregular distribution of insecticide is usually adequate to control jassids because of the mobility of the adults.

Disease control

A typical plant pathogen basically has four phases:

1 pre-penetration
2 penetration
3 post-invasion
4 sporulation and dispersal.

Ideally control is achieved before the pathogen can penetrate the host plant. It is possible that spores may arrive over a series of relatively short periods when conditions favour dispersal. Rapid penetration into the host plant then limits the time available for effective action by foliar-applied fungicides unless a systemic fungicide can curtail development of the post-invasion phase. In most cases a protectant fungicide has to be applied on several occasions to limit the spread of disease. Area-to-area and year-to-year variation in the onset of infection make it difficult to time sprays, so early application even before infection is usually recommended to avoid the possibility of delay in application when conditions favour an epidemic.

Where possible, farmers have preferred prophylactic seed treatment. Jeffs (1986) gives a general account of seed treatment. Seedling diseases such as damping-off caused by various species of *Pythium*, *Rhizoctonia* and *Fusarium* can be controlled by seed treatment, as can smuts and mildews on cereals and black arm *Xanthomonas malvacearum* on cotton. The amount of fungicide retained on the seed coat varies and most consistent results have been obtained with high loading rates. Free soil water is needed for uptake of a systemic so control can be inadequate if dry conditions prevail. Nevertheless, more consistent control with ethirimol, a systemic fungicide used for powdery mildew control in temperate cereals, was obtained by seed treatment rather than by using pelleted seed, spraying along the furrow, broadcasting granules or placement of granules alongside the seed. Foliar fungicide sprays on cereals have also increased in importance in Europe. Cereal farmers are not prepared to risk losing the considerable investment in their crops because of mildew or rust diseases.

Foliar sprays are widely used to control mildew and scab on deciduous fruit. Applications against mildew are difficult due to leaf expansion during the period of infection diluting the effect of the fungicide. To control post-harvest fruit rot a single spray of benomyl, a systemic fungicide, can be directed at the young apples in July instead of on the mature fruit.

Primary pod infection of black pod disease *Phytophthora palmivora* of cocoa in Nigeria is initiated primarily by inoculum from the soil which reaches the lower pods either by rain splash or movement of soil particles by insects (Gregory and Maddison 1981). Only about 2–3 per cent of the disease arises from cushion infection or sporulating trunk cankers in Nigeria. Removal of buds at the base of trees may reduce further spread of the disease, but if a fungicide with or without an insecticide is needed a restricted target at the base of trees and surrounding soil, rather than the canopy and individual pods, would be very much easier to spray. An alternative method is to have a reservoir of copper fungicide in a water permeable collar around the upper stem so that copper is leaked to pods lower down the tree (Shreenivasan et al 1990).

Weed control

Herbicides are particularly important in the early stages of crop growth, especially when removal of weeds by hand or mechanical cultivation disturbs the crop. Late in the season a herbicide application may be beneficial, even if there is an increase in yield when harvesting is easier in the absence of weeds.

Soil and foliar treatments may be applied (Fig. 2.9). The target for a herbicide can be

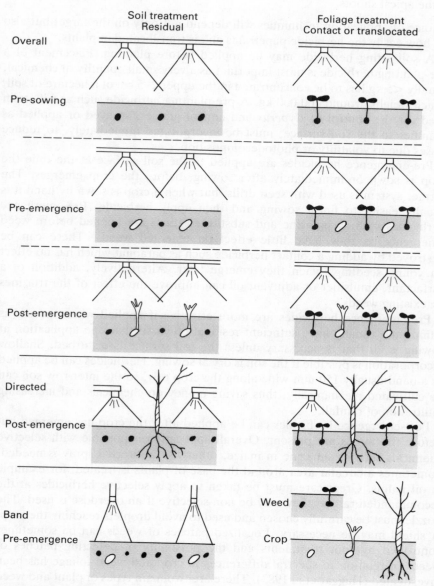

Fig. 2.9 Situations in which herbicides may be used for weed control in crops (after Fryer 1977)

1 the weed seed to prevent germination or kill the seedling immediately the seed germinates
2 the roots, rhizomes or other underground tissues
3 the stem, especially when applied to woody plants
4 the foliage
5 the apical shoots.

Choice of application techniques will depend not only on the target but also on how easily the herbicide penetrates and is translocated in plants.

A soil-acting herbicide may be applied before planting. Placement of a pre-planting herbicide is most important as a very small quantity of chemical, usually <5 kg has to be concentrated in the upper 2–5 cm of a hectare of soil, which weighs around 700 000 kg. A pre-planting herbicide such as trifluralin used for the control of *Cyperus* and annual grasses, sprayed or applied as granules to the soil surface, must be incorporated immediately, to reduce losses due to volatility or photodecomposition.

Pre-emergence herbicides are applied to the soil surface at the time the crop is sown or immediately after sowing, before the crop emerges. The former system is used with seed drills, but when a crop is sown by hand it is usually easier to finish sowing and then apply herbicide. Pre-emergence herbicides such as simazine and substituted ureas are applied before weed emergence as they have little effect on emerged weeds. These can be destroyed by adding a contact herbicide such as paraquat which has no effect on young seedlings when they emerge later. Alternatively, addition of a surfactant, emulsifier or adjuvant oil can improve the effect of the triazines on existing weeds.

Pre-emergence herbicides are more effective if applied when the soil is damp, and usually have sufficient residual effect so that one application at sowing is all that is necessary unless the soil surface is disturbed. Shallow incorporation is possible if the soil is dry at sowing. Herbicides can be applied as a band usually 150 mm wide along the crop row, so the inter-row soil can be cultivated mechanically, thus saving money on chemicals and increasing infiltration of rainfall.

Post-emergence herbicides can be applied after the crop has emerged but before the weeds are present. Overall application is possible with selective chemicals such as simazine in maize, otherwise a **directed** spray is needed. Sometimes a circular area around the base of plants is treated, for example in oil palms. Great care must be taken to apply selective herbicides at the recommended rate as they may be non-selective if an overdose is used. The nozzle must be carefully chosen and used to avoid droplets reaching the crop. A shield may be necessary. Localized patches of weeds can be sometimes controlled by spot treatments and the possibility of detecting patches of weeds in relation to spectral differences in crop and weed foliage has been investigated (Haggar et al 1983). There are two main types of plant and weed foliage to consider in relation to spray deposition: the narrow grass-like leaves of monocotyledons and the broad leaves of dicotyledons. As will be pointed out later, there are considerable differences in the detailed surface structure of leaves which affects the retention of spray droplets. Movement of deposits to leaf axils can improve the effect of some herbicides, thus spray volume, droplet size and formulation interactions need to be considered. The addition of a surfactant may improve retention by better penetration of the cuticle or affect redistribution over a plant surface. Chemical weed control often requires herbicides which selectively kill both monocotyledon and dicotyledon weeds within a cereal crop. Similarly in bean, potato and other broad-leaved crops, selective chemicals are needed to differentiate between the crop and a wide range of weed species.

When the roots of rhizomes are the main target, a herbicide such as glyphosate, which is translocated to these targets, may be applied to foliage. Many perennial weed and brush species are difficult and expensive to control with hand labour or mechanical means, but may be susceptible to herbicides.

There is also a potential for genetically engineered plants with resistance to certain herbicides, thus sowing a crop resistant to glyphosate allows this broad-spectrum herbicide to be used whereas previously the crop itself would be killed. However concern has been expressed that resistant crop plants could become weeds in subsequent seasons. Another approach has been to treat the crop seed with a safener, thus the rice weed can be controlled in rice fields with alachlor if the rice seed has been treated with 1.8 – naphthalic anhydride (AN) (Fryer 1977). Similarly other cereals, such as sorghum and maize can be protected from herbicides used to control annual grass weeds (Blair et al 1976).

Dispersion of herbicides makes it difficult to control aquatic weeds so slow release formulations (eg gels) have been used (Barrett 1978; Barrett and Logan 1982; Barrett and Murphy 1982).

Collection of droplets on targets

Droplets are collected on insects or plant surfaces by sedimentation and impaction (Johnstone 1985). Under still conditions even small droplets will eventually fall by gravity on to a horizontal surface. For example when fogging in a glasshouse only 0.5 per cent of a *Bacillus thuringiensis* treatment was recovered on the lower surface of leaves (Burges and Jarrett 1979). More important is the disposal of droplets in air currents in relation to different target surfaces. There is a complex interaction between the size of the droplet, the obstacle in its path, and their relative velocity (Langmuir and Blodgett 1946; E.G. Richardson 1960; May and Clifford 1967; Johnstone et al 1977). Collection efficiency of an obstacle in an airstream is defined as the ratio of the number of droplets striking the obstacle to the number which would strike it if the air was not deflected. In general, collection efficiency increases with droplet size and the velocity of the droplet relative to the obstacle, and decreases as the obstacle increases in size.

The sum of the cross-sectional area of the two airstreams passing on either side of an obstacle is only about 75 per cent of the original airstream, therefore the velocity of the deflected airstream is increased. Droplets tend to follow in the airstream and miss the obstacle unless the size of the droplet and its momentum are sufficient to penetrate the boundary layer of air around the obstacle. The distance (mm) over which a droplet can penetrate still air is

$$\frac{d^2 V \varrho_d}{18\eta},$$

where d = droplet diameter (m), V = the velocity of the droplet (m/s), ϱ_d = droplet density (kg/m³) and η = viscosity of air (Ns/m²). Even small droplets will impact if they are travelling at sufficient velocity to resist the change in direction of the airstream (Fig. 2.10). Collection efficiency on most flying insects is significantly less when droplet diameter is less than 40 μm (Spillman 1976) (Fig. 2.11), but it is these small droplets that remain airborne longer and are most likely to be filtered out by insects. The effect of terminal velocity and wind speed on collection efficiency of cylinders of different diameters is illustrated in Fig. 2.12 (Spillman, personal communication).

Most target surfaces are not smooth, and variations in the surface may cause local turbulence of the airflow. In this way, interception of a droplet or particle may occur if its path has been only partially altered. Impaction of droplets on leaves depends very much on the position of the leaf in relation to the path of the droplet. More droplets are collected on leaves which are 'fluttering' in turbulent conditions and thus present a changing target pattern. If wind velocity is too great – this often occurs when a high-speed air jet is used to transport droplets – the leaf may be turned to lie parallel with the

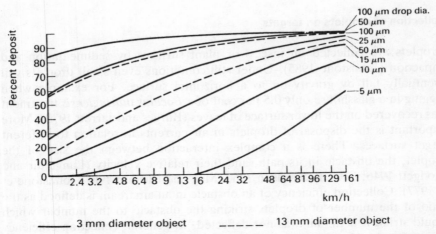

Fig. 2.10 Theoretical deposit achieved on objects of two sizes with several droplet sizes in airstreams of different velocity (after FAO 1974)

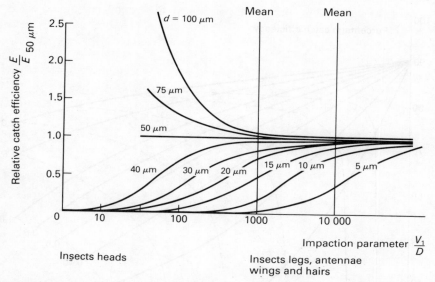

Fig. 2.11 Catch efficiency relative to 50μm droplets over a range of impaction parameters (after Spillman 1976)

airflow, so presenting the minimum area to intercept droplets. Morton (1977) noticed that the orientation of cotton leaves was affected by phototropism so that the maximum area was orientated towards the sun. He concluded that ideally when 'drift' spraying, two applications should be made. In the first application, spray droplets should be released when the wind is approximately from the direction of the sun and when the deposit will be predominantly on the upper surfaces of the exposed leaves. A second application is made with the wind towards the sun, so that more droplets are impacted on the undersurfaces of leaves. Such sprays can be applied during the early morning and repeated in the evening if the wind direction remains more or less constant. Similarly, sprays can be applied on two occasions with opposing wind directions, irrespective of the sun's position, in order to cover both sides of the plants. However the extent of the coverage needed will depend on the pest behaviour as discussed earlier. In practice two applications separated in time (two to three days) may be more important in providing treatment of new foliage.

Droplets which reach a surface may not be retained on it. Brunskill (1956) cited the familiar example of cabbage leaves in a rainstorm and pointed out that many surfaces reject liquids falling on them. Brunskill showed that retention of spray droplets on pea leaves could be increased by decreasing the surface tension of the spray, droplet diameter and the angle of incidence. His studies revealed that droplets which strike a surface such as a pea leaf become flattened, but the kinetic energy is such that the droplet then retracts

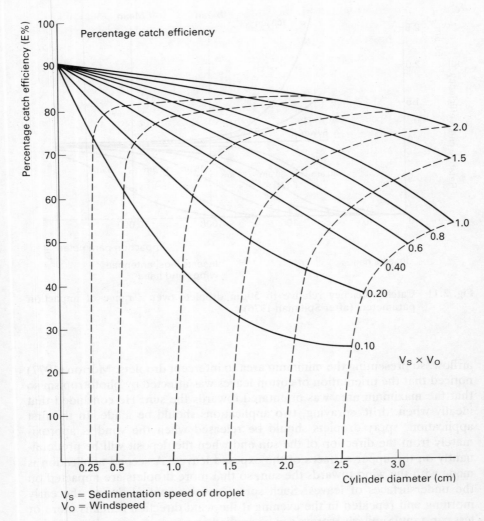

Fig. 2.12 Variation of catch efficiency with cylinder diameter, sedimentation speed and windspeed (Spillman personal communication)

and bounces away. Droplets below a certain size (<150 μm) have insufficient kinetic energy to overcome the surface energy and viscous changes and cannot bounce. Conversely, very large droplets (>200 μm) have so much kinetic energy that they shatter on impact. Bouncing from pea leaves is associated with the roughness of the surface.

Leaf roughness varies considerably between plants and also between upper and lower surfaces (Holloway 1970) and influences the spreading of spray droplets over the leaf surfaces (Boize et al 1976). Apart from conspicuous features caused by venation, the shape and size of the epidermal cells which may have flat, convex or hairy surfaces influence the topography of the leaf. The cuticle itself may develop a complex surface ornamentation. Various patterns of trichomes exist on leaves, but at the extremes 'open' patterns enhance the wetting of leaves, possibly due to capillary action, while 'closed' patterns are water-repellent. Holloway (1970) differentiated between the various types of superficial wax deposits on cuticle surfaces. A 'bloom' on a leaf surface occurs when these deposits are crystalline, for example when rodlets and threads are present.

When assessing wetting of leaves, there are two main groups of leaf surface, depending on whether the angle of contact (Fig. 2.13) is either above or below 90° (Table 2.3). With the latter group, superficial wax is not a feature, but on leaf surfaces with a contact angle above 90°, wax significantly

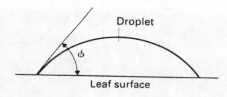

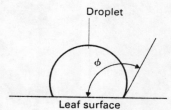

Fig. 2.13 Angles of contact

Table 2.3 Contact angles of water on some leaf surfaces (*after* Holloway 1970)

| | Leaf surface | |
	Upper	Lower
Eucalyptus globulus	170°	
Narcissus pseudonarcissus	142° 54′	
Clarkia elegans	124° 8′	159° 15′
Saponaria officinalis	100° 6′	106° 26′
Prunus laurocerasus	90° 50′	93° 32′
Rhododendron ponticum	70° 22′	43° 21′
Senecio squalidus	90° 10′	90° 15′
Rumex obtusifolius	39°	40° 5′
Plantago lanceolata	74° 23′	39° 32′

affects wettability. Contact angles of 90–110° occur on leaves with a smooth layer of superficial wax. Above 110°, the contact angles depend on the roughness of the surface. There is a generalization that leaf roughness is less important when the droplet size is below 150 μm, particularly as pesticides are formulated with surface-active agents. Ideally the advancing contact angle must be kept as high as possible and the receding angle as small as possible. Surface-active agents (surfactants or wetters) behave differently, depending on the leaf surface, so it is not possible to formulate optimally for all uses of the pesticide. The effect of a surfactant on droplets on a leaf surface is shown in Fig. 2.14. Surfactants affect retention more on leaf surfaces, such as pea leaves which are difficult to wet. Anderson et al (1987) pointed out that retention was also determined by the dynamic surface tension of the spray rather than the equilibrium surface tension.

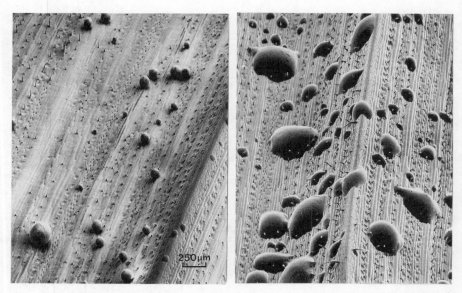

Fig. 2.14 Cryo-scanning electron micrographs of the abaxial surface of glasshouse-grown wheat showing spray droplets of the plant growth regulator paclobutrazol (0.5 g/l) as 'Cultar' to show effect of adding surfactant (1 g/l 'Synperonic NP8') (from Hart and Young 1987)

Spray coverage

When sprays are applied at high volume (HV), the aim is complete coverage of the crop, although in practice this can seldom be achieved, especially in a dense canopy. Reduction in the volume of spray has necessitated application of discrete droplets and, except in a few cases, control has been as good as that obtained with larger volumes. When discrete droplets are applied, the pesticide applicator needs to know the density of droplets required and the distribution of these droplets on the target. Microvariations in distribution of droplets have less effect on the control when a systemic pesticide is used or the chemical is translocated to the site of action. Systemic insecticides applied in large droplets to avoid drift have been effective against certain pests, such as aphids, due to redistribution of the active ingredient by the plants. Distribution is more important when activity is by contact.

Insect control of mobile pests such as jassids can be readily achieved without complete coverage, but more uniform coverage is needed to control scale insects and leaf miners. Johnstone et al (1972) used 1 droplet/mm^2 so that the deposited droplets (100 μm) should be sufficiently close to give a high probability of a direct hit on the small insects. Difficulties can arise, particularly with larger droplets, when the insect is capable of avoiding individual droplets. Polles and Vinson (1969) reported higher mortality of tobacco budworm larvae with 100 μm droplets of ULV malathion than with larger, more widely spaced droplets (300–700 μm) which the larvae were able to detect and avoid.

Almeida (1967) has reported that *Plodia interpunctella* larvae could avoid oil droplets containing natural pyrethrin. Avoidance was evident for oils of low boiling point even without the insecticide, whereas non-volatile oils behaved similarly to controls.

Later Munthali and Scopes (1982) using a micro-tip nozzle (Coggins and Baker 1983) applied uniform-sized droplets of the acaricide dicofol to the adaxial surface of bean leaves (*Phaseolus vulgaris*) infested with known numbers of a sessile pest, the egg stage of the red spider mite (*Tetranychus urticae*). The dicofol was mixed with a fluorescent tracer so that the number of droplets on leaves could be recorded. Few eggs were hit directly by droplets but subsequent redistribution of the oil-based formulation revealed that the amount of active ingredient per cm^2 required to kill 50 per cent of the eggs was least with 20 μm diameter droplets containing 1–2 per cent dicofol. At a lower concentration, the volume and hence dosage applied had to be increased. Similarly mortality was not increased at higher concentrations (Munthali 1984; Munthali and Wyatt 1986). Munthali referred to the area over which the pesticide spread as the 'biocidal area', although this term had been used much earlier in relation to fungicide deposits by Courshee et al (1954). M.G. Ford and Salt (1987) discussed Munthali's data and defined biocidal efficacy as the inverse of the LD$_{50}$, ie cm^2/μg (Fig. 2.15). They suggested that effective spreading of the active ingredient from the initial deposit may involve a diffusion-controlled process. Thus the concentration of active ingredient on the leaf will decrease radially from the centre of the initial deposit. Gradually more of the active ingredient will spread over an increasing area, but the rate of diffusion will progressively decrease. Using these data Sharkey et al (1987) used a simulation model to examine the response to discrete foliar insecticide deposits.

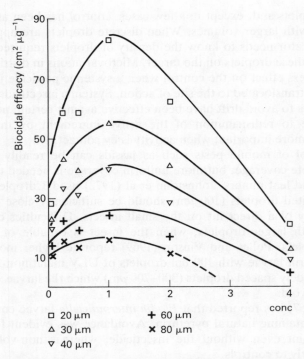

Fig. 2.15 The effect of spray concentration (& w/v) and in-flight droplet diameter on the biocidal efficacy (cm².ug⁻¹) of ULV formulations of dicofol applied to tobacco leaves for the control of red spider mites (*Tetranychus urticae*) (from M. G. Ford and Salt 1987)

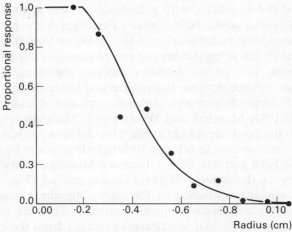

Fig. 2.16 Mortality of whitefly larvae as a function of the radial distance from the centre of a ULV deposit containing 10 per cent w/v experimental formulation of permethrin applied to the surface of an infested tobacco leafsurface. Time after application four days; in-flight droplet diameter 114μm, diameter spread factor 1.7

This approach indicates a maximum concentration to achieve control of a pest, but in practice a higher concentration may be required to ensure persistence of the deposits.

These results were substantiated by experiments with different formulations of permethrin against glasshouse whitefly (*Trialeurodes vaporarium*) scales (Fig. 2.16) (Abdalla 1984; Wyatt et al 1985; Adams et al 1987). Alm et al (1987) also reported that 120 μm droplets of bifenthrin were more efficient than 200 μm droplets against *T. urticae*.

In addition to the studies with sessile insects, M.G. Ford et al (1977) examined the response of *Spodoptera littoralis* larvae to bioresmethrin applied in 80 μm droplets. They suggested that there was a critical mass of insecticide per deposit otherwise transfer of the active ingredient from the leaf to the insect was inadequate. Similar results were obtained by Reay and Ford (1977) using permethrin. Control of newly hatched larvae could be achieved with different combinations of droplet size and numbers/cm² provided a minimum dosage had been applied. However, Omar and Matthews (1987) using diamond back moth (*Plutella xylostella*) larvae showed that efficiency of transfer could be increased by using small droplets. Crease et al (1987) also showed the importance of a high viscosity oil to enhance the effect of small droplets applied at ultra low volume. Small droplets of permethrin in vegetable oil were more effective than larger droplets against *Heliothis virescens*, but droplet size was not important with water-based sprays; however the latter were applied with a cone nozzle which would produce a wider spectrum of droplet sizes (Wofford et al 1987).

While such laboratory studies may indicate the ideal form of insecticide deposit on a leaf surface, the application of small droplets (< 100 μm) under field conditions is difficult as it is these droplets that are most vulnerable to downwind drift. This is discussed again later in Chapter 4. The relative size of an aphid tarsus, leaf surface and a spray deposit is indicated in Fig. 2.17.

As far as fungicide application is concerned, it might also seem impossible to achieve control of a disease unless there is complete coverage, since hyphae can penetrate plants at the site of spore deposition when suitable conditions occur. However, each speck of fungicide from a droplet has a zone of fungicidal influence. As mentioned earlier, Courshee, et al (1954) referred to this as the biocidal range of a droplet. They postulated that the maximum ratio between the effective fungicidal cover and the actual cover by the deposit of the fungicide residue of a droplet on drying when the droplet size is minimal. Initial infection of a disease such as potato blight (*Phytophthora infestans*) occurs usually in wet weather when most spores are collected on the upper surfaces of leaves (Beaumont 1947). The spores follow the movement of raindrops to the edges of leaves where the symptoms of blight attack first occur. Also a high proportion of pesticide spray deposit is redistributed over the leaf surface by rain, so however uniformly the initial deposit is applied, control can be maintained by the small proportion of deposit retained at the same sites as the spores (Courshee 1967).

Redistribution of fungicide over the surface of plants is very important

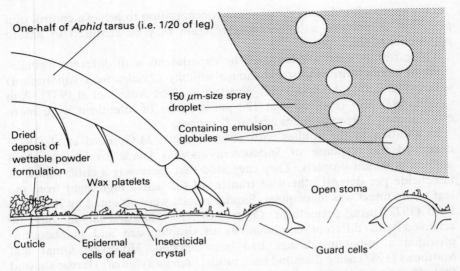

One-half of *Aphid* tarsus (i.e. 1/20 of leg)

150 μm-size spray droplet

Containing emulsion globules

Dried deposit of wettable powder formulation

Wax platelets

Open stoma

Cuticle

Epidermal cells of leaf

Insecticidal crystal

Guard cells

Fig. 2.17 Relative size of an aphid tarsus, spray droplet and leaf surface (from Hartley and Graham-Bryce 1980)

with other diseases. Coffee berry disease control has been achieved by spraying over the tops of trees with either ground (Pereira and Mapother 1972) or aerial equipment (Pereira, 1970). Chemical has been redistributed by rain to more or less the same areas of the plant as spores. This type of application is relatively simple for the farmer but as a high proportion of the chemical applied is wasted, the aim should be to improve distribution of the smaller quantities of pesticide in a suitable formulation so that it is retained and biologically active at those sites where control is needed.

What volume of spray is required?

Ideally the actual volume of liquid to be sprayed should be given in recommendations for each pesticide and crop pest situation. In practice a number of terms have been used, namely high, medium, low, very low and ultra low volume, usually abbreviated to HV, MV, LV, VLV and ULV. These terms have acquired different meanings for field and tree crops separately (Table 2.4). The trend has been to decrease the total volume of application, and thus reduce the cost of carting diluent (usually water) and the time required for application. Ultra low volume is defined as the minimum volume per unit area required to achieve economic control (Anon 1971). The actual volume will depend on the size of the target.

This chapter illustrates the wide diversity of targets for pesticides. As

Table 2.4 Volume rates of different crops (litres/ha)

	Field crops	Trees and bushes
High volume	>600	>1 000
Medium volume	200–600	500–1 000
Low volume	50–200	200–500
Very low volume	5–50	50–200
Ultra low volume	<5	<50

pesticides are biologically very active, application efficiency can be improved only if, instead of attempting to wet the whole target, the optimum droplet size is selected to increase the proportion of spray which adheres to the targets. More research is needed to define which is the optimum droplet size range collected by particular targets, but certain generalizations can be made. These generalizations are shown in Table 2.5. The trend with herbicides has been to apply large droplets (>250 μm) to reduce the risk of drift, but smaller droplets can be more effective.

Table 2.5 Optimum droplet size ranges for selected targets

Targets	Droplet sizes	
	Diameter (mμ)	Volume (picolitres)
Flying insects	10–50	0.5–65
Insects of foliage	30–50	14–65
Foliage	40–100	33–524
Soil (and avoidance of drift)	>200	>4 189

When a suitable droplet size has been selected, provided an estimate is made of the coverage (droplets/unit area) required, the volume of spray required can be calculated (Fig. 2.18) (Johnstone 1973b). For example, if a spray with 100 μm diameter droplets is applied and 50 droplets/cm^2 is required, then the minimum volume required is 2.5 litres/treated ha. If even-sized droplets could be produced the minimum volume required to achieve a droplet pattern with a density of 1 mm^2 is shown in Table 2.6 (Bals 1975b). Theoretically, very small volumes of spray per hectare are needed when applying droplets less than 100 μm in diameter (ie less than 524 picolitres per droplet). The application of more uniform droplet sizes is considered further in Chapter 8. Use of the appropriate droplet size and droplet density according to the target in this way is referred to as controlled droplet application (CDA) – CDA stresses the need for a narrow droplet size range (Matthews 1977 1978).

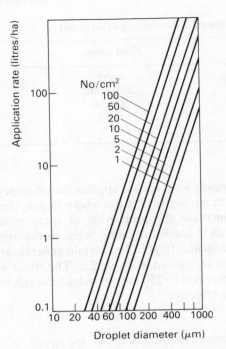

Fig. 2.18 Relation between droplet numbers, diameter and volume application rate
(after Johnstone 1972)

The target area requiring treatment may be much greater than the ground area, although most recommendations in the past have referred to ground area only. Martin (1958) and Way et al (1958) emphasize the importance of defining dosage per unit of plant surface ($\mu g/cm^2$) rather than dosage per ground area of crop. Some attempts to relate spray volume to target area have been made; thus Morgan (1964) referred to selection of spray volume in relation to the size of trees in orchards. Tunstall et al (1961) increased the volume of spray applied on cotton according to plant height (see p. 139). For plants less than 30 cm in height, 56 litres/ha was applied, and for each subsequent increase in height of 30 cm, the volume was increased by 56 litres/ha up to a maximum of 280 litres/ha. Similarly, Matthews (1971) recommended a reduction in swath width as cotton plants grew. Where spray coverage of leaves is required, ideally an estimate of the leaf area index (LAI) is required. The LAI is the ratio of leaf area to ground area and will vary with different crops according to the stage of plant growth. The LAI seldom exceeds about 6–7, as leaves without adequate light are usually shed. Thus, if the LAI is 3 and 2.5 litres is needed per hectare of foliage, then the total volume per ground hectare applied to that crop will be 7.5 litres.

Obviously such calculations of spray volume assume uniform distribution

Table 2.6 Minimum spray volumes

Droplet diameter (μm)	Spray liquid required (litres/ha) for density of 1 droplet/mm2 applied evenly to a flat surface
10	0.005
20	0.042
30	0.141
40	0.335
50	0.655
60	1.131
70	1.797
80	2.682
90	3.818
100	5.238
200	41.905
500	654.687

of the spray liquid to achieve the required droplet density. If distribution is uneven, the volume may have to be increased, but an increase in volume does not necessarily result in improved coverage. If a sprayer is used with a particular set of nozzles and spray volume is increased, most of the deposit will be increased in the most exposed target areas and there will be little improvement at other, more concealed, sites. Courshee (1967) illustrated this by plotting the cumulative percentage of targets (leaves) with different deposit densities (Fig. 2.19, line A). Doubling of spray volume or mass application rate could be expected to double the deposit density of each leaf (line B) but in practice, distribution curves illustrated in line C are more usually obtained. Three volume and mass application rates were examined on cotton using a knapsack sprayer with up to eight cone nozzles per row on a tail-boom. Doubling the mass application rate (ie spray concentration increased from 0.5 to 1.0 per cent) or increasing the volume by 50 per cent, failed to increase the yield significantly. In contrast, significant decreases occurred with a 0.1 per cent spray or if the volume was reduced to 67 per cent of the recommended volume (Table 2.7) (Matthews and Tunstall 1966). The lower dosage was inadequate due to the effects of weathering and dilution of the spray deposit by plant growth (Matthews 1966).

An increase in the number of points of emission in relation to the target by using more nozzles and/or narrower swaths can achieve more uniform distribution. In experiments on cotton with a spinning disc nozzle held above the crop, a high spray concentration (25 per cent) on 4.5 m swaths even at the dosage rate of 0.5 kg ai (active ingredient)/ha gave less control of the bollworm *Helicoverpa* than lower concentrations (5 per cent) applied on narrower swaths (0.9 m), although the volume applied did not exceed 10 litres/ha (Matthews 1973). Using such small volumes of liquid does not imply

Table 2.7 Effect of varying spray volume and concentration on yields of seed cotton.

	Concentration %	Volume litres/ha	Yield kg/ha
Recommended spray	0.5	56–227	3123
Doubling concentration	1.0	56–227	3221
Increasing volume by 50%	0.5	84–340	3138
Reducing concentration by 80%	0.1	56–227	2228
Reducing volume by 33%	0.5	37–150	2819

the need to spray concentrated formulations. If the LD_{50} contained in a single droplet can be determined, the concentration of the spray required in controlled droplet application can also be calculated (Fig. 2.20). When a high proportion of the spray reaches the target, a reduction in the total dosage compared with recommended dosages per unit of ground area may often be possible. This has been shown in particular with the 'Electrodyn' sprayer (see

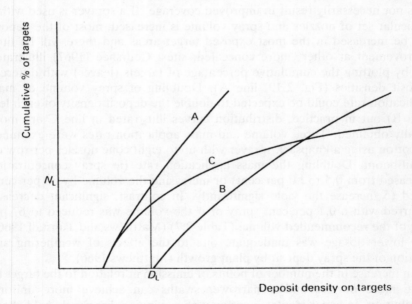

Fig. 2.19 Hypothetical deposit distribution curves on foliage. A typical distribution of doses on targets shown by curve A. If deposit on each target were doubled by doubling the application rate, curve B would be obtained, but in practice the heavy deposits are increased while many leaves continue to receive an inadequate deposit – curve C. The minimum deposit needed may be that indicated at point D (from Courshee 1967)

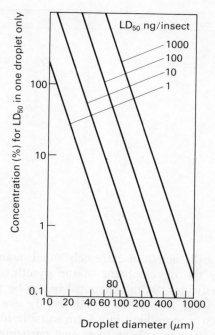

Fig. 2.20 Relation between toxicity, droplet diameter and concentration of active ingredient for one droplet to contain the LD_{50} (from Johnstone 1972)

Chapter 9) where the dosage of some insecticides can be significantly reduced compared with hydraulic sprays (R. Smith 1989). In general, the concentration of active ingredient in sprays for CDA should seldom exceed 10–20 per cent. The application of 1 kg/ha is equivalent to 100 nanograms/mm². The lethal dose for some insects may be as little as 1 nanogram of certain insecticides; thus in some circumstances there could be 100 times overkill.

3

Formulations

Pesticides are biologically active in extremely small quantities, and this has been accentuated by the development of the pyrethroid insecticides and other more active pesticides, so the chemical has to be prepared in a form that is convenient to use and to distribute evenly over large areas. The preparation of the active ingredient in a form suitable for use is referred to as 'formulation'. Manufacturers have their own particular skills in formulation, details of which are a closely guarded secret because of competition from rival companies. Flanagan (1983) gives a general overview of the principles of formulation.

Most pesticides are formulated for dilution in water, and as the measuring of the product and transfer to the sprayer brings the user into the closest contact with the active ingredient, improvements in packaging, including the use of water-soluble plastic containers, have aimed at reducing the risk of spillage and operator contamination. Changes in equipment have also made it easier to load spray liquids into the tank, by using closed filling systems or providing low level mixing facilities. Some formulations can be applied directly without dilution at ultra low volumes, but toxicological, technical and economic considerations limit the number of chemicals which can be used in this way. Many pesticides have been marketed as wettable powders, owing to the high cost of suitable solvents for the active ingredient, but with concern about the risk of inhalation hazards, research has led to new formulations using dispersible grains or suspension concentrates (Seaman 1990).

Types of formulation

A range of different formulations is usually available for each active ingredient to suit individual crop-pest and regional marketing requirements.

Differences between formulated products of one manufacturer may be due to the availability of solvents, emulsifiers or other ingredients at a particular formulation plant. Registration requirements also influence the availability of certain formulations.

Dry formulations

Dust

Dust is a general term applied to fine dry particles usually less than 30 μm diameter. Most dust formulations contain between 0.5 and 10 per cent of active ingredient. Transport of large quantities of inert filler is expensive so a manufacturer may ship more concentrated dusts that are diluted before use in the country importing them. Sulphur dust is applied against some pathogens without dilution.

The concentrate is prepared by impregnating or coating highly sorptive particles with a solution of the pesticide. Alternatively, it may be made by mixing and grinding together the pesticide and a diluent in a suitable mill. The concentrate is then mixed usually with the same diluent to the strength required in the field (Table 3.1). Diluent fillers with high surface acidity,

Table 3.1 Examples of dust formulations

	%
(a) **Dust concentrate**	
50% Sevin (Carbaryl) dust concentrate	
Technical carbaryl	50.5
a Montmorillonite clay, e.g. Pikes Peak clay	49.5
	100.0
(b) **Field strength dust (by dilution)**	
5% Sevin dust	
50% dust concentrate	10.2
a Montmorillonite clay, e.g. Pikes Peak clay	9.8
a Kaolinite clay, e.g. Barden A.G. clay	80.0
	100.0

alkalinity or a high oil absorption index need to be avoided as the formulation would be unstable. Suitable materials for the diluent or carrier are various clay minerals such as attapulgite, often referred to as fuller's earth, montmorillonite or kaolinite (Watkins and Norton 1955). Forms of silica or almost pure silica such as diatomite, perlite, pumice or talc are also used. Diatomite is composed of the skeletons of diatoms and, like all the other materials mentioned above, except talc, it is highly abrasive to the insect cuticle, and can have an insecticidal effect (David and Gardiner 1950). In the tropics, road dust drifting into hedges and fields is often very noticeable and can upset the balance of insect pests and their natural enemies. Dusts have been used to protect stored grain without an insecticide, but mortality is less as the moisture content is increased (Le Patourel 1986).

The use of dusts has declined as the extremely small particles are prone to drift downwind, and winnowing of independent particles of active ingredient and diluent can also occur (W.E. Ripper 1955; Eaton 1959). Often only 10–20 per cent or less of a dust is deposited on the target (Courshee 1960), so most dusts are now used to treat seeds, to protect horticultural crops grown in long narrow polythene tunnels, where the water in sprays can exacerbate fungal diseases, and in farm stores. Seed can be treated centrally by seed merchants, but the product used in the treatment should contain a warning colour and bitter ingredient to prevent such seeds being eaten by humans, birds or farm animals.

Most dust is removed from foliage by rain, although the very small particles can adhere very effectively to plant surfaces. In some cases redistribution by rain can be advantageous, thus dusts applied to the 'funnel' of maize plants may be washed to where stalk-borer larvae penetrate the stem. However most farmers now prefer to use larger microgranules to control this pest.

Dusts with a low content of active ingredient (0.5 per cent ai) with a short persistence such as malathion and pirimiphos methyl, have been mixed with grain (D.W. Hall 1970), but there is concern about residues, even though surface deposits are removed by washing before the grain is ground into flour and cooked.

Granules

Large, discrete dry particles or granules are used to overcome the problem of drift, although care is essential during application to avoid fracture or grinding of the granules to a fine dust, which could be dangerous if inhaled or touched. This is particularly important as pesticides which are too hazardous to apply as sprays, such as aldicarb, are formulated as granules. These granules may be coated with a polymer and graphite to improve the flow characteristics and reduce the risk of operator contamination. Granules are prepared by dissolving the pesticide in a suitable solvent and impregnating this on to a carrier which is similar to those used in dust formulations, namely attapulgite or kaolin. Other materials which have been used include vermiculite, coal dust, coarse sand and lignin (G.G. Allen et al 1973; Wilkins 1990). Sometimes a powder is made and the granule formed by aggregation (D. Whitehead 1976).

The choice depending on the sorptivity of the material, its hardness, bulk density (Table 3.2) (Elvy 1976). Some granules are designed to disintegrate rapidly on contact with water. Bulk density is especially important in relation to the volume of the product that has to be transported. Like dusts, the concentration of active ingredient is usually less than 15 per cent, so transport

Table 3.2 Properties of dust diluents and granule carriers

			Bulk density (g/dm3)	Specific gravity	pH
Oxides	Silicon	Diatomite	144–176	2.0–2.3	5–8
		Graded sands			
	Calcium	Hydrated lime	448–512	2.1–2.2	12–13
Sulphates		Gypsum	784–913	2.3	7–8
Carbonates		Calcite	769–1073	2.7	8–9
Silicates	1. Talc		480–833	2.7–2.8	6–10
	2. Pyrophyllite		448	2.7–2.9	6–7
	3. Clays	(a) Montmorillonite	608–705	2.2–2.8	6–10
		(b) Kaolinite	480–561	2.6	5–6
		(c) Attapulgite	432–496	2.6	7

costs per unit of active ingredient are high. The rate of release of a pesticide from the granules will depend on the properties of the pesticide, solvent and carrier, but the period of effectiveness is often longer than that obtained with a single spray application. The coating of a granule and the thickness of it can be selected to control the rate of release to increase persistence.

When an infestation can be predicted, a prophyllactic application of granules may be more effective than a spray, especially if weather conditions prevent sprays being applied at the most appropriate time. Uptake of a pesticide by a plant may be negligible if the soil is dry and movement of chemical to the roots is limited, so granules of certain pesticides are more suitable on irrigated land where soil moisture can be guaranteed. Conversely, there may be phytotoxicity under very wet conditions. Granules have been used extensively in rice cultivation where they are broadcast but the main advantage of granules is that they can be placed very precisely, so less active ingredient may be required and there is less hazard to beneficial insects. They are often placed alongside seeds or seedlings at planting, but spot treatment of individual plants is possible later in the season. In Kenya, control of the stalk-borer of maize has been achieved with a 'pinch' of granules dropped down each maize 'funnel' (P. Walker 1976). Banana plants may be treated with granules to control borrowing nematodes. Granules are often applied by hand, but this should be discouraged even if the person wears gloves. Simple equipment with an accurate metering device is available for both placement and broadcast treatments (see Chapter 12). With more precise placement, there is also less hazard to beneficial insects.

Despite the advantage of not mixing the pesticide on the farm, there has been a rather slow acceptance of granule application. One main drawback is that equipment required for granule application is more specialized than a sprayer and, with a smaller range of products available in granule form, farmers are reluctant to purchase a machine with a limited use. Granules are often applied at sowing in which case the applicator has to be designed to operate in conjunction with a seed-drill or planter. Second, development of suitable equipment has been hampered by a lack of research to determine the best means of distributing granules to maximize their effectiveness, especially with herbicides where uniform distribution is essential. Variation in the quality of granules has also caused difficulties in calibrating equipment. Granules have been categorized by mesh size, the numbers indicating the coarsest sieve through which all the granules pass, and on which the granules are retained (Table 3.3), but similar samples may have quite different particle size spectra (Table 3.4) (D. Whitehead 1976). The Agriculture (Poisonous Substance) Regulations in the UK require that not more than 4 per cent by weight shall pass a 250 μm sieve, and 1 per cent by weight through a 150 μm sieve when the more toxic pesticides are formulated as granules (Crozier 1976). The size range affects the number of particles per unit area of target (Table 3.4).

Larger granules (8/15) which fall easily, even through foliage, are used principally for application in the soil or to water surfaces, for example to control mosquito larvae or aquatic weeds. Granules, including soft ones such as bentonite which release the toxicant quickly, have been widely used to control various pests in paddy fields where they can be broadcast by hand. Movement of insecticide is partly by systemic action but also some chemical

Table 3.3 Sieve analysis of two samples described as 8/22 mesh granules

Pass mesh no.	Retained by mesh no.	Percentage of granules in sample	
		A	B
8	12	2	10
12	16	36	60
16	22	42	30

Table 3.4 Estimated number of attapulgite granules per unit area

Mesh size	Particle size (μm)	Calculated no of particles */m² applying 1 kg/ha
8/15	2360–1080	32
15/30	1080– 540	253
20/40	830– 400	817
30/60	540– 246	2712
80/120 (microgranule)	200– 80	78 125

* The number of granules per kg will depend on whether dried or calcined granules of attapulgite are used.
Number of granules per plant can be calculated knowing the plant density.

is carried by capillary action between stems and the leaf bases. With some pesticides, there may also be a localized fumigant effect. Chemical is lost if granules are carried out of the fields in irrigation water, so smaller microgranules (80–250 μm particle size) which adhere to foliage are used for application to rice plants. Size 30/60 granules are normally used for stalk-borer control on maize.

Impregnating fertilizer granules with pesticides has been considered to eliminate the cost of an inert carrier and save time and labour during application. Apart from the possible breakdown of the pesticides when combined with fertilizer, there may be different requirements for timing and placement of toxicants and nutrients, although a broadcast application of certain herbicides plus fertilizer ('herbiliser') has been used with success (Ogborn 1977).

P. Walker (1971) has reviewed the subject of residues following the application of granules.

Dry baits

Pesticides are sometimes mixed with edible products or sometimes with inert materials, usually to form dry pellets which are attractive to pests. Cutworms and locust hoppers have been controlled by using bran as a bait, and banana bait has been used in cockroach control. Baits have also been used to control leaf-cutting ants (Lewis 1972; Phillips and Lewis 1973) and slugs. Maize and rodenticides have been mixed in wax blocks for rat control in palm plantations. Peregrine (1973) has reviewed the use of toxic baits. A major problem with pelleted baits is that they can be eaten by domesticated animals and disintegrate readily in wet weather and are then ignored by the pests. Non-pelleted baits go mouldy very rapidly, but a silicone waterproofing agent can be added to delay mould development. For invertebrate pests such as ants, the bait can be dispersed in the infested area, but mammalian pests may develop 'bait shyness', especially if dead animals are left near a bait station. Pre-baiting or a mixture of poisoned and unpoisoned baits reduces this.

Dry fumigants

Aluminium phosphide compressed in small, hard tablets with ammonium carbonate on exposure to moisture releases the fumigant phosphine, together with aluminium hydroxide, ammonia and carbon dioxide. The tablets can be distributed evenly throughout a mass of grain in stores. Normally, no appreciable evolution of the fumigant occurs immediately, and respirators are not required if application is completed in less than one hour. The exposure period for treatment is usually three days or longer, so precautions must be taken to avoid personnel becoming affected.

Other fumigants such as methyl bromide are supplied as liquefied gases under pressure in special containers, and their use is described in Chapter 14.

Formulations for application as sprays

A few pesticides dissolve readily in water and can be applied as solutions. Examples are the sodium, potassium or amine salts of MCPA and 2,4-D. Owing to insolubility in water, many require formulating with surface-active agents or special solvents.

Wettable powders (WP)

These formulations, sometimes called dispersible or sprayable powders, consist of finely divided pesticide particles, together with surface-active agents that enable the powder to be mixed with water to form a stable homogeneous suspension. Wettable powders frequently contain 50 per cent active ingredient, but some contain higher concentrations. The upper limit is usually determined by the amount of inert material such as synthetic silica (HiSil) required to prevent particles of the active ingredient fusing together during processing in a hammer or fluid energy mill ('micronizer'). This is influenced by the melting point of the active ingredient, but an inert filler is also needed to prevent the formulated product from caking or aggregating during storage. The amount of synthetic silica needs to be kept to a minimum as this material is very abrasive. Apart from wear on the formulating plant, the nozzle orifice on sprayers is liable to erosion, thus increasing application rates.

Wettable powders have a high proportion of particles less than 5 μm and all the particles should pass a 44 μm screen. Ideally, the amount of surface-active agents should be sufficient to allow the spray droplets to wet and spread over the target surface, but the particles should not be easily washed off by rain.

Wettable powders should flow easily to facilitate measuring into the mixing container. Like dusts, they have some extremely small-sized particles, so care must be taken to avoid the powder concentrate puffing up into the spray operator's face. To overcome this problem, the powder can be granulated with a highly water soluble or water absorbing material and binding agent to form dispersible grains or granules (Fig. 3.1) (Wright and Ibrahim 1984) (Table 3.5) which disintegrate rapidly when mixed with water. Bell (1989) describes five methods of forming such granules: extrusion granulation, pan granulation, spray drying, fluid bed granulation and agglomeration.

Most wettable powders are white, so to avoid the risk of confusing powder from partly opened containers with foods like sugar or flour, small packs containing sufficient formulation for one knapsack sprayer load have been introduced in some developing countries (Gower and Matthews 1971). Water-soluble plastic sachets of some products have also been introduced: the whole package is added to the spray tank. The wettable powder should disperse and wet easily when mixed with water and not form lumps. To

Table 3.5 Example of a wettable powder formulation

75% Wettable powder	% wt
Technical	76.5
Inert filler: HiSil 233 (hydrated silicon dioxide)	21.0
Wetting agent, e.g. Igepon T.77	1.5
Dispersing agent, e.g. Marasperse N	1.0
	100.0

Fig. 3.1 Dispersible granule formulation on mixing with water (courtesy Ciba Geigy Agrochemicals)

ensure good mixing some pesticides should be pre-mixed with about 5 per cent of the final volume of water and creamed to a thin paste. When added to the remaining water the pre-mix should disperse easily with stirring and remain suspended for a reasonable period. The surface-active or dispersing agent should prevent the particles from aggregating and settling out in the application tank. The rate of sedimentation in the spray tank is directly proportional to the size and density of the particles (see p. 78). Suspensibility is particularly important when wettable powders are used in equipment without proper agitation. Many knapsack sprayers have no agitator. Suspensibility of a wettable powder suspension is checked by keeping a sample of the suspension in an undisturbed graduated cylinder at a controlled temperature (World Health Organization 1973). After 30 min. a sample is withdrawn halfway down the cylinder and analysed. The sample should contain at least 50 per cent of the pesticide.

Methods of preparing wettable powders have been described by Polon (1973). Some wettable powders contain too much surface-active agent and foam when air is mixed in the spray liquid. Foam within the spray rig may cause intermittent application, and is prevented by keeping air out of the spray system. No more than 10 ml of foam should remain in a 100 ml graduated cylinder 5 min after mixing a sample of spray at field strength. Foam can be dispersed by silicones such as Silcolapse.

Wettable powders should retain their fluidity, dispersibility and suspensibility even after prolonged storage. Containers should be designed to that even if wettable powder is stored in stacks, the particles are not affected by pressure and excessive heat, which may cause agglomeration. The World Health Organization requires tests for dispersibility and suspensibility after the wettable powder concentrate has been exposed to tropical storage conditions. Poor quality wettable powders are difficult to mix and readily clog filters in the spray equipment.

Normally, wettable powder formulations are not compatible with other types of formulation, although some have been specially formulated to mix with emulsions. Mixing wettable powders with an emulsion frequently causes flocculation or sedimentation, owing to a reaction with the surface-active agents in the emulsifiable concentrate formulation. Sometimes a small quantity of an emulsifiable concentrate can be added to a wettable powder already diluted to field strength, but compatibility should always be checked before mixing in the field.

Suspension concentrates (SC)

Farmers generally prefer to use a liquid formulation as it is easier to measure out small quantities for use in closed systems. Furthermore some environmental authorities have restricted the use of certain solvents and surfactants. These factors have led to more interest in suspension concentrates in which a particulate is pre-mixed with a liquid. Initially these colloidal suspensions had a short shelf life as the pesticides sedimented to form a clay deposit which was not easily resuspended. Advances in milling of particles (L.A. Dombrowski and Schieritz 1984) and improved dispersing agents (Heath et al 1984; Tadros 1989) have significantly enhanced the shelf life of aqueous based suspension concentrates, which are often referred to as 'flowables' (Fraley 1984). Mixtures of two pesticides which are not easily coformulated in an emulsifiable concentrate can be formulated as a suspension concentrate.

Emulsifiable concentrates (EC)

An important component in these formulations is the emulsifier, a surface-active agent which is partly hydrophilic and partly lipophilic. A pesticide dissolved in a suitable organic solvent such as xylene or cyclohexanone cannot be mixed with water, since the two liquids form separate layers. The addition of an emulsifier enables the formation of a homogeneous and stable dispersion of small globules, usually less than 10 μm in size, of the solvent in water. The small globules of suspended liquid are referred to as the disperse phase, and the liquid in which they are suspended is the continuous phase. The concentration of many emulsifiable concentrate formulations is usually 25 per cent w/v active ingredient. One of the lowest concentrations available commercially is 8 per cent w/v tetradifon, but manufacturers prefer to use the highest concentration possible, depending on the solubility of the pesticide in a particular solvent. Some pesticides, such as carbaryl, cannot be formulated economically as an emulsifiable concentrate because the solvents in which the active ingredient is soluble are too expensive for field use.

Van Valkenburg (1973) has discussed the factors which affect the stability of an emulsion which involves a complex dynamic equilibrium in the disperse phase-interface-continuous phase system. The stability of an emulsion is improved by a mixture of surfactants as the anionics increase in solubility at higher temperatures, whereas the reverse is true of non-ionic surfactants (van Valkenburg 1973). Becher (1973) lists a number of emulsifiers, together with a numerical value for the hydrophile–lipophile balance (HLB). An unstable emulsion 'breaks' if the disperse phase separates and forms a 'cream' on the surface, or the globules coalesce to form a separate layer. Creaming is due to differences in specific gravity between the two phases, and can cause uneven application.

Agitation of the spray mix normally prevents creaming. Breaking of an emulsion after the spray droplets reach a target is partly due to evaporation of the continuous phase, usually water, and leaves the pesticide in a film which may readily penetrate the surface of the target. The stability of emulsions is affected by the hardness and pH of water used when mixing for spraying and also conditions under which the concentrate is stored. High temperatures and frost can adversely affect a formulation.

Choice of solvent may also be influenced by its flash point so as to reduce possible risks of fire during transportation and use, especially with aerial application. For example, naphthenes are too inflammable for use as insecticide solvents. Emulsifiable concentrates have been applied without mixing in water, but their use as a ULV formulation is not advisable owing to the high volatility of the solvent.

Emulsions pre-mixed with a small quantity of water to form a mayonnaise-type formulation deteriorate in storage so are not used.

Miscible oil formulations are similar to emulsifiable concentrates, but contain an oil in place of, or in addition to, the organic solvent. These products are less volatile and more suitable for applications in hot, dry climates.

Invert emulsions

Use of a viscous invert (water in oil) emulsion has been considered for aerial application of herbicides to minimize spray drift (Pearson and Masheder 1969) but has not been accepted due to the need for specially designed equipment.

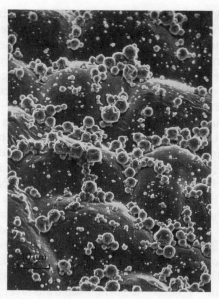

Fig. 3.2 A microencapsulated formulation of a pheromone on the adaxial surface of a cotton leaf (photo ICI Agrochemicals)

Encapsulated pesticides

Microencapsulated formulations have been developed primarily for volatile chemicals, eg pheromones (Fig. 3.2) (D. Hall and Marr 1989) and for controlled release of the pesticide. There are three basic processes:

1 a physical method of covering a core with a wall material
2 a phase separation in which microcapsules are formed by emulsifyng or dispersing the core material in an immiscible continuous phase, or
3 by using the second process followed by an interfacial polymerization reaction at the surface of the core.

These processes are discussed in detail by Marrs and Scher (1990) and Tsuji (1990). The persistence of a deposit can be controlled by varying the wall thickness and type of polymer as well as the size of the microcapsule. Special materials to screen the effect of UV light can be incorporated in the capsules. Microcapsules less than 10 μm diameter have been sprayed very effectively, but the wall thickness relative to the actual capsule size needs to be optimized for specific pesticides and their intended use. Beneficial insects are less exposed (Dahl and Lowell 1984), although it has been argued that bees can collect capsules as their size is similar to pollen grains. Specificity can be increased especially if a suitable attractant is used with a stomach poison, for example in leaf-cutting ant control (Markin et al 1972). Retention of microcapsules applied in suspension in water with stickers such as Acronal

4D is good on foliage even after rain (Phillips and Gillham 1973). In practice slow-release characteristics of microcapsules are particularly useful for application of chemicals which affect the behaviour of insects (Campion 1976). Application of the pheromone disparlure was reported by Beroza et al (1974), dicastalure by Marks (1976) and gossyplure by Campion et al (1989). W.H. Evans (1984) has described a soluble acrylic system for applying the pheromone gossyplure.

Some surfactants solubilized in the aqueous phase of an insecticide microcapsule dispersion can provide a second barrier to the microcapsule wall. This enables the dermal toxicity of a product to be reduced without affecting the release rate of insecticide to such an extent that its activity is reduced to an unacceptable level.

Ultra-low-volume formulations

When small spray droplets are used to achieve effective coverage, the evaporation of the droplets needs to be minimized. The decrease in size of droplets between the nozzle and the target as a result of evaporation is discussed in relation to meteorological factors in Chapter 4.

As mentioned earlier, a few insecticides such as malathion can be applied as the technical material without any formulation, although there is no need for such high concentrations of active ingredients. The number of such pesticides is very restricted so a suitable solvent has to be used. Barlow and Hadaway (1974) investigated a number of solvents to determine which were sufficiently non-volatile for a spray deposit to remain liquid for days or weeks rather than minutes or hours (Table 3.6).

Although meteorological factors considerably influence rates of evaporation, they concluded that a suitable solvent should have a boiling point of at least 300°C at atmospheric pressure.

In addition to low volatility, a solvent suitable for ULV application should have a low viscosity index, ie the same viscosity at different temperatures, should be compatible with a range of chemicals and not be phytotoxic. The specific gravity should be high to increase the terminal velocity of small droplets, and pesticides should readily dissolve in it. Viscosity is particularly

Table 3.6 Volatility of single compounds from cellulose papers at 25°c (from Barlow and Hadaway 1974)

Compound	Boiling point (at 760 mm)	Volatility (g/m²/day)
n–decane	174	2 030
Isophorone	215	290
n–hexadecane	287	2.7
Dibutyl phthalate	340	0.05

Table 3.7 Physical properties of solvents (from Maas 1971) (italic type signifies undesirable characteristics)

		Dissolving power	Volatility	Viscosity	Phytotoxicity
I	Low boiling aromatic hydrocarbons, eg xylene and solvent naphtha	Good	*High*	Low	Low
II	High boiling aromatic hydrocarbons, eg Iranolin KEB	Good	Low	Low	*High*
III	Aliphatic hydrocarbons, eg white spirit, kerosene	*Poor*	*Medium*	Low	Low
IV	High boiling alcohols, eg nonanol	*Medium*	Low	Low	*High*
V	Ketones, eg cyclohexanone	Good	*High*	Low	*Medium*
VI	Special solvents, eg pine oil and tetralin	Good	Low	Low	*High*
VII	Vegetable oils eg cotton-seed oil and castor oil	*Poor*	Low	*High*	Low
VIII	Glycolethers and glycols	*Medium*	Low	Low	Low
Ideal ULV solvent		Good	Low	Low	Low

important in relation to flow rate of liquid to the nozzle. The risk of phytotoxicity is reduced with small droplets. Solvents with all these characteristics are not available (Table 3.7) so a mixture of solvents may be used which overcomes to some extent, a compromise, between persistence and the need for the spray droplet to spread and penetrate an insect cuticle or plant surface. If droplet size is increased to allow for the volatility of one component of a mixture, fewer droplets can be sprayed from a given volume, reducing the coverage of the target, for example because of the cube relationship between diameter and volume of a droplet, doubling the diameter from 75 to 150 μm reduces the number of droplets to one-eighth. Data on the volatility of certain commercial formulations used in aerial spraying of cotton in Swaziland are given by Johnstone and Johnstone (1977). Solvents used in ULV formulations should have no detrimental effects on the application equipment and fabric of aircraft.

Several different mixtures of a low volatility oil and a more volatile solvent have been tried (Coutts and Parish, 1967; Johnstone and Watts 1966). Vegetable oils such as cotton seed and soya bean oil have been in some ULV formulations (Scher 1984).

Special solution formulations of carbaryl, DDT and dimethoate (Maas 1971) were applied successfully on cotton at 2.5 litres/ha, using a sprayer with a spinning disc nozzle (Matthews 1973) and also with aerial application (Mowlam 1974) but phytotoxicity was evident on foliage if droplets were too large. The high cost of these special formulations has limited their use, so products diluted in water have been used at 10–15 litres/ha, ie very low volume (VLV) on narrow swaths with the spinning disc held close to the crop (Mowlam et al 1975). In some areas, molasses has been added as an anti-evaporant and the volume of water reduced to 5 l/ha (Gledhill 1975).

Farmers using these VLV techniques still have to prepare the spray as with conventional hydraulic spraying, but are using a more concentrated spray. While this is suitable for some pesticides, the original concept of using ULV treatments was to eliminate mixing on the farm. In some situations pre-packaged products are available to fit directly on the sprayer without any further mixing. Progress in this direction requires closer collaboration between the equipment and chemical manufacturers to match the pesticide product with the sprayer. Products suitable for electrostatic spraying are referred to in Chapter 9. Other products, including several herbicides, are marketed for spinning disc sprayers. These products usually contain either a vegetable oil or refined mineral oil. The latter are selected with a minimum of unsulfonated residue (UR) of 92 per cent to reduce phytotoxicity. A light mineral oil alone or with a fungicide is used at very low volume (VLV) 10–30 litres/ha to arrest development of banana leaf spot 'Sigatoka' (*Mycosphaerella musicola*) (Klein 1961). Copper fungicide in a heavy alkylate oil was more resistant to field weathering when applied to control angular leaf spot of cucumber (Mabbett and Phelps 1976). Superior deposition was achieved with ULV copper sprays for control of early blight of tomato (Mabbett and Phelps 1974). Carbendazin fungicide mixed with a high-grade paraffinic oil was successfully applied to groundnuts for *Cercospora* control (Mercer 1976). In general, ULV formulations should be checked for phytotoxicity at the proposed field application rate and also at double the rate, using the correct droplet size. Multiple applications may be required to detect any undesirable symptoms which do not show after a single application. Ideally studies on phytotoxicity should include measurements of photosynthesis and respiration in the crop. Excessive rates of application or too large a droplet are likely to have an adverse effect on plants. The lower surface tension of oils allows greater penetration through stomata of certain leaves, and also through lipoidal leaf cuticles. As these oils are such poor solvents, a suitable solvent and cosolvent to dissolve sufficient active ingredient may have to be added. An advantage of mineral oils is that spraying equipment is not corroded or otherwise affected.

Specially formulated oils such as rape seed oil plus emulsifier may be added to other formulations mixed with water, specially for controlled droplet application (G. Barnett 1990). The proportion of oil in the final spray will depend on the volume applied; usually only 1–2 litres of oil per hectare can be used economically. Further evaluation of this type of carrier is needed with a range of chemicals on different crops to establish optimum concentrations and application rates. Less active ingredient may be required against some pests or weeds when formulations based on oils are used because the chemical is spread more effectively on the target and is less likely to be washed off plant surfaces.

Uptake of the active ingredient may be reduced if high concentrations of active ingredient are applied in minimal volumes when users attempt to apply the same dose per unit areas as used in HV sprays. This may be caused by localized toxic effects preventing further absorption of the active ingredient.

Fog formulations

In thermal fogging machines, an oil solution of insecticide is normally used. Kerosene or diesel oil is a suitable solvent provided the solution is clear, and no sludge is formed. If a sludge is present a cosolvent, such as heavy aromatic naphtha (HAN) or other aromatic solvent, with a flash point in excess of 65°C should be used. Consideration of flash point is particularly important to avoid the hot gases igniting the fog. Wettable powder formulations have been used, but are normally mixed with a suitable carrier. Certain carriers are based on methylene chloride and a mixture containing methanol. Pre-mixing the powder with some water is advisable, especially with certain wettable powder formulations, so that a clod-free suspension is added to the carrier. Care must be taken to ensure that the viscosity of the fogging solution allows an even flow, and that powder formulations remain in suspension, as the spray tank on fogging machines is not equipped with an agitator. Pesticides such as pirimiphos methyl which have a fumigant effect are ideally applied as an aerosol spray or fog provided the appropriate concentration is retained for a sufficient time.

Smokes

The pesticide is mixed with an oxidant and combustible material which generates a large amount of hot gas. Water vapour with carbon dioxide and a small quantity of carbon monoxide is produced when a mixture of sodium chlorate and a solid carbohydrate (eg sucrose) is used with a retarding agent such as ammonium chloride. The pesticide is not oxidized, as sugars are very reactive with chlorate. Care has to be taken in the design and filling of smoke generators to avoid an explosion and to control the rate of burning. The high velocity of the hot gas emitted from the generator causes the pesticide to be mixed with air, before condensation produces a fine smoke. The period of high temperature is so short that breakdown of the active ingredient is minimal. Smokes have been used in glasshouses and in warehouses and ships' holds. Care must be taken to avoid the smoke diffusing into nearby offices or living quarters, which should be evacuated during treatment.

A special form of smoke generator is the mosquito coil. The coils are made from an extruded ribbon of wood dust, starch and various other additives and colouring matter, often green, together with natural pyrethrins or allethrin. MacIver (1963, 1964a, 1964b) gives a general description of the coils and their biological activity. Each coil is usually at least 12 g in weight and should burn continuously in a room without draughts for not less than 7.5 hours. Chadwick (1975) suggested that the sequence of effects of smoke from a coil on a mosquito entering a room, is deterrency, expellency, interference with host finding, bite inhibition, knock-down and eventually death. The coils provide a relatively cheap way of alleviating the nuisance from mosquitoes during the night.

Other formulations

Pressure packs

The pesticide is dissolved in a suitable solvent and propellent, in pressure packs described more fully in Chapter 5. Droplet size decreases rapidly after formation of droplets at the nozzle when a very volatile solvent such as xylene is used.

Tar distillates

These contain a mixture of aromatic compounds such as benzene, naphthalene and anthracene, tar bases such as pyradine and tar acids including phenol and cresol. This mixture starts to boil at 210°C, but less than 65 per cent of the volume has boiled at 350°C. Tar distillates have been used as dormant sprays.

Solubilized formulation

Solubilization refers to the mixing of a water-soluble pesticide with oil by using suitable surfactants as cosolvents. The aim is to produce a formulation which increases penetration of the bark or leaf cuticle, but permits translocation of the active ingredient in the plant. The technique has been tried with some herbicides (D.J. Turner and Loader 1974). The effect of leaf-applied glyphosate was improved compared with an aqueous spray, but the cost of oil and solvents was too great.

Banding materials

Localized application of pesticide to the trunk of trees can be achieved by banding. Grease bands have been used to trap insects climbing trees.

Plastic strips

Some volatile insecticides such as dichlorvos have been impregnated on to plastic strips. When hung in a closed room or room with minimal ventilation the concentration of vapour released from the strip is sufficient to control some insects.

Choice of formulation

Formulations have usually been selected on the basis of convenience to the user. Farmers who have large tractor-mounted sprayers fitted with hydraulic agitation prefer liquid concentrates which can be poured into the tank or transferred straight from the container as a volume of concentrate is much easier to measure than to weigh out a powder. Nevertheless, in many parts of the world, the less expensive wettable powder has been used extensively despite the need to pre-mix the powder to form a thin paste before diluting to the correct strength. Pre-packaging selected weights of wettable powder to knapsack or tractor equipment has helped reduce the problem of weighing powders on the farm. Wettable powders have a rather better shelf life than emulsifiable concentrates – an important factor when it is difficult to forecast requirements accurately. Greater use of dispersible granules is expected to replace wettable powders.

Barlow and Hadaway (1947) showed that a particulate deposit was more readily available to larvae walking on sprayed leaves than an emulsifiable concentrate, perhaps because the emulsion was absorbed more readily by the leaf. Large-scale trials on cotton confirmed that by using a wettable powder instead of an emulsifiable concentrate formulation against bollworm, farmers obtained a higher yield at less cost (Matthews and Tunstall 1966). Similarly, DDT wettable powder was recommended for residual deposits on walls of dwellings for mosquito control. With powder formulations, particle size is important, especially with some pesticides such as insect growth regulators. In general, micronization of a formulation provides finer particles which are more effective for contact pesticides than when coarse particles are present. When stomach poisons are applied, surface deposits are effective against leaf-chewing insects, but less so against borers which often do not ingest their first few bites of plant tissue. The effectiveness of stomach poison can be improved by the addition of a feeding stimulant, such as molasses, to these sprays. Carbaryl wettable powder is relatively ineffective against the noctuid *Helicoverpa armigera*, but up to 20 per cent molasses added to the spray gave improved larval mortality and also considerable mortality of moths feeding on the first three nights following application.

Wettable powder formulations of systemic insecticides and fungicides and many herbicides are used, but uptake is sometimes increased by a suitable solvent or oil carrier which can penetrate leaf and seed cuticles more rapidly. Refined white oils have also been added to emulsifiable concentrate sprays applied at HV to improve penetration of the toxicant where the cuticle is particularly resistant to uptake of water-based sprays. Control of scale insects on citrus and other crops is a good example of this, where the addition of a suitable oil improves the effectiveness of an insecticide. Addition of an emulsifier also enhances wetting of certain types of foliage and redistribution of pesticides. This may be very important with some fungicides, and herbicides such as difenzoquat.

Choice of formulation has often been dictated by the availability of equipment in developing countries. Low percentage concentration dusts and granules can often be applied by hand or shaken from a tin with a few holes punched in it, when a sprayer is not available. On the other hand, farmers may be reluctant to use granules where neither labour nor specialized equipment is readily available. Shortage of water in many areas has dictated the use of dusts or granules, but high transport costs have favoured the use of highly concentrated formulations as these are less bulky.

Reduction but not elimination of drift, particularly with aerial application, is assisted by formulations containing thickening agents. Adjuvants of this type have included a polysaccharide gum with thixotropic properties, alginate derivatives, hydroxethyl cellulose and various polymers. The addition of some of these has created mixing problems and increased costs. Drift has been more effectively reduced by changing the spray nozzles to provide a coarser spray preferably with a narrow range of droplet sizes or applying granules.

Choice of formulation may be determined by phytotoxicity – as some plants, or indeed individual varieties, are susceptible to certain solvents and other ingredients – or merely by impurities due to the use of cheap solvents. Phytotoxic effects may be caused by chemical burning, physically by droplets on the plant surface acting as lenses which focus the sun's rays on the plant tissue, or by subsequent effects on plant growth. The persistence of a formulation can be improved by adding 'stickers', but care must be taken to avoid protecting the deposit too much that the availability of it to a pest is reduced. Persistence to rain-washing was improved by formulating wettable powders with amine stearate (Phillips and Gillham 1971). Such formulations were effective on foliage that was difficult to wet – for example cabbage leaves (Amsden 1962) – and were also useful where new growth was insufficient to justify repeat applications. Apart from the use of various additives, including oils, improved rain-fastness can also be achieved with fine particles which are not readily washed off by rain. Even a tropical thunderstorm fails to remove all the dust from surfaces, as electrostatic forces hold the particles to a surface. Advantages of small size and slow release of a pesticide are combined with microencapsulated formulations.

The use of formulations of low concentration reduces the toxic hazard when measuring and mixing the concentrate material with diluent or applying it undiluted. This hazard can be reduced by properly designed and standardized containers which allow the concentrate to be either poured easily or pumped directly into the spray tank. The greatest danger occurs when the spray operator does not wear protective clothing, especially when only a small quantity is required from a large container. Spillage may occur over the operator's hands or feet or a splash may contaminate the eyes or skin. Water-soluble plastic sachets eliminated handling of concentrate formulations, but strong outer packaging is needed to keep the sachets dry until required.

Fig. 3.3 Litre container incorporating measure to dispense small volumes of liquid pesticide formulations

Some liquid products are now packed in containers which incorporate a measure (Fig. 3.3).

Usually the choice of formulation is decided by what is readily available, and the price. In general, the cheapest in terms of active ingredient are wettable powder formulations and those with the highest amount of active ingredient per unit weight of formulation. When assessing costs the whole application technique needs to be considered, since the use of a particular formulation may affect the labour required, the equipment and spraying time. Whichever formulation is chosen, users must read the instructions with great care before opening the container. Manufacturers attempt to provide clear instructions on the label of each container, but limitations of pack size may restrict information on the label, in which case an information leaflet is usually attached to the container. Care must be taken to avoid premature loss of labels, and important information can also be easily obliterated by damage under field conditions. If in doubt about the correct dosage rate and method of use, before starting check alternative sources of information such as the appropriate pesticide manual or crop pest handbook.

4

Droplets

Importance of droplet size in pest management

Sprays contain a large number of droplets, that is very small spheres of liquid, mostly less than 0.5 mm in diameter. Droplet size is highly important if pesticides are to be applied efficiently with the minimum contamination of the environment. Pesticidal sprays are generally classified according to droplet size (Table 4.1). In the UK spray quality is based on assessment of the droplet size spectrum (Doble et al 1985). Certain nozzles are used as standard reference nozzles to indicate each category (Fig. 2.1a). The whole spectrum is considered rather than one parameter, such as the vmd indicated in Table 4.1. With concern about spray drift, which is indicated for coarse, medium and fine sprays in Figure 2.1b, further consideration is being given to nozzle classification by measuring the spray droplets at a set distance from the nozzle mounted in a wind tunnel and at specified wind speeds (P.C.H. Miller et al 1989).

Aerosol sprays are used principally for drift spraying against flying insects. Some aerosols (30–50 μm) and mists are ideal for treating foliage with VLV or ULV rates of application. When drift must be minimized a medium or

Table 4.1 Classification of sprays according to droplet sizes

Volume median diameter of droplets (μm)	Droplet size classification	
<25	Fine aerosol	
26–50	Coarse aerosol	Very fine spray
51–100	Mist	
101–200	Fine spray	
201–300	Medium spray	
>300	Coarse spray	

coarse spray is required, irrespective of volume applied although care is needed when large volumes of a coarse spray from hydraulic nozzles are applied as an appreciable volume of fine droplets can be produced. A fine spray is used when compromise between reduced drift and good coverage is needed. An electrostatic charge on droplets can improve deposition and thus reduce the risk of drift, but small charged droplets can move downwind, especially when a cloud of droplets expands due to the interaction of droplets of the same charge, and then some of the spray gets carried by the wind or thermal upcurrents.

The most widely used parameter of droplet size is the volume median diameter (vmd) measured in micrometres (μm). A representative sample of droplets of a spray is divided into two equal parts by volume so that one half of the volume contains droplets smaller than a droplet whose diameter is the vmd, and the other half of the volume contains larger droplets (Fig. 4.1). A

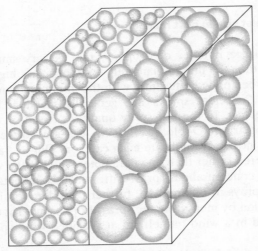

Fig. 4.1 Diagrammatic representation of the vmd – half of the volume of spray contains droplets larger than the vmd while the other half has smaller droplets

few large droplets can account for a large proportion of the spray and so can increase the value of vmd, which on its own does not indicate the range of droplet sizes. The number median diameter (nmd) divides droplets into two equal parts by number without reference to this volume, thus emphasizing the small droplets. Because the measurement of vmd and nmd is affected by the proportion of large and small droplets, respectively, the ratio between these parameters is often an indication of the range of sizes, thus the more uniform the size of droplet the nearer is the ratio to 1 (Table 4.2). Sometimes the range of droplet size is indicated by the 'span' in which the difference in

Table 4.2 Some examples of spray droplet size data for different nozzles

Nozzle	Flow rate	Spray liquids used	rpm or pressure	vmd	nmd	vmd/nmd
Vortical	35 ml/min	Deodorized kerosene	0.2 bar	12		
Electrodynamic	6 ml/min	ED blank	25 kV	48	46	1.04
Spinning cup 52 mm dia.	30 ml/min	ULV	15 000 rpm	70	42	1.67
Fan 80°	200 ml/min	water + wetting agent	300 kPa	99	22	4.5
Fan 80°	800 ml/min	"	300kPa	145	13	11.2
Cone D4/25	1.47 l/min	"	500 kPa	228	42	5.4
Spinning disc 90 mm dia.	60 ml/min	"	2 000 rpm	260		
Airshear	50 ml/min	Risella oil	85 m/s	132	25	5.3
Airshear	400 ml/min	water + wetting agent	85 m/s	282	35	8.1
Airshear	480 ml/min	"	100 m/s	90	24	3.7

Notes: These centrifugal-energy nozzles are described in Chapters 5 and 8. Ideally for CDA the vmd/nmd ratio should be less than 2.0, when sampling droplets with a laser light diffraction (Malvern Particle Size Analyzer)

the diameter for 90 and 10 per cent of the spray by volume is divided by the vmd.

$$\text{Span} = \frac{D_{0.9} - D_{0.1}}{D_{0.5}}$$

Another single parameter, originally favoured by Maas (1971) is the vad (volume average diameter) which can be readily converted to the average droplet volume (adv) expressed in picolitres (Bateman 1989). Dividing 10^{12} by the adv also gives the estimated number of droplets of uniform size that can be obtained from 1 litre of liquid. Other measures of droplet size and dispersion used mainly in relation to nozzles used in industry are described by Lefebvre (1989).

When choosing a given droplet size for a particular target, consideration must be given to the movement of spray droplets or particles from the application equipment towards the target. The magnitude of the effects of gravitational, meteorological and electrostatic forces on the movement of droplets is influenced by the size of the droplets. The size of individual droplets has not always been considered in the past, as most nozzles produce a range of droplet sizes, and when sprays are applied in HV, droplets coalesce to provide a continuous film of liquid on the target. However, there are new nozzles commercially available which enable to the user to select a particular droplet size range. The theoretical droplet density obtained if uniform droplets were distributed evenly over a flat surface is given in Table 4.3.

Table 4.3 Theoretical droplet density when spraying 1 litre evenly over 1 ha

Droplet diameter (μm)	Number of droplets/cm²
10	19 099
20	2 387
50	153
100	19
200	2.4
400	0.298
1 000	0.019

The number of droplets available from a given volume of liquid is inversely related to the cube of the diameter, thus the mean number falling on a square centimetre n, of a flat surface is calculated from

$$n = \frac{60}{\pi}\left(\frac{100}{d}\right)^3 Q,$$

where d = droplet diameter (μm), Q = litres per hectare.

Movement of droplets

Effect of evaporation

The surface area of the spray liquid is increased enormously when broken into small droplets, especially when the diameter of the droplets is less than 50 μm (Fig. 4.2). A droplet will lose any volatile liquid over this surface area. The rate of evaporation decreases as the evaporation from a droplet saturates the surrounding air. Changes in the concentration of spray due to non-volatile ingredients may depress the vapour pressure of the solvent. The disadvantage of water, the principal diluent of pesticidal sprays, is that it is volatile. Many of the organic solvents used in emulsifiable concentrates are also highly volatile.

Studies of freely falling water droplets (>150 μm) have indicated their diameter decreased linearly with time, but below 150 μm, the rate at which the diameter decreased, increased by about 27 per cent (Spillman 1984). It has been shown that the change occurs when the fall Reynolds number is about 4, below which evaporation occurs from the whole surface. At higher Reynolds numbers a toroidal vortex of trapped air can become fully saturated, thus reducing the rate of evaporation from part of the surface (Fig. 4.3).

Batchelor (1967) had shown that the liquid in a falling droplet would follow certain streamlines (Fig. 4.4) from which Spillman (1984) postulated that as volatile liquid evaporated from the surface, the concentration of any involatile component will increase, and if it has a much greater viscosity, the

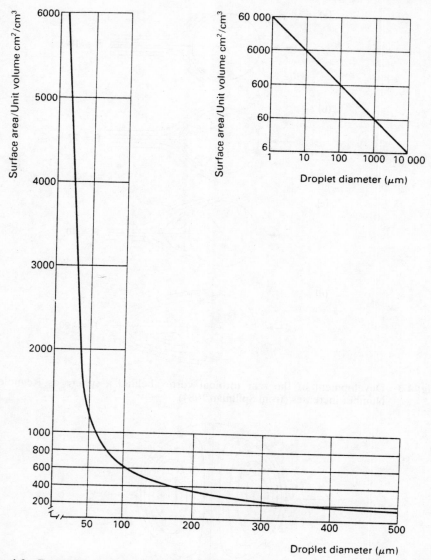

Fig. 4.2 Rate of increase of specific surface or reduction of droplet diameter (after Fraser 1958)

surface velocity will decrease. This can result in a new rigid skin of involatile material over the surface. Studies with molasses suggested the thickness of the skin was 1.5–3 μm for 70–100 μm droplets, containing 10–20 per cent molasses. In-flight encapsulation has also been used with other polymers and oils, thus Wodagenah and Matthews (1981a) reported an increase in oil content of droplets produced with a spinning disc caused a decrease in their

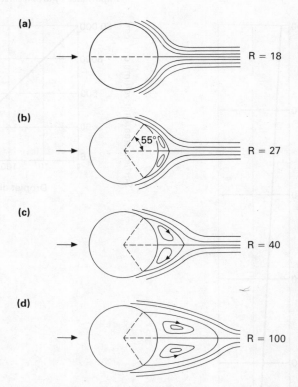

Fig. 4.3 Development of the rear toroidal vortex behind a sphere as Reynolds Number increases (from Spillman 1984)

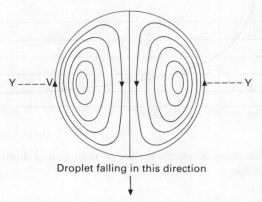

Fig. 4.4 Streamlines of the flow induced by surface friction on a falling droplet (from Batchelor 1967)

initial size due to a lower flow rate with increased viscosity, but the percentage change in their volume while moving 2 m from the nozzle was much less.

The simplest equation to indicate the lifetime of water droplet measured in seconds (Amsden 1962) is as follows:

$$t = \frac{d^2}{80\Delta T},$$

where d = droplet diameter (μm) ΔT = difference in temperature °C between wet and dry thermometers.

Table 4.4 Lifetime and fall distance of water droplets at different temperatures and humidities

| | Temperature (°C) | 20 | | 30 | |
| | ΔT (°C) | 2.2 | | 7.7 | |
	RH(%)	80		50	
Initial droplet size (μm)	Lifetime to extinction		Fall distance	Lifetime to extinction	Fall distance
50	14 s		0.5 m	4 s	0.15 m
100	57 s		8.5 m	16 s	2.4 m
200	227 s		136.4 m	65 s	39 m

This shows clearly that the size of small droplets of water-based sprays decreases rapidly, leaving an aerosol droplet of involatile material or a particle even in temperate conditions of 20°C with 80 per cent relative humidity (Table 4.4). The theoretical distance a droplet of water will fall under the force of gravity (Fig. 4.5) before all the water has evaporated is given by

$$\frac{1.5 \times 10^{-3}d^4}{80\Delta T}\text{cm}$$

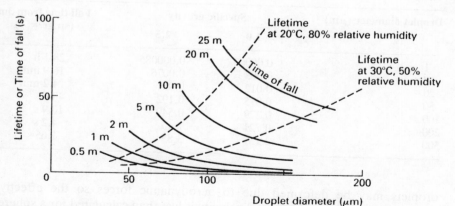

Fig. 4.5 Time of fall from indicated heights (unbroken curves) and lifetime at two ambient states (broken curves) related to droplet diameter (after Johnstone 1971)

The speed at which droplet size decreases is faster, under tropical conditions at a higher temperature and lower humidity. Thus, Johnstone and Johnstone (1977) recommend that spraying water-based formulations at 20–50 litres/ha with 200–250 μm vmd droplets should cease if ΔT exceeds 8°C or the dry-bulb temperature exceeds 36°C, but at lower application rates (10–15 litres/ha) and with smaller droplets (150–175μm vmd), the criteria for ceasing spraying are reduced to ΔT 4.5°C and dry-bulb temperature to 32°C.

Effect of gravity

A droplet released in still air will accelerate downwards under the force of gravity until the gravitational force is counterbalanced by aerodynamic drag forces when the fall will continue at a constant **terminal velocity**. Terminal velocity is normally reached in less than 25 mm by droplets smaller than 100 μm diameter, and 70 cm for a 500 μm droplet. The size, density and shape of the droplet, and the density and viscosity of the air all affect terminal velocity. Thus

$$V_t = \frac{gd^2 \varrho_d}{18\eta}$$

where V_t = terminal velocity (m/s), d = diameter of droplet (m), ϱ_d = density of droplet (kg/m^3), $\mathbf{g}$ = gravitational acceleration (m/s^2), η = viscosity of air in newton seconds per square metre (1 Ns/m^2 = 10 P (poise)) = 181 μP at 20°C. This equation is usually referred to as Stokes' Law.

The most important factor affecting terminal velocity is droplet size. The terminal velocity for a range of sizes of spheres is given in Table 4.5 and is approximately the same for liquid droplets within this range, but larger

Table 4.5 Terminal velocity (m/s) of spheres and fall time in still air

Droplet diameter (μm)	Specific gravity		Fall time from 3m (sp.gr = 1)
	1.0	**2.5**	
1	0.00003	0.000085	28.1 h
10	0.003	0.0076	16.9 min
20	0.012	0.031	4.2 min
50	0.075	0.192	40.5 s
100	0.279	0.549	10.9 s
200	0.721	1.40	4.2 s
500	2.139	3.81	1.65 s

droplets may be deformed due to aerodynamic forces so the effective diameter is reduced and terminal velocity is less than calculated for a sphere.

Owing to their low terminal velocities, droplets less than 30 μm diameter will take several minutes to fall in still air. Examples of the time to fall to ground level when released from a height of 3 m above the target, as in some aerial applications, are given in Table 4.5. Small droplets are thus exposed for a longer period to the influence of air movements. In a light breeze for example a constant wind velocity of 1.3 m/s parallel to the ground, a 1 μm droplet released from 3 m can theoretically travel over 150 km downwind before settling out. In contrast, a 200 μm droplet can settle in less than 6 m downwind if the droplet remains the same size.

If air moved smoothly over flat ground (laminar flow), the distance (S) that droplets travel downwind could be predicted from the equation $S = HU/V_t$ where U is the wind speed, H is height of release and V_t is the terminal velocity. Adjustment of height inversely with wind speed to keep HU constant has been used to deposit droplets of a given size at a fixed distance downwind of the source. This method was originally described by Gunn et al (1948) and is known as the Porton method of spraying.

In practice the airflow is not laminar. Even when spraying over a flat field resistance to air movement increases close to the ground so that wind speed is zero at ground level due to surface friction.

The extent of crop friction is influenced by the topography of the ground and variability in crops. More irregular surfaces in interplanted crops such as maize and groundnuts will have more air turbulence than a field of wheat when all the plants are of similar height.

Johnstone (1971) calculated the proportion of spray deposited at various distances downwind for a range of vmd and HU values (Fig. 4.6). The intersections with broken lines on Figure 4.6 indicate the loss due to evaporation from aqueous sprays at 30°C and 50 per cent relative humidity for several indicated wind speeds and heights. Clearly, drift is minimized with droplets larger than 200 μm. The majority of droplets less than 70 μm are still travelling horizontally with laminar flow when over 100 m downwind. As such droplets are more readily collected on vertical targets, 70–90 μm droplets were selected for drift over the swath and deposit on vegetation in front of locust hoppers to achieve stomach poisoning (Courshee 1959). The distance which droplets travel downwind depends very much on the type of vegetation or other obstacles in their path. Drift may occur over long distances across a flat field, whereas foliage of a field crop can filter out most droplets (Payne and Schaefer 1986). Drift can also be reduced by using an electrostatically charged spray, as described in Chapter 9.

Effect of meteorological factors

The proportion of spray which reaches the target is greatly influenced by local climatic conditions, so an understanding of the meteorological factors affecting the movement of droplets necessitates information on the climate close to the ground. The basic factors are temperature, wind velocity, wind direction and relative humidity.

Air temperature is affected by atmospheric pressure which decreases with height above ground level, so that if a mass of air rises without adding or

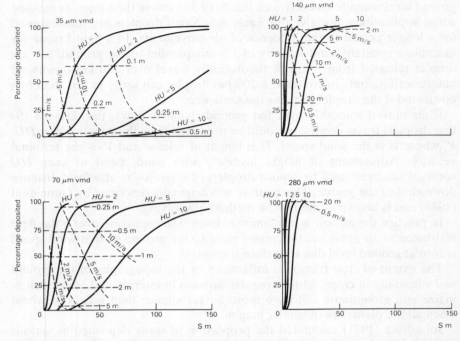

Fig. 4.6 Percentage of spray deposited at different distances downwind for various *HU* values and droplet sizes (unbroken curves). The intersections of the broken and unbroken curves define evaporation loss (read as 100 − percentage deposited) for aqueous spray at 30°C, 50 per cent relative humidity, for several wind speeds (broken curves) and heights (broken lines) (Johnstone 1971)

removing heat, it expands and cools. A decrease in temperature of approximately 1°C for 100 m in dry air is referred to as the adiabatic lapse rate. If the temperature decreases more rapidly, a super-adiabatic lapse rate exists. Under these conditions a mass of air which is close to the ground, warmed by radiation from the sun, will start to rise and continue to do so while it remains hotter and lighter than surrounding air. These convective movements of air result in an unstable atmosphere, and thus turbulent conditions, such as those associated with the formation of thunderstorms when large changes in wind speed and direction often occur (Fig. 4.7b). Turbulence occurs at night under monsoonal conditions.

A temperature decrease less than the adiabatic lapse rate, inhibits upward movement of air so the atmosphere is stable. When the ground loses heat by radiation and cools more rapidly than the air above it, air temperature increases with height and an inversion condition exists (Fig. 4.7a). Inversions typically occur in the evening when there is a clear sky following a hot sunny day and may persist until after dawn and until the sun heats up the ground.

(a)

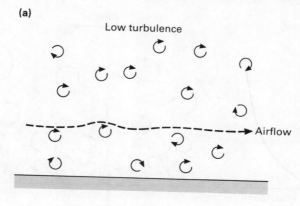

(b)

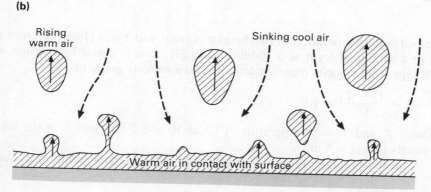

Fig. 4.7 (a) Stable inversion conditions (b) Air turbulence caused by surface heating – super adiabatic lapse rate conditions

Fog or early morning mist occurs during inversion conditions, when wind velocity is low and airflow approaches a smooth or laminar state, that is there is very little turbulence. Irregularities in the ground surface cause masses of air to be mixed by friction and eddies develop (Fig. 4.8). These cause rapidly fluctuating gusts, lulls and changes in wind direction to occur. This mixing of air may destroy an inversion or it may persist at a higher level. Therefore the stability of the atmosphere is affected by the movement of masses of air from convection caused by thermal gradients, and surface friction determined by local topography. Roughness of vegetation, causing resistance to airflow, is one of the factors influencing surface friction. Detailed accounts of micro-meteorology are given by Sutton (1953) and Pasquill (1974).

When studying the dispersal of spray droplets, the most established parameter to express meteorological factors is the dimensionless Richardson number (L.F. Richardson 1920), which combines measurements of wind

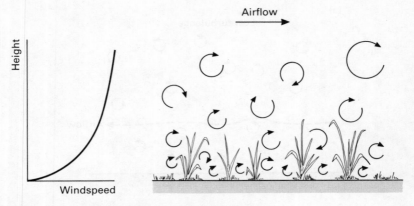

Fig. 4.8 Air turbulence caused by surface friction

velocity and temperature at two heights. Coutts and Yates (1968) referred to it in a simplified form as a stability ratio SR, as it is easier to measure an average wind velocity more accurately than a velocity gradient

$$SR = \frac{T_2 - T_1}{U^2} \times 10^5,$$

where T_2 and T_1 = temperatures (°C) at 10 and 2.5 m and U = the wind velocity (cm/s) at 5 m.

A positive stability ratio indicates temperature-inversion conditions. Such

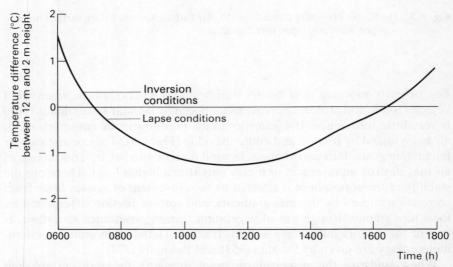

Fig. 4.9 Diurnal variation of temperature lapse between 12m and 2m height – data collected at Makoka Research Station, Malawi (adapted from Johnstone 1972)

conditions are ideal for applying aerosol sprays directed against tsetse flies, so spraying is often done at night. A negative stability ratio occurs when there is turbulent mixing. Normal lapse rate and mild mixing conditions prevail at near-zero values when the stability ratio is rather insensitive. Another limitation is that the gustiness of the wind is not considered.

Suitable apparatus to record the temperature and wind-velocity gradient has been described by Huntington and Johnstone (1973) and Johnstone (1972). The diurnal variation in temperature gradient is shown in Figure 4.9. On cloudy days there may be less convection.

Droplet dispersal

A droplet will follow the resultant direction (V^r) depending on the combined effects of gravity (V^f) mean wind velocity (V^x) and turbulence (V^z) which can be upward when convection forces prevail (Fig. 4.10). This can often be

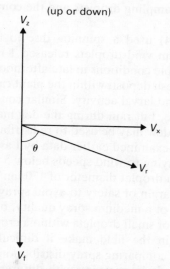

Fig. 4.10 Resultant direction of a droplet (V_r) depending on the magnitude of effects of gravity, wind and convection air movement (after Johnstone et al 1974b)

clearly seen when spray is drifted over several rows during unstable conditions. A large proportion of the spray may not be collected within the crop being treated. Bache and Sayer (1975) found that peak deposition of small droplets downwind was proportional to the height of the nozzle and inversely proportional to the intensity of turbulence, whereas larger droplets are relatively unaffected by turbulence and sediment according to the *HU/V* relationship. Thus, upward movement of small droplets ($<60\ \mu m$) is counter-balanced by downdraughts which return the droplets elsewhere. When relatively small areas are involved (hectares rather than square kilometres) the downdraughts may deposit droplets on a totally different area, contaminating other crops or pastures.

Evidence of this has been clearly demonstrated when an untreated crop, susceptible to a particular pesticide, shows distinctive symptoms of damage; thus 'strap leaf' on cotton due to 2,4-D spraying can be detected often at considerable distances from the site of application, especially if the volatile ester was sprayed. Losses due to evaporation may be so great, with prolonged transport in the air, that the droplets become too small to impact and continue drifting over long distances. Nevertheless, drift techniques can be used to apply small droplets by diffusion over large areas with a non-volatile formulation to avoid droplets getting too small. The choice of chemical is also important, as small quantities of a non-persistent chemical break down rapidly to insignificant residues.

Study of droplet dispersal has been hampered principally by the lack of monodisperse sprays (ie sprays having uniform-sized droplets) in the field, as well as by problems of sampling droplets and the complexity of meteorological factors.

Johnstone et al (1974) used a spinning disc to apply a relatively non-volatile spray with 90 μm vmd droplets released 1 m above a cotton crop. They concluded that stable conditions in late afternoon or early evening were preferable to create a fresh deposit within the plant canopy immediately prior to the period of moth and larval activity. Similar conditions at dawn are also suitable, to spray foliage, but rain during the day may wash off some of the deposit. Early morning dew may be used to redistribute fungicides.

Bache et al (1988) re-examined earlier data by Lawson and Uk (1979) and in terms of limiting spraying to wind speeds below 5 m/s at a height of 10 m, they suggest a minimum droplet diameter of 170 μm for a non-volatile liquid or 220 μm to allow a margin of safety to avoid spray drift. This corresponds to the recommendation of a medium spray quality, but does not relate to the biological effectiveness of small droplets within a crop canopy.

Variable conditions in the field make it difficult to compare different sprays, so some studies comparing spray distribution achieved with different machines have been made by spraying with different-coloured dyes in each machine tested almost simultaneously (Johnstone and Huntington 1977). Parkin et al (1985) used two food dyes – red erythrosine and water blue to ULV and LV aerial sprays. Cayley et al (1987) have suggested using a series of chlorinated esters as tracers, while Babcock et al (1990) used the ninhydrin reaction to quantify deposits of an amino acid.

Other studies on air movement within a crop have been made in a wind tunnel, using scale models of sprayers and trees (Hale 1975). The pattern of airflow in the laboratories simulated field conditions, but such models are unable to scale down droplet size. The technique can examine effects of air volume and velocity produced by different fans in relation to the density of trees, their shape and size. Studies on orchard spraying confirmed that better penetration of the tree foliage was achieved by moving a large volume of air at low velocity than by using less volume in a high-velocity air jet. Laboratory studies may be more suitable for examining the aerodynamical properties of smaller plants. Thus, rice plants wave in the wind much more than maize plants, and may be more efficient in collecting small droplets due to differences in plant form and elasticity.

Spray distribution

The distribution of spray on foliage and other targets can be assessed by adding a fluorescent tracer to the spray and examining deposits under UV light (Fig. 4.11). Insoluble micronized powders such as zinc sulphide, Saturn yellow or Lumogen have been widely used as tracers, but require careful mixing with a suitable surfactant before the spray is diluted to the correct volume. Uvitex 2B has been widely used as it retains sensitivity at low concentrations (0.5 w/v) over extended periods (30–50 days) (G.M. Hunt and Baker 1987). Fluorescent oil-soluble dyes have been added to spray oils to observe spray coverage (Dean et al 1961).

Samples of the target, often leaves, are collected and sorted into various arbitrary categories depending on the amount of cover achieved (Staniland

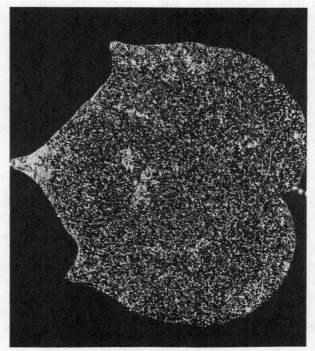

Fig. 4.11 Fluorescent spray deposit on cotton leaf (photo: ICI Agrochemicals)

1959, 1960; Pereira 1967). Spray coverage has also been assessed by measuring the emission of visible light with a fluorometer (Yates and Akesson 1963). Such data are best presented as deposit frequency tables which permit an estimation of the relative areas which are overdosed or underdosed (Stafford et al 1970). Quantitative assessments have been used extensively in the study of spray coverage in orchards by Sharp (1973) who developed methods for handling large numbers of samples, Sharp (1974) has also used Saturn yellow to trace the initial placement of soil-applied herbicides. Uk and Parkin (1983) described a portable surface fluorimeter to assess deposits on leaves in the field but a thick layer of fluorescent material is masked by the surface layer (Cowell et al 1988 a, 1988b). Patterson (1963) examined the effectiveness of different concentrations of tracer (Saturn yellow) in relation to application rate. As a compromise between constant weight per unit area and constant concentration he proposed that the concentration should be altered in proportion to the square root of the spray volume. In metric units,

$$\text{per cent concentration} = 0.25 \sqrt{\frac{561}{Q}},$$

where Q = litres/ha, thus at 100 litres/ha the concentration is 0.6 per cent, which Patterson regarded as an upper limit. Higher concentrations of pigment have been used when less than 100 litres/ha have been sprayed.

Courshee and Ireson (1961) combined subjective assessment of a large number of samples from the field with chemical measurement of deposits on small subsamples taken from each class of spray deposit used in the subjective assessment. This technique was used to assess spray deposits on cotton (Matthews and Johnstone 1968).

Qualitative assessment of the coverage obtained in farmers' fields with wettable powder formulations has been achieved by placing table-tennis balls coated with blackboard paint or pieces of black polythene in the target area. The white deposit readily shows up on a black background. Similarly, dyed spray liquids have been collected on ordinary unpainted table-tennis balls or paper targets.

Detailed qualitative studies of individual droplets on leaf surfaces is possible using scanning electron microscopy and cathodoluminescence (Fig. 4.12) (Hart, 1979; Hart and Young 1987). Detailed spatial distribution of a pesticide can be provided by elemental mapping with a scanning electron microscope with energy dispersive X-ray (EDX) analysis. Fluorescence microscopy and auto-radiography, have also been used to trace deposits (G.M. Hunt and Baker 1987).

Dobson et al (1983) used neutron activation analysis to determine the level of dysprosum in spray deposits within a crop and up to 100 m downwind. The technique has the advantage of sensitivity and speed of analysis at reasonably low cost when compared with quantitative techniques which require expensive solvents for extraction.

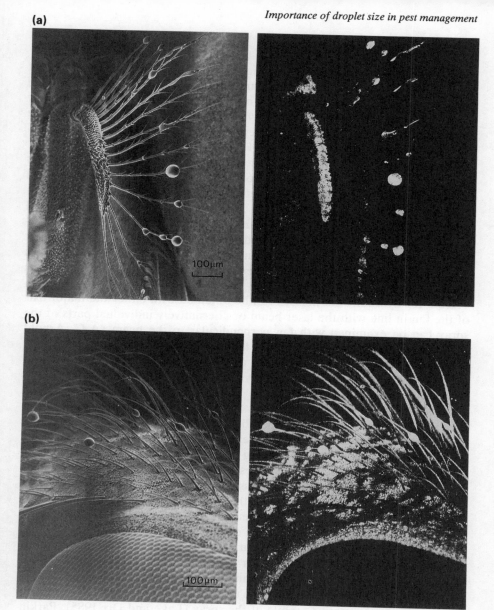

Fig. 4.12 Cathodoluminescence image of 'Ulvapron' oil containing 'Uvitex OB' and Brilliant Yellow R (photo ICI Agrochemicals). Spray droplets on tsetse fly: (a) arista (b) near compound eye

Determination of droplet size

Measurement of spray droplets in flight is now possible with a number of instruments which use a laser beam and sensitive light detectors to provide information which is analysed rapidly by a computer. Two of the most widely used analysers have been the Malvern Particle Size Analyser and the Knollenburg Particle Measuring System (PMS), which use spatial and temporal sampling methods respectively (Fig. 4.13).

Using the Malvern, spray is projected through a laser beam so that light diffracted by the droplets is focused on a special photodetector in the focal plane of the lens (Fig. 4.14). The detector consists of 30 concentric, semi-circular, photosensitive rings which convert the light into an electrical energy signal processed by the computer using a Rosin-Rammler (Swithenbank et al 1977) or a model-independent distribution to calculate the 'best fit' of the measured data. The volume of spray in different size ranges is calculated, the size range being dependent on the focal length of the lens used.

Data for hydraulic fan nozzles can be obtained with either the major axis of the fan in line with the laser beam or alternatively individual parts of the spray can be examined with fan perpendicular to the beam (A.C. Arnold 1983). A standardized positioning of the nozzle relative to the laser beam is essential especially with cone nozzles as the position of the beam relative to the spray cone can very significantly affect the droplet spectra measured (Combellack and Matthews 1981a).

Data can be obtained very easily with the equipment providing useful comparisons between nozzles, operating pressures and formulations (A.C. Arnold, 1983; Combellack and Matthews, 1981b).

The PMS developed by Knollenberg (1970, 1971 1976) to measure particles in clouds was first used to assess a pesticide spray from an air blast orchard sprayer (Reichard et al 1977). The instrument has a laser and lens system that enables droplets passing thrugh the focal plane to form a shadow on a photodiode array (Fig. 4.15). Droplet size is a function of the number of elements obscured during the passage of the droplet. Any droplet that is not in the correct plane produces an out-of-focus pattern so the computer is programmed to accept or reject data depending on the shadow produced. Droplets passing through a small sampling volume are measured over a period of time (temporal sampling) so the nozzle (or probe) has to be moved to obtain a representative sample of the spray (Lake and Dix 1985). Parkin et al (1980) reported on the use of the PMS with particular reference to aerial atomizers mounted in a wind tunnel, and the unit has also been fitted to an aircraft (Yates et al 1982).

Samples of different nozzles were measured by both the Malvern and PMS as well as other systems to provide droplet spectra data to assess spray quality (Doble et al 1985) and provide a nozzle classification system described in the next Chapter 5. A.C. Arnold (1987) has also reported on direct comparisons between the two measuring systems. In general the vmd is significantly

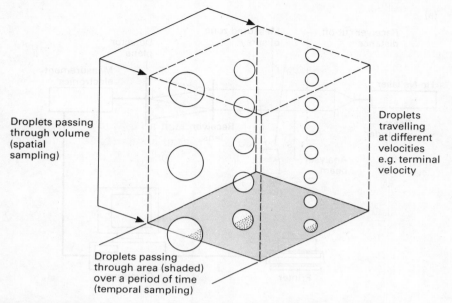

Droplets passing
through volume
(spatial
sampling)

Droplets
travelling
at different
velocities
e.g. terminal
velocity

Droplets passing
through area (shaded)
over a period of time
(temporal sampling)

Fig. 4.13   Diagrammatic representation of spatial and temporal sampling

greater with the PMS, except at very low flow rates. The difference between the systems is due to the effects of droplet velocity (Frost and Lake 1981) thus with a monodisperse spray, both measuring systems should give similar results.

In-line holography can also be used to measure spray droplets in flight (Dunn and Walls 1978) but the reconstruction of the holograms and their analysis is slow compared with other laser techniques.

Another instrument is the Aerometrics Phase/Doppler Particle Analyser (PDPA) which uses an interference pattern formed as a droplet passes through the intersection of two laser beams to measure droplet size. The velocity of the droplet can also be measured using laser/doppler velocimetering (Bachalo et al 1987). As with the PMS, the volume sampled is small so the nozzle has to be tracked through the sampling area to provide a representative sample of the total spray (Western and Woodley 1987).

Another system of measuring aerosol droplets has been used – the principle of a hot-wire anemometer in which the size of the droplets is calculated from measurements of the cooling effects due to evaporation. A 5 μm platinum sensor (Fig. 4.16) is heated to a uniform temperature, and the temperature drop over a length of wire which is related to the droplet diameter, is measured by the change in resistance (Mahler and Magnus 1986). The instrument is portable so can be used to measure spray quality within a crop canopy (Adams et al 1989).

(a)

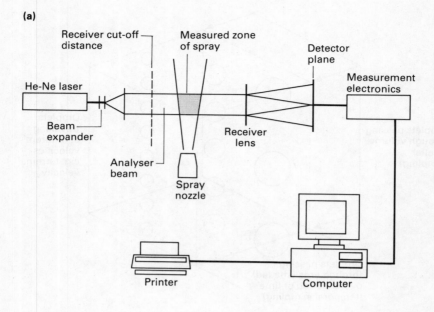

(b)

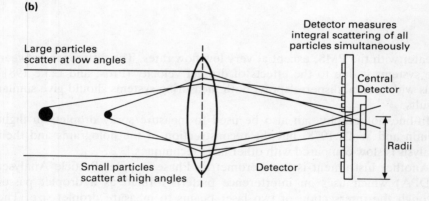

Fig. 4.14 (a) Layout of a Malvern light diffraction particle size analyser (b) Light diffraction of two different sized droplets (c) Photo detector (d) Fourier transform property of receiver lens

(c)

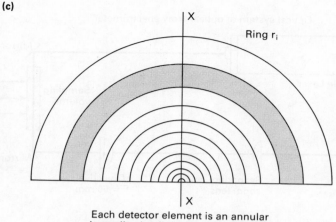

X

Ring r_i

X

Each detector element is an annular
ring collecting light scattered between
two solid angles of scatter w_1 and w_u

(d)

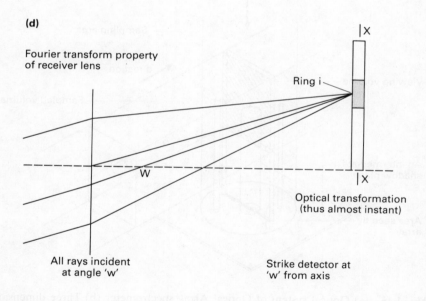

Fourier transform property
of receiver lens

Ring i

W

X

X

Optical transformation
(thus almost instant)

All rays incident
at angle 'w'

Strike detector at
'w' from axis

Sampling surface

If a laser system is not available or droplets are sampled in the field it is
usually necessary to collect droplets on a suitable surface on which a mark,
crater or stain is left by their impact. A standard surface is magnesium oxide,
obtained by burning two strips of magnesium ribbon, each 10 cm in length,
below a glass slide so that only the central area is coated uniformly. The slide
should be in contact with a metal stand to prevent unequal heating of the

(a)

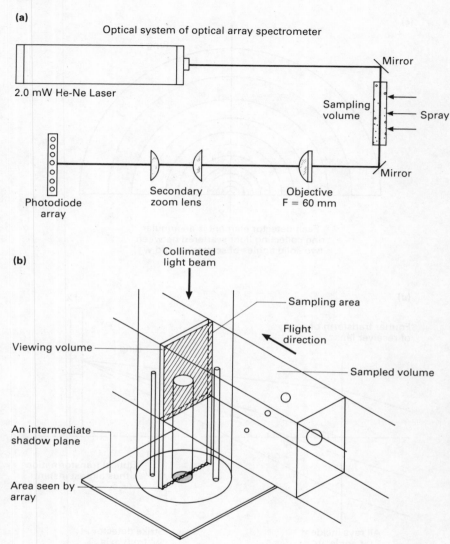

Fig. 4.15 (a) Optical system of Optical Array spectrometer (b) Three dimensional diagram to show shadow effect on detector

glass. On impact with the magnesium oxide a droplet (20–200 μm diameter) forms a crater which is 1.15 times larger than the true droplet size (K.R. May 1950). The difference in size between the crater and true size is the spread factor. The reciprocal of the spread factor is used to convert the measurements of craters (or stains) to the true size; thus for magnesium oxide this factor is 0.86. The factor is reduced to 0.8 and 0.75 for measuring droplets

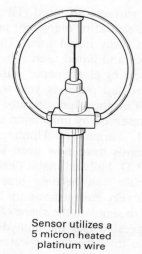

Sensor utilizes a
5 micron heated
platinum wire

Fig. 4.16 Hot-wire droplet sensor

between 15 and 20 μm and 10 and 15 μm diameter, respectively. The magnesium oxide surface is less satisfactory for smaller droplets, and those above 200 μm may shatter on impact. Droplets below 100 μm may bounce unless they impinge at greater than terminal velocity.

Glass slides waved through a spray cloud are not efficient collectors of droplets less than 40 μm in diameter (Mount and Pierce 1972), so sampling of airborne aerosol droplets is either with a cascade impactor which requires a vacuum pump (K.R. May 1945) or by rotating thin rods (0.5–1.6 mm) (C.W. Lee 1974) coated with magnesium oxide, or microscope slides (Thornhill 1979). 'Teflon' coated slides have been used to collect droplets of a relatively involatile spray such as technical malathion (Carroll and Bourg 1979). Alternatively, sampling surfaces can be placed on the floor of a settling chamber, but sufficient time must be allowed for all the small droplets released into the chamber to sediment.

Droplets as small as 5 μm diameter in an involatile spray can also be sampled using a 'harp' in which extremely fine tungsten wires are stretched taut across a frame (McDaniel and Himel, 1977) but the wires are extremely delicate. A.C. Arnold (1980) demonstrated an electrostatic sampler for collecting aerosol droplets in the laboratory.

Water droplets can be collected on a grease matrix, but the droplets must be covered quickly with oil to prevent evaporation reducing their size. A suitable matrix has one part of petroleum jelly and two parts of a light oil (risella oil or medicinal paraffin). No spread factor is needed as the droplets resume their original spherical shape on the surface of the matrix.

The surfaces described so far are difficult to use in the field, so glossy paper such as Kromekote card (A.H. Higgins 1967) or photographic paper

(Johnstone 1960) or other surfaces are used. The spray is coloured with a water-soluble (eg lissamine scarlet or nigrosine) or oil-soluble (eg waxoline red) dye according to the spray liquid used. The spread factor must be determined for each surface and liquid used. The addition of a dye in the spray can be avoided by treating glossy paper with a water sensitive dye, eg bromophenol blue. The treated paper is yellow while it is kept dry but aqueous droplets deposited on the paper produce blue stains of the ionized dye. The papers can be preserved by immersion in ethyl acetate to remove the non-ionized dye (C.R. Turner and Huntington 1970). Commercially available water-sensitive cards have now been widely used to detect the presence of the spray (eg B.D. Hill and Inaba 1989), but unless the cards are kept dry, the whole surface can become blue quite rapidly in humid conditions and the surface very easily shows up blue fingerprints. Tu et al (1986) cut the paper strips to simulate the shape of rice leaves. Babcock et al (1990), however, have suggested that the ninhydrin colour reaction can be used and is more definitive than the use of water sensitive papers, but the samples have to be taken back to the laboratory for processing.

Artificial surfaces are vastly different from natural leaf (Fig. 4.17) or insect

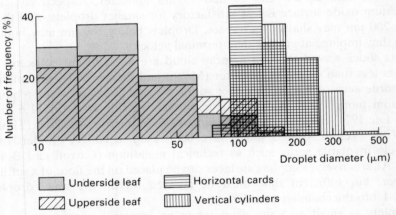

Fig. 4.17 Frequency of droplets of different sizes collected on leaves and artificial surfaces (adapted from Uk 1977)

surfaces (Uk 1977) so, ideally, measurements of droplet size are needed on particular targets. The approximate size of droplets on natural surfaces has been measured by adding a known quantity of fluorescent particles (FPs) to the spray (Himel 1969a). The method is suitable for droplets in the range 20–70 μm if the spray contains a uniform suspension of 2×10^8 FPs/ml, but counting individual particles is very tedious as a doubling of droplet diameter means an eightfold increase in the number of particles (Fig. 4.18).

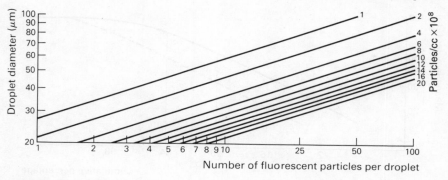

Fig. 4.18 Number of fluorescent patricles (FPs) in droplets of different size according to the concentration of FPs in the spray (after Himel 1969a)

Measurements of droplets

One method is to view the sample of droplets with a microscope fitted with a graticule such as a Porton G12 graticule (Fig. 4.19). The microscope must have a mechanical stage to line up the stains or craters on the sampling surface with a series of lines on the graticule. The distance between these lines from the base line Z increases by a $\sqrt{2}$ progression. A stage micrometer is needed to calibrate the graticule. Use of the graticule is

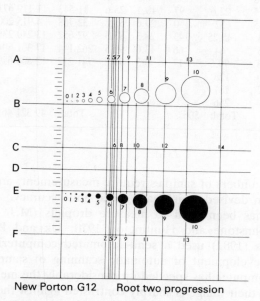

Fig. 4.19 Porton G12 Graticule (courtesy Graticules Ltd)

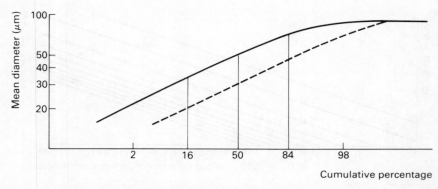

Fig. 4.20 Graph of cumulative percentage against mean droplet size to calculate the volume (solid curve) and number (broken curve) median diameter of a spray

Table 4.6 An example of the calculations required to determine the nmd and vmd (from Matthews 1975a)

Graticule no.	Upper class size (D)	True size upper limit (d)	Mean size (dm)	Number in class (N)	%N	Σ %N	dm³	Ndm³	% Ndm³	Σ % Ndm³
4		13.2								
5		18.8	16	33	6.5	6.5	4 096	135 168	0.3	0.3
6		26.5	22.6	97	19.1	25.6	11 543	1 119 671	2.3	2.6
7		37.5	32	150	29.6	55.2	32 768	4 915 200	10.0	12.6
8		53	45.25	143	28.3	83.5	92 652	13 249 236	26.9	39.5
9		75	64	66	13.0	96.5	262 144	17 301 504	35.1	74.6
10		106	90.5	17	3.3	99.8	741 217	12 600 689	25.5	100.1
11		150	128							
12		212	181							
13	780	300	256							
			Total:	506			Total:	49 321 468		

laborious if large numbers of samples require measurement, and alternative methods have been devised to increase speed and accuracy. Apparatus to shear the image has been used to measure droplets (M.J. Barnett and Timbrell 1962; Johnstone and Huntington 1970; Uk and Parkin 1983). Nguyen and Jarvis (1982) used a semi-automated computerized digitizer system, but the development of automatic scanning of samples using an image-analysing computer has speeded up considerably the measurement of droplets, provided their image is sharply contrasted against the background (C. Fisher 1971; Bond 1974; P.C. Jepson et al 1987; Last et al 1987).

Calculation of number and volume median diameter

An example of calculation of nmd and vmd is shown in Table 4.6. The number of droplets in each class is recorded in column N. The percentage of droplets is then calculated and the cumulative percentage plotted on log probability paper against the mean diameter (Fig. 4.20), which is chosen because not all droplets in any particular size range are as large as the maximum for a given class. As the volume of a sphere is $\pi d^3/6$, and $\pi/6$ is a common factor, the cube of the mean diameter is calculated and multiplied by the number of droplets in that class (Ndm^3). These figures are then expressed as percentages of the total volume of the sample and the cumulative percentages plotted on the same graph (Fig. 4.20). The nmd and vmd are then read at the 50 per cent intersect, and in the example are 30 and 50 μm respectively. A computer program (J.F. Cooper 1991) can be used for calculation of the droplet parameters, but for those without access to a calculator or computer, a graphical method of calculating vmd was published by Amsden (1975).

5

Nozzles

All sprayers have three features in common. Spray liquid is held in a container (spray tank) from which it is moved by pumps, pressure or a gravity-feed system to one or more outlets called nozzles. A nozzle is strictly the end of pipe through which liquid can emerge as a jet. In this book, the term 'nozzle' is used in the wider sense of any device through which spray liquid is emitted, broken up into droplets and dispersed at least over a short distance. Further distribution of spray droplets is influenced principally by natural air movements, although on certain sprayers an airstream is used to direct droplets towards the appropriate target.

Usually the amount of liquid sprayed is metered at the nozzle, but on some sprayers, such as motorized knapsack mistblowers, the metering of liquid is by means of a separate restrictor. The nozzle is therefore one of the most important parts of a sprayer, yet it is often neglected and seldom checked to ensure that expensive chemicals are being applied at the correct rate. Energy is needed to break up the liquid into droplets, and nozzles are generally classified according to the energy used, namely hydraulic, gaseous, centrifugal, kinetic, thermal and electrical (Table 5.1). Electrical energy has been used to produce electrostatically charged droplets, and manipulate their trajectories in relation to the combustion of fuels (Thong and Weinberg 1971), and for the application of pesticides (Coffee, 1979; Matthews 1989). Some nozzles, especially those using centrifugal energy, are often referred to as atomizers, and the process of producing sprays as 'atomization'. A detailed description of atomization and sprays is given by Lefebvre (1989), with special reference to the requirements in combustion technology. There is no universal nozzle, different designs being used to achieve the appropriate droplet spectrum.

Types of Nozzle

Hydraulic energy nozzle

Production of droplets

A large range of hydraulic nozzles has been designed in which liquid under pressure is forced through a small opening or orifice so that there is sufficient velocity energy to spread out the liquid, usually in a thin sheet which becomes unstable and disintegrates into droplets of different sizes. Pressure properties of the liquid such as surface tension, density and viscosity, and ambient air condition, all influence the development of the sheet. A minimum pressure is essential to provide sufficient velocity to overcome the contracting force of surface tension and to obtain full development of the spray pattern. The minimum pressure for most nozzles is at least 1 bar (14 psi) but usually 2–3 bar is required. An increase in pressure opens the sheet and also increases the flow rate in proportion to the square root of the pressure. Flow rate divided by the square root of the pressure differential is equal to a constant, commonly termed the flow number (FN).

Fraser (1958) pointed out that there are three distinct modes of disintegra-

Table 5.1 Different types of nozzle and their main use

Energy	Type	Uses
Hydraulic	Deflector	Low-pressure nozzle with coarse spray; large orifice not liable to clog. Mainly used for herbicide applications.
	Fan*	Spraying flat surfaces, e.g. soil and walls†
	Even-spray fan	Band sprays
	Cone*	Foliage sprays
	Solid stream	Spot treatments
Gaseous	Air blast	Foliage sprays, especially tree crops and bushes
	Vortical	Aerosol space sprays
Centrifugal	Spinning disc or cage*	Application of minimal volumes with controlled droplet size. Slow rotational speeds: large droplets for placement sprays. Fast rotational speeds: mist/aerosols for drift and space spays. Also to fit air-carrier sprayers.
Kinetic		Coarse sprays, especially herbicides
Thermal	Fogging*	Space treatments, especially inside buildings and forests
Electrostatic	Annular Linear*	Ultra low volume electrostatically charged sprays

Notes: * Types of nozzle on aircraft
  † Volume of spray depends on surface, ie run-off occurs at approximately 25 ml/m²

tion of the sheet – perforated, rim, and wavy-sheet disintegration (Fig. 5.1) – but only one mechanism of disintegration in which separate filaments of liquid break up into droplets.

Perforated sheet disintegration occurs when holes develop in the sheet and, as they expand, their boundaries form unstable filaments which eventually break into droplets. In rim disintegration, surface tension contracts the edge of the sheet and forms rims which produce large droplets at low pressure, but at higher pressures threads of liquid are thrown from the edge of the sheet. Rim disintegration is similar to droplet formation from ligaments thrown from a centrifugal energy nozzle. Whereas in perforated sheet and rim disintegration droplets are formed at the free edge of the sheet, wavy-sheet disintegration occurs when whole sections of the sheet are torn away

(a)

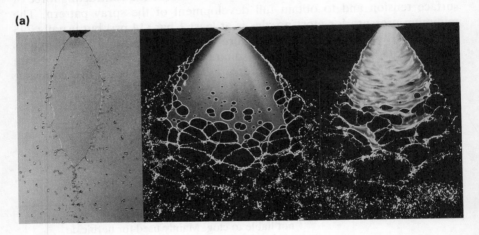

(b)

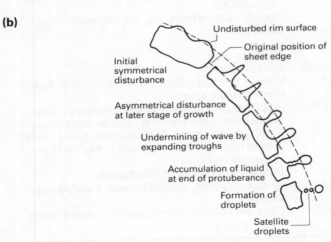

Undisturbed rim surface

Original position of sheet edge

Initial symmetrical disturbance

Asymmetrical disturbance at later stage of growth

Undermining of wave by expanding troughs

Accumulation of liquid at end of protuberance

Formation of droplets

Satellite droplets

Fig. 5.1 (a) Rim, perforated and wavy-sheet disintegration (photos N. Dombrowski, Leeds University) (b) Diagram showing rim disintegration. *Note:* formation of satellite droplets

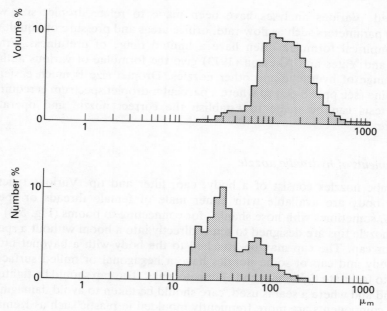

Fig. 5.2 Example of droplet distribution from a fan nozzle from Malvern Particle Size analyser

before reaching the free edge (Clark and Dombrowski 1972). The droplets formed vary considerably in size (Fig. 5.2) in the range 10–1 000 μm, owing to the irregular break-up, so the volume of the largest droplets is a million times that of the smallest. Their average size decreases with an increase in pressure and increases with a larger orifice. The range of sizes is less at the higher pressures, especially in excess of 15 bar. During forward movement of the sprayer, inwardly curling vortices are formed on either side of a flat-fan nozzle so that small droplets are carried in a low energy trailing plume and subsequently more vulnerable to drift away from the intended target (B.W. Young 1991). Reduction in the number of small droplets liable to drift can be achieved by using minimum pressures and a large orifice. Shrouding the orifice with hot gases causes the sheet to break up closer to the nozzle (N. Dombrowski et al 1989) but so far this has not proved to be a practical system for applying pesticides. An increase in surface tension and viscosity of the spray liquid will also result in larger droplets, so various additives such as thickeners, gels and foaming agents have been used to reduce the number of small droplets liable to drift. However, an increase in viscosity with low surface tension may result in a considerable number of small droplets (Yates and Akesson 1973). Viscosity should be less than 15 centistokes for adequate droplet formation. At higher viscosities, the angle of spray emerging from the orifice is decreased, possibly to such an extent that a straight jet is produced. Various analyses have been made to relate droplet size with various parameters such as flow-rate, orifice areas and pressure of liquid, but such empirical formulae often have a limited range of usefulness. Fraser (1958) and Yates and Akesson (1973) give the formulae of various authors for a range of hydraulic and other nozzles. Droplet size is much easier to determine (see pp. 88–97) so where a particular droplet spectrum is required, initial tests can be made to establish the correct nozzle and operating pressure.

Components of hydraulic nozzle

Hydraulic nozzles consist of a body, cap, filter and tip. Various types of nozzle body are available with either male of female threads or special clamps, sometimes with hose shanks, for connecting to booms (Fig. 5.3) and some nozzle tips are designed to screw directly into a boom without a special body or cap. The cap may be attached to the body with a bayonet fitting. The body and cap of some nozzles have a hexagonal or milled surface or wings to facilitate tightening and eliminate leaks. The cap should be tightened by hand and where a seal is used, care should be taken to avoid damaging it. These components are more frequently moulded in plastic such as Kematal. Some nozzles are not provided with a filter, but as spray liquid is readily contaminated with dust or other foreign matter which can block the nozzle tip, a suitable filter should be used in the nozzle body. A 50-mesh filter is usually adequate, except for very small orifice tips when an 80-, 100- or 200-mesh filter may be needed. A coarse strainer normally equivalent to 25-mesh, may be used with large orifice nozzles (Fig. 5.4). The filter may incorporate a small spring and ball valve as an anti-drip device. This type of valve may be affected by chemical corrosion and, as it falls out as soon as the nozzle cap and tip are removed, the spray operator can be contaminated by spray liquid. An anti-drip device, consisting of a diaphragm check valve held in place by a separate cap, is preferred since the nozzle tip can be changed without losing spray liquid from the boom (Fig. 5.5). The diaphragm check valve, which was designed originally for equipment on aircraft to prevent spray liquid dripping outside the treated area, is now widely used on tractor and other spray equipment. The diaphragm is usually made from a synthetic rubber which may be affected by certain chemicals, so a polytetrafluoroethylene (PTFE) disc can be inserted to protect the diaphragm. The efficiency of these valves should also be checked, using a wettable powder suspension in case particles prevent proper seating of the diaphragm. It is also possible to incorporate a spray management valve that will provide a constant pressure as well as act as a check valve (Fig. 5.6).

Most nozzles were manufactured from brass, which is unaffected by a wide range of chemicals and can be readily machined. However, brass is easily abraded by particles. Alternative nozzle tips are made in hardened stainless steel, ceramics and certain plastics such as Kematal. Plastic tips are sometimes more resistant to abrasion than metal tips because moulded tips have a

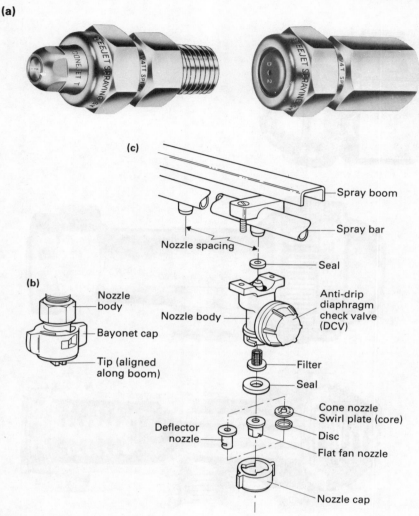

(a)

(c)

Spray boom

Spray bar

Nozzle spacing

Seal

(b)

Nozzle body

Nozzle body

Anti-drip diaphragm check valve (DCV)

Bayonet cap

Tip (aligned along boom)

Filter

Seal

Cone nozzle
Swirl plate (core)

Deflector nozzle

Disc

Flat fan nozzle

Nozzle cap

Fig. 5.3 (a) Hydraulic nozzles – male and female nozzle body (Spraying Systems Co.) (b) plastic nozzle with bayonet cap and diaphragm check valve (c) exploded view of nozzle (Lurmark Ltd.)

Fig. 5.4 Strainer, 50-mesh and 100-mesh filters (Spraying Systems Co.)

(a)

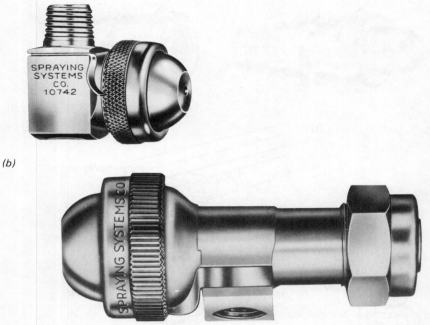

(b)

Fig. 5.5 (a) Diaphragm check valve (Spraying Systems Co.) (b) Diaphragm check valve incorporated into nozzle body (Spraying Systems Co.)

Table 5.2 The BCPC Nozzle Code

Nozzle type	Spray angle	Nozzle output	Rated pressure
F (Fan) HC (Hollow cone) D (Deflector) FE (Even spray) FLP (Low pressure)	Given in degrees if known	Given in litres per minute	Normally 3 bar but 1 bar for deflector and low pressure nozzles

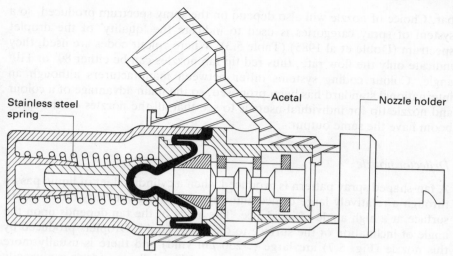

Fig. 5.6 Spray Management Valve (SMV) (Fluid Technology)

smoother finish. The surface of metal tips has microscopic grooves as a result of machining and drilling the orifice; the rough finish presumably causes turbulence and enhances the abrasive action of particles suspended in a spray liquid. The threads of some nozzle bodies and caps manufactured in plastic are easily damaged by constant use, especially if they are over-tightened with a spanner. Various hydraulic nozzle tips are manufactured to provide differences in throughput, spray angle and pattern. The tip and cap of some nozzles are integrated.

Each manufacturer has its own system of identifying different nozzles, including colour coding, so an independent code has been introduced to be recommended without referring to an individual manufacturer. The code uses four parameters to describe a nozzle: the nozzle type, spray angle at a standard pressure, flow rate and the rated pressure (Table 5.2). As an example, F110/1.6/3 refers to 110 degree fan nozzle, 1.6 l/min output at 3

Table 5.3 Spray quality for agricultural nozzles in the UK

Spray quality	Retention on difficult leaf surfaces	Used for	Drift hazard
Very fine	Good	Exceptional circumstances	High
Fine	Good	Good cover	Medium
Medium	Good	Most products	Low
Coarse	Moderate	Soil herbicides	Very low
Very coarse	Poor	Liquid fertilizers	

bar. Choice of nozzle will also depend on the spray spectrum produced, so a system of spray categories is used to indicate the 'quality' of the droplet spectrum (Doble et al 1985) (Table 5.3). Where colour codes are used, they indicate only the flow rate, thus red tip fan nozzles can be either 80° or 110° angle. Colour coding systems differ between manufacturers although an international standard has been proposed, so the main advantage of a colour and nozzle tip for individual users is to see that all the nozzles on the spray boom have the same output.

Deflector nozzle

A fan-shaped spray pattern is produced when a cylindrical jet of liquid passes through a relatively large orifice and impinges at high velocity on a smooth surface at a high angle of incidence. The angle of the fan depends upon the angle of inclination of the surface to the jet of liquid. Droplets produced by this nozzle (Fig. 5.7) are large (>250 μm vmd) and there is usually more spray at the edges of the fan (spray 'horns'). The deflector nozzle is normally operated at low pressures and has been widely used for herbicide application to reduce the number of small droplets liable to drift. When applying herbicides, the spray is normally directed downwards, but the lance can be inverted to direct spray sideways under low branches. The effect of nozzle orientation on the spray pattern has been reported by Krishnan et al (1989). Deflector nozzles have also been used where blockages could occur if a smaller elliptical fan nozzle orifice were used, and also where a wide swath is required with the minimum number of nozzles. They are sometimes referred to as flooding, anvil or impact nozzles. This type of nozzle has been produced in plastic, colour-coded according to the size of orifice. A full circular pattern can be obtained if the side of the nozzle is not shrouded. Deflector nozzles have been used on fixed pipes in citrus orchards to apply nematicides, herbicides and systemic insecticides, metered into the irrigation water, around the base of individual trees.

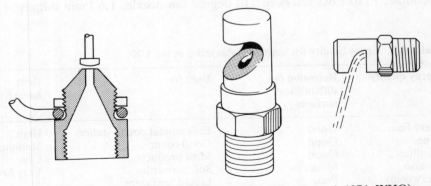

Fig. 5.7 Deflector nozzles (after Equipment for Vector Control, 1974, WHO)

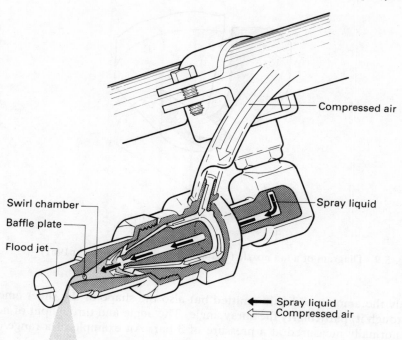

Compressed air

Swirl chamber

Baffle plate

Flood jet

Spray liquid

⬅ Spray liquid
⇐ Compressed air

Fig. 5.8 Deflector nozzle with air assisted atomization (*Airtec Nozzle*, courtesy: Cleanacres Ltd)

A deflector nozzle has also been incorporated into a twin-fluid nozzle so that droplet formation and dispersal are affected by combinations of liquid and air pressure (Fig. 5.8) (Cowell and Lavers 1987). Some of the air is entrapped by the spray liquid to form aerated droplets. Their main use has been in the application of low volumes without too small an orifice liable to blockages. Spray drift from this nozzle was significantly lower than that obtained from flat fan nozzles operated at 100 litres per hectare (Rutherford et al 1989) provided the nozzle is not used at too high an air pressure (>10 bar) or very low flow rates (<0.5 l/min/nozzle) (Western el al 1989), otherwise drift could be exacerbated (Cooke and Hislop 1987).

Fan nozzle

If two jets of liquid strike each other at an angle greater than 90°, a thin sheet is produced in a plane perpendicular to the plane of the jets. The internal shape of a fan nozzle (Fig. 5.9) is made to cause liquid from a single direction to curve inwards so that two streams of liquid meet at a lenticular or elliptical orifice. The shape of the orifice is particularly important in determining not

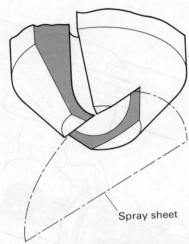

Spray sheet

Fig. 5.9 Diagram of a fan nozzle (adapted from N. Dombrowski 1961)

only the amount of liquid emitted but also the shape of the sheet emerging through it, particularly the spray angle. The angle and throughput of nozzles is normally measured at a pressure of 3 bar. An example of a range of fan nozzles is shown in Table 5.4.

Some farmers prefer to use 110° rather than 80° or 65° nozzles to reduce the number required on a boom or to lower the boom to reduce the effect of drift although droplets are on average smaller with the wider angle. Boom height is very important and computer simulations have predicted more drift from 80° angle nozzles 50 cm above the crop compared to 110° nozzles at 35 cm height (Hobson et al 1990). Boom height can also be reduced by directing the spray forwards instead of directly down into the crop. The spray pattern usually has a tapered edge with the lenticular shape of orifice (Fig. 5.10a), and these nozzles may be offset at 5° to the boom to separate overlapping spray patterns and avoid droplets coalescing between the nozzle and target. Great care must be taken to ensure that all the nozzles along a boom are the

Table 5.4 Examples of flat fan nozzles

Nozzle code	Flow rate (l/min) at			Spray quality at 3 bar	Volume Rate (l/ha) at 3 bar with 50 cm space between nozzles at 8 km/h forward speed
	2 bar	3 bar	4 bar		
F110/0.40/3	0.326	0.40	0.46	fine	60
F110/0.80/3	0.653	0.80	0.924	fine	120
F110/1.20/3	0.980	1.20	1.385	fine	180
F110/2.00/3	1.632	2.00	2.309	medium	300
F110/3.20/3	2.612	3.20	3.693	coarse	480

same and to ensure that they are spaced to provide the correct overlap according to the boom height and the crop which is being sprayed. Details of the position of nozzles and boom height on tractor sprayers are given in Chapter 7. Fan nozzles are ideal for spraying 'flat' surfaces such as the soil surface and walls. They have been widely used on conventional tractor and aerial spray booms and on compression sprayers for spraying huts to control mosquitoes (Gratz and Dawson 1963). Low-pressure fan nozzles provide the same throughput and angle of a conventional fan tip, but at a pressure of 1 bar instead of 3 bar (Bouse et al 1976). A narrow band of spray requires a rectangular spray pattern when herbicides are applied to avoid under-dosing the edges of the band, so a fan nozzle with an 'even-spray' pattern is required (Fig. 5.10b), especially with pre-emergence herbicides. A new series of 'Lo-

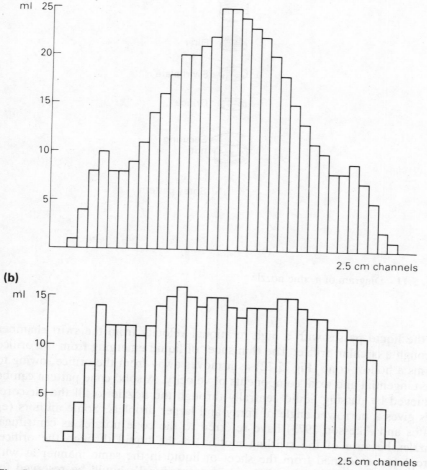

Fig. 5.10   (a) Spray pattern with a fan nozzle. (b) Spray pattern with an even-spray nozzle

drift' fan nozzles has an insert on the input side which, with a slight change in design of the actual tip, reduces the percentage of spray volume in droplets smaller than 100 μm and the volume median diameter is increased.

Nozzles with an offset orifice can also be used, for example at the end of a boom to increase its effective width.

Cone nozzle

Liquid is forced through a swirl plate, having one or more tangential of helical slots or holes, into a swirl chamber (Fig. 5.11). An air core is formed

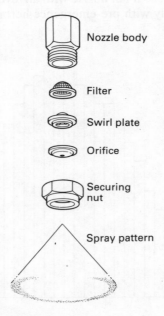

Nozzle body

Filter

Swirl plate

Orifice

Securing nut

Spray pattern

Fig. 5.11 Diagram of a cone nozzle

as the liquid passes with a high rotational velocity from the swirl chamber through a circular orifice. The thin sheet of liquid emerging from the orifice forms a hollow cone (Fig. 5.12) as it moves away from the orifice, owing to the tangential and axial components of velocity. A solid cone pattern can be achieved by passing liquid centrally through the nozzle to fill the air core; this gives a narrower angle of spray and larger droplets. Some authors (eg Yates and Akesson 1973) have referred to the cone nozzles as centrifugal nozzles because of the swirling motion of the liquid through the orifice. Droplets are formed from the sheet of liquid in the same manner as with other hydraulic nozzles, and the term 'centrifugal' should be reserved for those nozzles with a rotating surface (spinning disc).

(a)
(b)

(c)

Fig. 5.12 (a) Solid cone nozzle, (b) Hollow cone nozzle – disc type, (c) Hollow cone – 'ConeJet' type (Spraying Systems Co.)

A wide range of throughputs, spray angles and droplet sizes can be obtained with various combinations of orifice size, number of slots or holes in the swirl plate, depth of the swirl chamber and the pressure of liquid. Some manufacturers designate orifice sizes in sixty-fourths of an inch; thus D2 and D3 discs have orifice diameters of 2/64 in (0.8 mm) and 3/64 in (1.2 mm), respectively.

Reducing the orifice diameter, with the same swirl plate and pressure, diminishes the spray angle and throughput (Table 5.5). The smaller the openings are on the swirl plate, the greater the spin given to the spray. Also

Table 5.5 Effect on throughput and spray angle of certain combinations of disc and swirl plate of hollow–cone nozzles

Orifice	Orifice diameter (mm)	Swirl plate	Pressure (bar) 1.03		Pressure (bar) 2.8	
			Throughput (litres/min)	Angle (°)	Throughput (litres/min)	Angle (°)
D2	1.04	13	0.22	41	0.30	67
		25	0.38	32	0.61	51
		45	0.49	26	0.76	46
D4	1.60	13	0.31	64	0.45	79
		25	0.68	63	1.10	74
		45	0.83	59	1.36	69
D6	2.39	25	1.06	77	1.67	85
		45	1.32	70	2.20	79

a wider cone and finer spray is produced with a smaller swirl opening. An increase in pressure for a given combination of nozzle and swirl plate increases the spray angle and throughput. On some cone nozzles the swirl slots are cut on the back of the orifice disc, but usually the orifice disc and swirl plate are separate parts. The depth of the swirl chamber between the swirl plate and orifice disc can be increased with a washer to decrease the angle of the cone and increase droplet size.

Flow rate of the nozzle can also be adjusted if some of the liquid in the swirl chamber is allowed to return to the spray tank or pump suction. The use of these by-pass nozzles has been investigated by Bode et al (1979) and Ahmad et al (1980 1981).

For a given output and pressure, a cone nozzle produces a finer spray than the equivalent fan nozzle.

Variable-cone nozzles are available in which the depth of the swirl chamber can be adjusted during spraying, but this type of nozzle is suitable only when a straight jet or wide cone is needed at fairly short intervals as intermediate positions cannot be easily duplicated.

Cone nozzles have been used widely for spraying foliage because droplets approach leaves from more directions than in the single plane produced by a flat fan. A second swirl chamber can be positioned immediately after the orifice (Fig. 5.13) to reduce the proportion of small droplets produced by

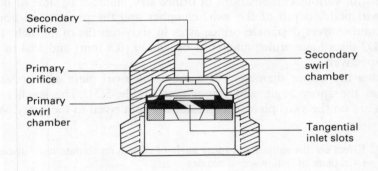

Fig. 5.13 Diagram of 'Raindrop' nozzle

cone nozzles and thus reduce drift. Air is drawn into this second swirl chamber and mixes with swirling liquid, the next result of which is the production of larger, aerated droplets. The additional swirl chamber on a nozzle operated at 2.8 bar can reduce the proportion of droplets less than 100 μm diameter from over 15 per cent to less than 1 per cent (Brandenburg, 1974; Ware et al 1975). This type of nozzle is used for application of herbicides.

Plain jet or solid stream nozzle

This nozzle is similar to a cone nozzle but without a swirl chamber, and sometimes may have more than one orifice. It is used for various purposes including the spot treatment of weeds, young shrubs or trees with herbicide, and has been used to project spray to pods high in the canopy of cacao trees. This type of nozzle is used to apply molluscicides to control vectors of schistosomiasis to ponds and at intervals along canals where there is insufficient flow of water to redistribute chemical from a point source at the head of the canal. A long thin plastic tube attached to a solid stream nozzle has been used to inject pesticides into cracks and crevices for cockroach control.

Foam or air-aspirating nozzle

Air is drawn into and mixed with the spray liquid before it is delivered through the nozzle orifice (Fig. 5.14). Foam is produced, provided a suitable

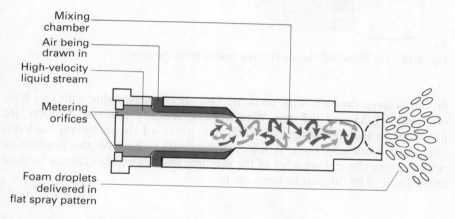

Fig. 5.14 Diagram of a foam nozzle

chemical foaming agent has been added to the spray liquid. Foam nozzles, to produce fan- and cone-shaped patterns, and special aircraft nozzles are available and are easily interchanged with other hydraulic energy nozzles. Up to 60 per cent reduction in spray drift has been claimed, but the technique does not eliminate small droplets and has not been widely accepted. The main advantage is that foam droplets can be seen easily, hence their use in swath marking (see pp. 173–5). An air-aspirating nozzle producing spray without foaming adjuvant resulted in significantly greater spray recovery than a low-pressure fan nozzle (Bouse et al 1976).

Microfoil nozzle

A needle nozzle (Fig. 5.15) for use on aircraft, particularly helicopters,

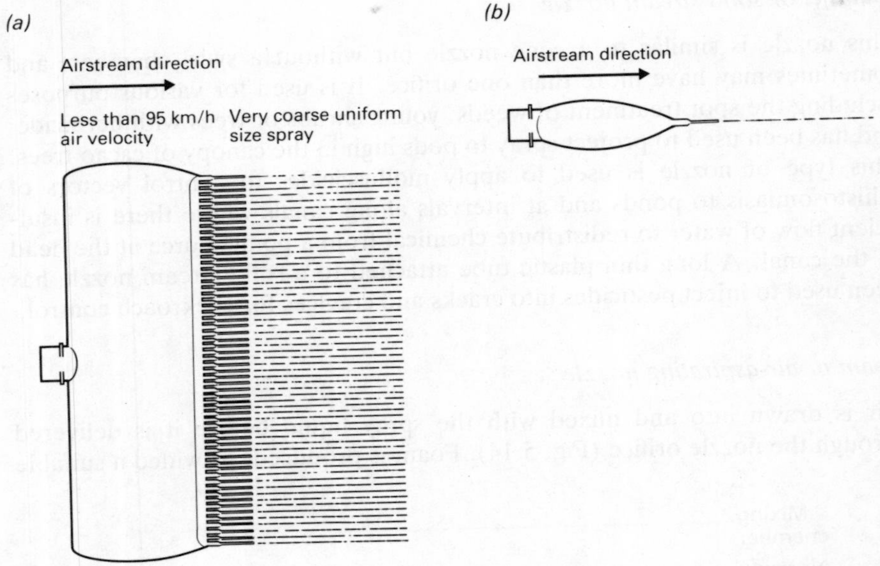

(a)   (b)

Airstream direction   Airstream direction

Less than 95 km/h   Very coarse uniform
air velocity   size spray

Fig. 5.15   (a) Microfoil Nozzle (b) The nozzle from the side

produces large droplets with no drift. The nozzle is available with two sizes of needle (0.33 and 0.7 mm diameter). Practically uniform droplets are produced in the size range 800–1000 μm, provided the airstream does not exceed 95 km/h and there is a low-pressure (0.14 bar) on the needle-like orifices along the trailing edge of the nozzles. Smaller needles cannot be used as they would be difficult to keep open.

Gaseous energy nozzle ('twin-fluid')

Disintegration of liquids into droplets can be achieved by impacting one fluid, the liquid containing pesticide, with another fluid, usually an airstream. One of the simplest types of twin-fluid nozzle was used in the 'Flit gun' (Fig. 5.16) now largely replaced by the pressure pack or 'aerosol' can. The flow of liquid is independent of the airflow so liquids with viscosities as high as 50 centistokes can be applied with this type of nozzle. This is particularly important for certain formulations such as technical malathion. Many twin-fluid nozzles have been designed specially for industrial uses, for example spray-drying of milk and other products and paint spraying. Twin-fluid nozzles may be classified as internal or external mixers, depending on whether the gas and liquid meet inside or outside the nozzle body but, with internal mixing, application is more difficult to control as the pressures of the two liquids react with each other. Some twin-fluid nozzles used for applying pesticides have a straight jet of air at high velocity over an orifice, as used on

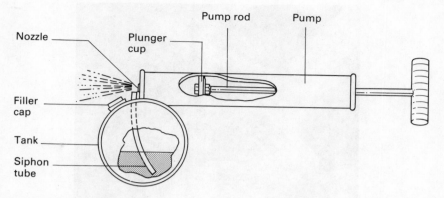

Fig. 5.16   Flit gun

motorized knapsack mistblowers (Figs. 5.17 and 5.18). A negative pressure at the orifice allows liquid to flow into the airstream owing to a venturi effect, and the jet of liquid emerging through the orifice is then sheared into droplets. Liquid is normally fed at low pressure (approximately 0.2 bar) to the nozzle through a suitable restrictor, instead of relying entirely on the venturi effect. The most important variable for droplet size control is the air/liquid ratio, thus larger droplets are obtained with increased liquid flow (Table 5.6) and a reduction in air velocity. Liquid should be spread into a sheet to maximize the effect of airflow and obtain efficient break-up.

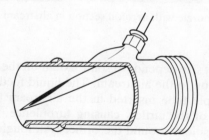

Fig. 5.17   Motorized mistblower nozzle showing simple tube inserted in airstream

Table 5.6   Example of the variation in droplet size with increased flow rate, obtained with a motorized mistblower

Flow rate (l/min)	Droplet size (vmd)
0.7	200
1.6	242
2.0	285

Fig. 5.18 Mistblower nozzle with aerofoil section in airstream

Variation in droplet size depends to some extent on the design and position of the orifice in relation to the airstream. The liquid feed on many knapsack mistblowers is a simple tube inserted in the airstream (Fig. 5.17), but the tube obstructs some of the airflow causing turbulence, so that air velocity varies at different positions around the orifice. Although air over the orifice is moving at a high velocity (eg 50 m/s), the emergent airstream velocity decreases rapidly owing to the drag of the atmosphere. This is discussed more fully in relation to sprayer design in Chapter 10.

In other twin-fluid nozzles, an annular airstream is separated from the liquid orifice to create a turbulent vortex of air into which the liquid is spread as a thin sheet. The vortex is often produced by a series of fixed vanes mounted around the liquid orifice (Fig. 5.19). Droplet formation of viscous liquid is improved if the shearing action is doubled by having an initial break-up when liquid is fed into a straight air jet which meets a vortex of air which shatters the initial droplets to produce very small droplets. This double shear nozzle is used in equipment to produce aerosols and has been widely used to

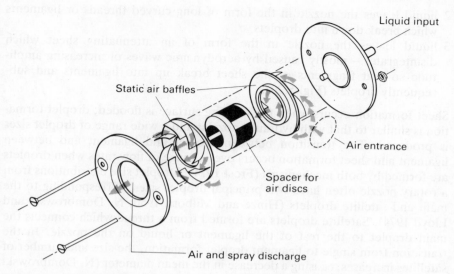

Fig. 5.19 Vortical nozzle – liquid fed into airstream, droplets fed into air vortex

apply ULV formulations such as technical malathion for adult mosquito control (see pp. 241–5), since it was developed originally for military use (Morrill et al 1955).

Droplet formation is achieved at relatively low air pressures (less than 0.3 bar), but when aerosol droplets are required, the volume of air must be increased and also the difference in velocity of the air/liquid must also be greatly increased.

Centrifugal-energy nozzle (eg spinning discs)

Hydraulic nozzles have been widely used for the application of pesticides as they are relatively inexpensive to manufacture, but as they produced a wide range of droplet sizes, the need for an alternative nozzle has been long recognized (Fraser 1958). Centrifugal-energy nozzles have proved valuable in the laboratory as a means of obtaining a narrow spectrum of droplets, but early attempts to use them in the field were not successful, since they were operated under 'flooded' conditions in an attempt to apply the same volumes of liquid as used with hydraulic nozzles.

Liquid is fed near the centre of a rotating surface so that centrifugal force spreads the liquid to the edge at or near which the droplets are formed. Fraser et al (1963) have defined three methods of droplet formation as the liquid flow rate is increased. These are:

1 single droplets leave directly from the nozzle at low flow rates

2 liquid leaves the nozzle in the form of long curved threads or ligaments which break down into droplets

3 liquid leaves the nozzle in the form of an attenuating sheet which disintegrates – mostly caused by aerodynamic waves of increasing amplitude so that fragments of the sheet break up into ligaments and subsequently droplets (Fig. 5.20).

Sheet formation occurs when the rotating surface is flooded; droplet formation is similar to that with hydraulic nozzles and a wide range of droplet sizes is produced. The transition between droplet and ligament and between ligament and sheet formation occurs over a range of flow rates when droplets are formed by both mechanisms (Frost 1974). Droplet size distributions from a rotary nozzle often have two principal droplet sizes, corresponding to the main and satellite droplets (Hinze and Milborn 1950; N. Dombrowski and Lloyd 1974). Satellite droplets are formed from a thread which connects the main droplet to the rest of the ligament or liquid on the nozzle. In the transition from single to ligament droplet formation, the size and number of satellites increases, causing a decrease in the mean diameter (N. Dombrowski and Lloyd 1974).

The diameter of droplets produced singly by a rotary nozzle can be approximately calculated from the following equation (Walton and Prewett 1949):

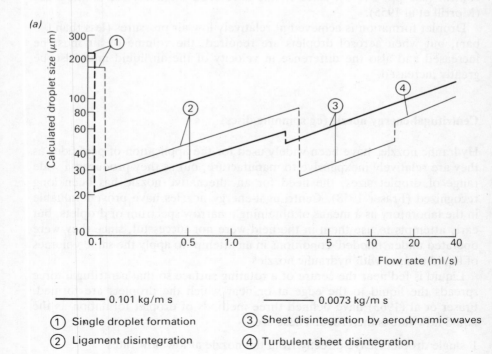

(a)

1 Single droplet formation
2 Ligament disintegration
3 Sheet disintegration by aerodynamic waves
4 Turbulent sheet disintegration

— 0.101 kg/m s   — 0.0073 kg/m s

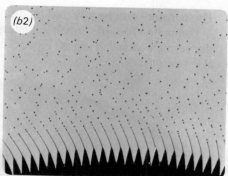

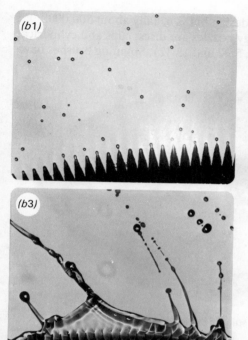

Fig. 5.20 (a) Variation of droplet size with flow rate on a single spinning disc (after N. Dombrowski and Lloyd 1974). (b) Single droplet, ligament and sheet formation from a spinning disc: (1) Herbi disc 2 000 rpm, 60 ml/min; (2) 2 500 rpm 100ml/min; (3) 1000rpm 800ml/min. (photos Micron Sprayers Ltd.)

$$d = K\frac{1}{w}\sqrt{\frac{\gamma}{D\varrho}},$$

where:

d = droplet diameter (μm)

w = angular velocity (rad/s)

D = diameter of disc or cup (mm)

γ = surface tension of liquid (mN/m)

ϱ = density of liquid (g/cm³)

K = constant which has been found experimentally to average 3.76 (Fraser 1958).

This can be written as

$$d = \frac{\text{constant}}{\text{rpm}}$$

119

The constant will be affected by disc design but is usually about 500 000.

The main types of centrifugal energy nozzles are discs, cups and cylindrical sleeves or wire mesh cages (Figs 5.21, 5.22 and 13.12). Spinning brushes have also been used.

Fig. 5.21 Example of grooved toothed spinning disc

(a)

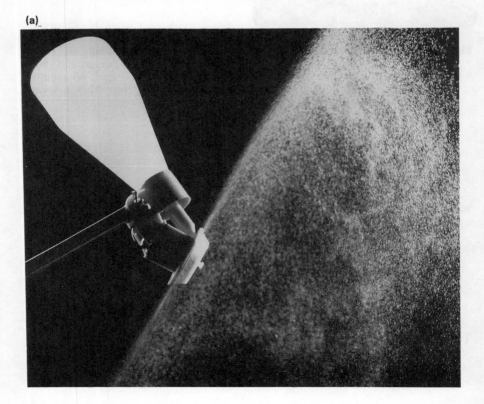

(b)

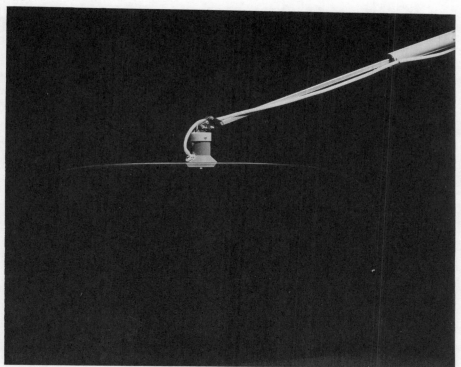

Fig. 5.22 (a) Disc operating at high speed for small droplets of insecticide (b) Slow speed disc with herbicide application

Studies of disc design have concentrated mainly on smooth-edged discs, but Fraser (1958) reduced droplet size by 13 per cent with a 45° chamber around the edge. Bals (1970) made discs with serrated edges to provide 360 half-pyramidical 'zero issuing points' to reduce the force required to overcome surface tension and breakaway droplets of a given size or, for a given force, produce smaller droplets. Bals (1976) introduced discs with a grooved inner surface to provide a reservoir of liquid to feed 'ligaments' of spray liquid to individual issuing points around the periphery. A very narrow range of droplet sizes is produced with discs having both grooves and teeth (Fig. 5.23), hence their suitability for controlled droplet application. There is an optimum flow rate for a given rotational speed; this decreases with increased speed. Application of a higher flow rate was possible when the larger cup-shaped disc was used with grooves to each of the peripheral teeth (Heijne 1978). Similarly a series of stacked discs on the rotary cages allows higher flow rates to be applied with sprayers on vehicles or aircraft.

Spinning discs, cups or cages are less liable to clog, but they are subject to different types of wear and breakdown compared with hydraulic nozzles.

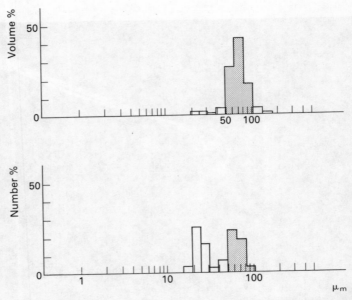

Fig. 5.23 Narrow droplet spectrum from a spinning disc 30 ml/min 120 rps (a) Volume distribution (b) Number distribution. (Malvern Particle Size Analyzer data)

Centrifugal-energy nozzles can be mounted in the airstream emitted from mistblowers (see Chapter 10). However, droplet spectra produced are affected by the interaction of centrifugal and air shear forces. Large droplets produced from the rotary nozzle will be sheared at high air velocities, thus with rotary sleeve nozzles, a higher air velocity produced a wider droplet spectrum when the nozzle rotated at only 50 revs/s (Hewitt 1991). The distance over which droplets are thrown from the periphery of a disc is important when droplets have to be entrained in such an airstream. According to Byass and Charlton (1968), the upper limit of droplet size from a nozzle mounted in an airstream into which the droplets have to be turned can be determined by an equation given by Prandtl (1952),

$$d = \frac{K\gamma}{1/2 \varrho V^2},$$

where

V = the velocity of the airstream (m/s)
d = diameter of the largest surviving droplet (μm)
K = a constant depending on the droplet size range
γ = surface tension (mN/m)
ϱ = density of air (g/cm³).

The actual distance a droplet is thrown depends largely on the effect of air resistance, which is reduced if more droplets are produced. Courshee and Ireson (1961) showed that within certain limits the distance (*S*) single droplets were projected in ambient air is approximately proportional to the square root of the product of droplet size (*d*) and disc diameter (*D*).

$$S = 1.3\sqrt{(dD)},$$

thus a 250 μm droplet produced on a 9 cm disc should travel 61.7 cm.

Kinetic-energy nozzle

A filament of liquid is formed when liquid is fed by gravity through a small hole, for example in the rose attachment fitted to a watering can, or the simple dribble bar which can be used for herbicide application. The liquid filament when shaken breaks into large droplets. This principle was used in the 'Vibrajet' in which an electrically powered unit oscillates a hollow tube through 25° at a frequency of 58 Hz. Liquid was fed by gravity through the shaft of the tube covered by a plastic sleeve which had an array of holes, the size of which affected droplet size in the range from approximately 500 μm (0.25 mm diameter hole) to 1 250 μm droplet (0.74 mm diameter hole) (Lake 1970). A hand-carried applicator was described by Clayphon and Thornhill (1974a) but they have not been used widely because the extra cost of each nozzle is not outweighed by any advantage to the farmer. In particular, volumes of spray are generally over 100 litres/ha.

Thermal-energy nozzle

Thermal-energy nozzles used to produce fogs are most effectively described in relation to specific fogging machines described in Chapter 11. In general, a stream of hot gas decreases the viscosity of an oil in which the pesticide is dissolved and shatters the liquid into small droplets. The droplets are vaporized and, on leaving the nozzle, condensation occurs to form a cloud of aerosol droplets, mostly less than 30 μm vmd, in sufficient quantity to reduce visibility. Actual droplet size depends on the characteristics of the pesticide solution and the flow rate. Increase in flow rate results in larger droplet. Normally, fogs are used for disinfestation without residual deposits, but some residual activity can be achieved with some insecticides if the maximum flow rate is used to produce a 'wet fog'. The hot gas may be obtained from the exhaust of an internal combustion engine or a small pulsejet engine, but other methods can be used.

Electrical-energy nozzle

Coffee (1979) described a nozzle in which a high voltage was applied to a semi-conducting liquid emitted through a narrow annulus or slit. The charged

Fig. 5.24 A spray from an electrodynamic nozzle

liquid forms ligaments (Fig. 5.24) which break up into electrostatically charged droplets with a very narrow size range (Fig. 5.25). More details are given in Chapter 9, together with descriptions of other nozzles adapted to produce an electrostatically charged spray.

Pressure-pack nozzle

A convenient but expensive means of producing aerosol droplets of a wide range of products such as hair lacquers, paints and deodorants in addition to pesticides, is a pressure pack, more commonly referred to as an aerosol or aerosol product. The pressure pack consists of a mixture of a propellant and a solution of active ingredient, packaged under pressure in a suitable container equipped with a valve (Fig. 5.26). The propellant is usually a liquefied gas. Fluorinated hydrocarbons (eg Freon 12) were chosen because

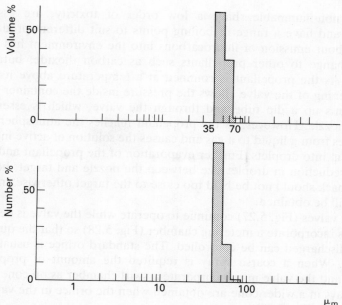

Fig. 5.25 Droplet spectrum from an electrodynamic nozzle at 25Kv 6ml/min (a) Volume distribution (b) Number distribution (Malvern Particle Size Analyzer data)

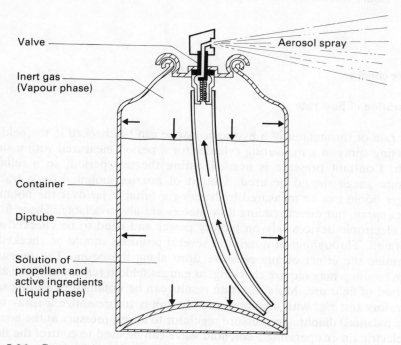

Fig. 5.26 Cross section of typical pressure pack

they are non-flammable, have a low order of toxicity, are essentially odourless and have a range of boiling points to suit different products but concern about emission of fluorocarbons into the environment has necessitated a change to other propellants such as carbon dioxide, butane and nitrogen. As the propellant is confined at a temperature above its boiling point, opening of the valve allows the pressure inside the container to force the contents up a dip tube and through the valve, which is essentially a pressure nozzle. However, as the propellant reaches the atmosphere, some of it flashes from a liquid to a gas and causes the solution of active ingredient to break up into droplets. Further evaporation of the propellant and solvent causes a reduction in droplet size between the nozzle and target, hence the pressure pack should not be held too close to the target otherwise an uneven deposit will be obtained.

Typical valves (Fig. 5.27) continue to operate while the valve is depressed, but others incorporate a metering chamber (Fig. 5.28) so that the quantity of product discharged can be controlled. The standard orifice is usually about 0.43 mm. When a coarse spray is required the amount of propellant is reduced, and the valve may incorporate a swirl chamber as on cone nozzles. Finer sprays in a wider cone are obtained when the orifice in the valve has a reverse taper (Fig. 5.29).

The problem for some of the alternative propellants such as compressed carbon dioxide is that the pressure decreases as the pressure pack empties. The pressure increases with temperature when a liquefied compressed gas is used.

Nozzle checks

Calibration of flow rate

Flow rate or throughput of a hydraulic nozzle can be checked in the field by collecting spray in a measuring cylinder for a period measured with a stopwatch. Constant pressure is needed during the test period, so a reliable pressure gauge should be used. Output of nozzles mounted on a tractor sprayer boom can be measured by hanging a suitable jar over the boom to collect spray, but direct reading flow meters are also available. Those fitted with electronic devices rely on battery power and need to be checked and calibrated. Throughput of nozzles at several positions should be checked to determine the effect of any pressure drop along the boom. The pressure-gauge readings may require checking, as gauges seldom remain accurate after a period of field use. More accurate results can be obtained by setting up a laboratory test rig, with a compressed-air supply to pressurize a spray tank and a balanced diaphragm pressure regulator to adjust pressure at the nozzle. An electric timer operating a solenoid valve can be used to control the flow. The test rig should have a large pressure gauge frequently checked against

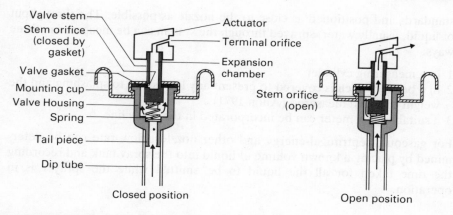

Valve stem
Stem orifice
(closed by
gasket)
Valve gasket
Mounting cup
Valve Housing
Spring
Tail piece
Dip tube

Actuator
Terminal orifice
Expansion
chamber

Stem orifice
(open)

Closed position

Open position

Fig. 5.27 Spray valve in closed and open position

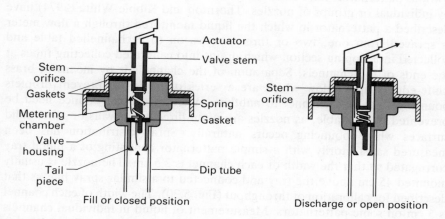

Stem
orifice
Gaskets
Metering
chamber
Valve
housing
Tail
piece

Actuator
Valve stem

Spring
Gasket

Stem
orifice

Dip tube

Fill or closed position

Discharge or open position

Fig. 5.28 Metering valve

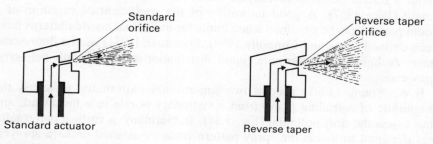

Standard
orifice

Reverse taper
orifice

Standard actuator

Reverse taper

Fig. 5.29 Reverse taper orifice

standards and positioned as close to the nozzle as possible. The throughput of liquid, usually water, sprayed through the system can be measured in three ways:

1 in a measuring cylinder
2 in a beaker which is covered to present any losses due to splashing and the weight of liquid measured (Anon 1971)
3 a suitable flow meter can be incorporated in the spray line.

For gaseous, centrifugal-energy and other nozzles, flow rate can be determined by placing a known volume of liquid into the spray tank and recording the time taken for all the liquid to be emitted while the sprayer is in operation.

Spray pattern – hydraulic nozzles

Various patternators have been designed to measure the distribution of liquid by individual or groups of nozzles. Thornton and Kibble-White (1974) have described a patternator in which the liquid monitored through a flow meter is sprayed from one, two or three nozzles on to a channelled table and collected in a sloping section which drains into calibrated collecting tubes at the ends of the channels. Separation of the channels is by means of brass knife-edge strips, below which are a series of baffles to prevent droplets bouncing from one channel to another. Whether droplet bounce need be prevented is debatable, as nozzles are often directed at walls or other solid surfaces where bouncing occurs naturally. Spray distribution has been measured satisfactorily with a simple patternator consisting of a metal tray corrugated so that the width of each channel is 2.5 cm. The nozzle is usually mounted 45 cm above the tray and connected to a similar spray line as that described for calibration of throughput (Fig. 5.30). The width of each channel is 5 cm on some patternators. Measurement of liquid in individual channels is better than a photographic record since the thickness of the collecting tubes may vary slightly. Considerable variation in the pattern can occur with successive runs with individual nozzles (R.G. Richardson et al 1986). Although normally used in the laboratory, patternators can be positioned under a tractor boom to investigate variation in spray distribution along its length (Rice 1967). A good indication of the coefficient of variation of a boom pattern can be obtained when limits for individual nozzle patterns have been defined (Rice and Connolly 1969). Speelman (1971) used a fluorescent tracer technique to determine liquid distribution achieved with field-crop sprayers.

B.W. Young (1991) used a two-dimensional patternator to assess the magnitude of a trailing plane from a stationary nozzle in a headwind, and thus assess the drift potential (Fig. 5.31). In Germany, a vertical patternator was designed to assess the spray pattern from air-assisted orchard sprayers (Fig. 5.32) (Kummel et al 1991).

Fig. 5.30 Patternator

Nozzle erosion

The orifice of the nozzle tip is enlarged during use by the combined effects of the spray liquid's chemical action and the abrasive effect of particles, which may be the inert filler in wettable powder formulations or, more frequently, foreign matter suspended in the spray. This is referred to as nozzle-tip erosion and results in an increase in liquid flow rate, an increase in droplet

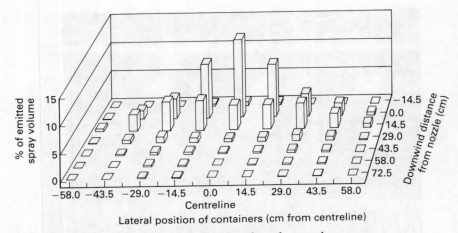

Fig. 5.31 Two-dimensional patternator data for a fan nozzle

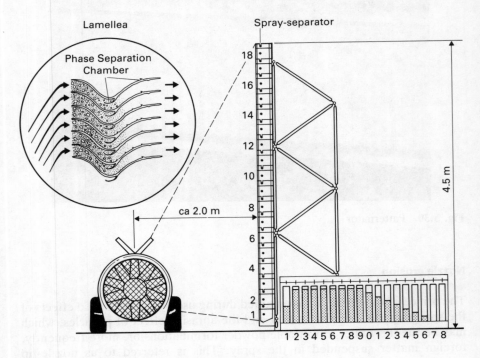

Fig. 5.32 Vertical patternator for an orchard sprayer

size and an alteration in spray pattern. Increase in flow rate can result in over-use of pesticides and increased costs, especially where large areas are involved, and throughput should be checked regularly and the tip replaced when the cost of the cumulative quantity of pesticide wasted equals the cost of a replacement nozzle tip, if the rate of erosion is fairly regular (Kao et al 1972). Rice (1970) reported increases in throughput of 49–63 per cent with brass nozzle tips after 300 h wear with a 1 per cent copper oxychloride suspension, whereas, with stainless steel, ceramic and plastic tips, throughput increased only by 0–9 per cent over the same period. Over 70 per cent of sprayers examined in a survey in the UK had at least one nozzle with an output which varied more than 10 per cent from the sprayer mean. In extreme cases the maximum output was three times that of the minimum throughput (Rutherford 1976). Beeden and Matthews (1975) showed a 10 per cent increase in throughput after 35 ha-sprays for cone nozzles used for spraying cotton in Malawi; therefore the average individual farmer should replace nozzle tips after three seasons. When water with a large amount of foreign matter in suspension is collected from streams or other sources, a farmer is advised to collect it on the day before spraying and allow it to settle overnight in a large drum. Nozzle tips should be removed after each spray application and carefully washed to reduce any detrimental effect of chemical residues. When cleaning a nozzle, a hard object such as a pin of knife should not be used, otherwise the orifice will be damaged (see Chapter 15).

Assessment of the effect of abrasion on a nozzle tip can be made in the laboratory by measuring the throughput before and after spraying a suspension of a suitable abrasive material. A suitable test is to spray 50 litres of a suspension containing 20 g of synthetic silica (HiSil 233) per litre (Jensen et al 1969; Anon 1990b) but other materials which have been used include white corundum powder which abrades nozzles similarly to HiSil but in one-third of the time.

6

Hand-operated hydraulic sprayers

Hydraulic-energy nozzles have been the most widely used in comparison with the other types described in Chapter 5, because considerable flexibility can be achieved by interchanging the tips in a standard nozzle body to provide a wide range of outputs and spray patterns at low cost. Hydraulic nozzles are used on a wide range of sprayers from a simple hand-syringe type to equipment mounted on aircraft. Hand-operated equipment is described in this chapter, power-operated equipment in Chapter 7, and aerial equipment in Chapter 13.

Sprayers with hydraulic pumps

Syringes

There are various types of simple syringe-type sprayers in which liquid is drawn from a reservoir into a pump cylinder by pulling out the plunger; the liquid is then forced out through a nozzle on the compression stroke. These sprayers have been mostly replaced by the double-acting slide pump or hand compression sprayer. A small syringe-type sprayer is useful for spot treatment, for example *Striga* can be killed in maize, using 20 g/ha of ametryne solution, applying 1 ml as a coarse spray to an area of 25 cm diameter (Ogborn 1972). They are also used to inject systemic insecticides into holes previously bored into trees.

The double-acting slide pump consists of a piston pump in which one valve is mounted at the inlet end of the cylinder, and a second valve is in a tube which is used as a piston. A handle grip is positioned on both the pump cylinder and piston, and the pump is operated by holding the piston handle grip firmly while directing the nozzle and moving the cylinder in and out continuously. On the first stroke, liquid from a separate container is drawn into the cylinder past the inlet valve, and on the return stroke, the inlet valve is closed and the liquid is forced past the second valve into the piston tube. The piston seal is usually a thick washer. If this is too tight in the cylinder, pumping effort will be excessive, and conversely, if too loose, pumping will be inefficient. An effective seal is also needed between the piston rod and end of the cylinder, otherwise spray liquid will leak over the operator's hands. A rubber stop between the cylinder and the piston handle grip cushions the end of the pump stroke. The piston handle grip is usually enlarged to contain a small pressure chamber to even out the variation in pressure between strokes of the pump. A knapsack container or ordinary bucket can be used to contain the liquid which must be kept stirred, especially if wettable powder suspensions are being applied. The delivery tube to the pump usually has a filter fitted at the inlet end.

Because of the narrow diameter of the pump cylinder (15–20 mm) and long stroke (25–40 cm), quite high pressures can be obtained with small nozzles. Output will, of course, vary with the length of pump stroke, and they are not suitable for precise application of pesticide. Various types of hydraulic nozzles can be used, but these sprayers are often fitted with an adjustable cone nozzle.

As both hands are required for operation, the pump is very tiring to use, so they are suitable only when spraying is over a short period. Many were used on small farms in tropical countries, for example for cotton spraying in Uganda (T.R. Jones 1966), but have been replaced by knapsack or spinning disc sprayers.

Stirrup pump

Another version of the double-acting pump is the bucket or stirrup pump. Two operators are normally required – one to work the pump while the other directs the nozzle. The lower end of the pump is immersed in the spray liquid in a bucket. The pump, which sometimes has a solid piston plunger, is steadied by a footrest or stirrup on the ground next to the bucket. Ideally the position of the stirrup can be adjusted to allow buckets of different depths to be used. Agitation is seldom provided, but on some sprayers there is a paddle agitator in the bucket which may cause splashing when the bucket is nearly empty. Agitation is also possible by recycling some of the liquid through a nozzle mounted at the lower end of the pump. The outlet of the pump is near the top and is connected to a hose, usually 6 m in length, with a lance and any type of hydraulic nozzle. Longer hoses are difficult to handle. Spraying is continuous because an air chamber is incorporated in the spray line, and on some models pressures up to 10 bar can be obtained fairly easily.

With stirrup pumps great care has to be taken to avoid spillage of toxic chemicals as the liquid is in an open container. These pumps are strongly constructed and should withstand considerable wear in the field if properly cleaned after use. The pump may be a simple plunger type, or a piston with a cup washer or gasket gland which should be smeared with grease, especially if it is made of leather. Stirrup pumps have been used to treat apple orchards in India, but preference is now given to motorized portable line systems. In vector control, larvicides can be applied to the surface of water and to spray a residual insecticide on the wall surface of dwellings but compression sprayers are now preferred. Specifications for a stirrup pump have been published by the World Health Organization (WHO/EQP/3.R3). They have been used to apply molluscicides to water and, by removing the nozzle, can also be used to transfer liquids from a container to a sprayer.

Knapsack sprayers – lever-operated

One of the most widely used small sprayers is the lever-operated knapsack sprayer, the design of which has changed very little since they were first manufactured for the application of fungicides in vineyards in the late nineteenth century (Lodeman 1896; Galloway 1891). A lever-operated sprayer consists of a tank which will stand erect on the ground and, when in use, fit comfortably on the operator's back like a knapsack, a hand-operated pump, a pressure chamber, and a lance with an on/off tap or trigger valve and one or more nozzles.

The tank is now usually made of polypropylene or alternative plastic, as it is less expensive to manufacture with these materials and they are extremely resistant to most of the agrochemicals used. A UV-light inhibitor is incorporated in the plastic. Some manufacturers may supply a tank made from brass or mild steel, either galvanized or coated with an epoxy-resin material. Stainless steel tanks are also used. The usual capacity of the tank is about 15 litres so that the total weight is not too excessive to be carried by the operator. Some sprayers with lighter plastic tanks have a larger capacity, but the total weight of the full sprayer should not exceed 20 kg. Smaller 10 litre sprayers are also available. Plastic tanks can be moulded to fit the operator's back more comfortably than was possible with metal tanks. The volume of spray in the tank is indicated by graduated marks, moulded in plastic tanks. It is also possible to have interchangeable polyethylene tanks with an externally mounted pump. These are useful when spraying a number of different products in one trial. To facilitate filling, the tank should have a large opening not less than 95 mm in diameter at the top. Also, a large opening permits operators to put their gloved hands inside the tank if necessary for cleaning. The tank opening should have a tight-fitting lid, in which any air vent is fitted with a valve to prevent spray liquid splashing out and down the operator's back. A 50-mesh filter, fitted with an air vent, should be placed at least 50 mm into the opening, to reduce the risk of liquid spilling over the tank during filling. A coarser filter is often used.

Lever-operated knapsack sprayers can be divided into those with an overarm lever or underarm lever, and second, those with a piston or diaphragm pump (Fig. 6.1). The piston pump is normally used where higher pressures at the nozzle are required, and the diaphragm pump is to be preferred when suspensions are being applied which are liable to cause

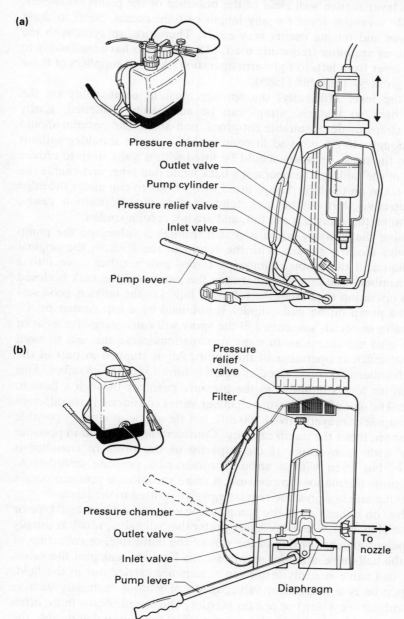

(a)

Pressure chamber

Outlet valve

Pump cylinder

Pressure relief valve

Inlet valve

Pump lever

(b)

Pressure relief valve

Filter

Pressure chamber

Outlet valve

Inlet valve

Pump lever

Diaphragm

To nozzle

Fig. 6.1 Lever-operated knapsack sprayers (a) Piston pump type (b) diaphragm pump type (from BCPC publication)

erosion of the piston chamber. The pump is connected by a system of linkage to a lever which is pivoted at some point on the side of the tank. The overarm lever is easier to operate when walking between plants that meet across the row, as the lever is then well clear of the branches of the plants. However, operating the overarm lever for any length of time causes blood to drain from the arm and fatigue occurs very easily. Therefore, sprayers with the underarm lever are more frequently used. Many sprayers have the facility to change the lever from left- to right-arm operation. A list of suppliers of these sprayers is given by Thornhill (1985).

To use the lever efficiently, the sprayer must fit comfortably on the operator's back so that the straps can be adequately tightened. Easily adjustable straps made of suitable rot-proof, non-absorbent material should be wide enough (40–50 mm) to fit comfortably over the shoulder without cutting into the neck. Sprayers should be fitted with a waist strap to reduce movement of the tank on the operator's back while pumping, and enable the load to be taken on the hips. Straps fitted with a hook to clip under the edge of the protective skirt of the tank tend to slide out of position easily, especially when the sprayer is not full, and are not recommended.

When using the sprayer, liquid is drawn through a valve into the pump chamber with the first stroke. With the return of the lever to the original position, liquid in the pump chamber is forced past another valve into a pressure chamber. The first valve between the pump and the tank is closed during this operation to prevent the return of liquid to the tank. A good seal between the pump piston and cylinder is obtained by a cup washer or 'O' ring. Abrasive materials suspended in the spray will cause excessive wear of the pump, also the chemicals in some formulations cause the seal to swell and prevent efficient operation of the pump. Air is trapped in part of the pressure chamber and compressed as liquid is forced into the chamber. This compressed air forces liquid from the pressure chamber through a hose to the nozzle. The size of the pressure chamber varies considerably on different types of knapsack sprayers (160-1 300 ml), but should be as large as possible and at least ten times the pump capacity. Considerable variations in pressure will occur with each stroke if the capacity of the pressure chamber is inadequate, but even with a strongly constructed pressure chamber to withstand these fluctuations in pressure, a small variation in pressure occurs while spraying unless a pressure regulating valve is fitted to the lance.

The valves on either side of the pump can be either of a diaphragm type or a ball valve (Fig. 6.2). Some operators prefer the ball valve, which is usually made of polypropylene. Pitting of the side of the ball valve, or collection of debris in the ball-valve chamber may cause the liquid to leak past the valve. Also, the ball valve is easily lost when repairs are carried out in the field. The alternative is a diaphragm valve, made of various materials such as synthetic rubber (eg Viton) or certain plastics. The chemicals, or more often the solvents, used in some formulations can affect the material and cause the valve to swell up. This causes the valve to stick and block the passage of liquid through the pump unless there is adequate space for the diaphragm

(a)

(b)

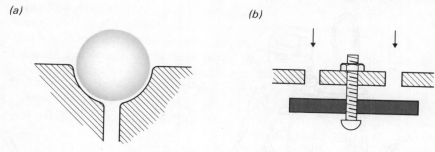

Fig. 6.2 Diagram of (a) ball-valve and (b) diaphragm valve

valve to move. With many knapsack sprayers an agitator or paddle is fitted to the lever mechanism, or directly to the pressure chamber, to agitate the spray liquid in the tank. On some sprayers hydraulic agitation is provided by a jet of liquid by circulating part of the pump's output back into the tank. Agitation is essential when spraying suspensions in order to reduce settling out, which can occur rapidly with some wettable powder formulations. The pressure chamber and pump are fitted outside the tank of some sprayers to facilitate maintenance, but they are more vulnerable to damage if the sprayer is dropped. The pressure chamber may be fitted with a relief valve so that the operator cannot over-pressurize it. This should not be used as a pressure control valve and should be touched only when the tank is empty and cleaned.

When using lever-operated knapsacks, the operator moves the lever up and down several times with the tap closed so that pressure is built up in the pressure chamber. The tap is opened and the operator continues to pump steadily with one hand while spraying. Ideally a pressure control valve is also

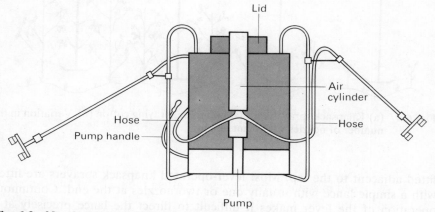

Fig. 6.3 Nozzles mounted on rear of knapsack sprayer

(a)

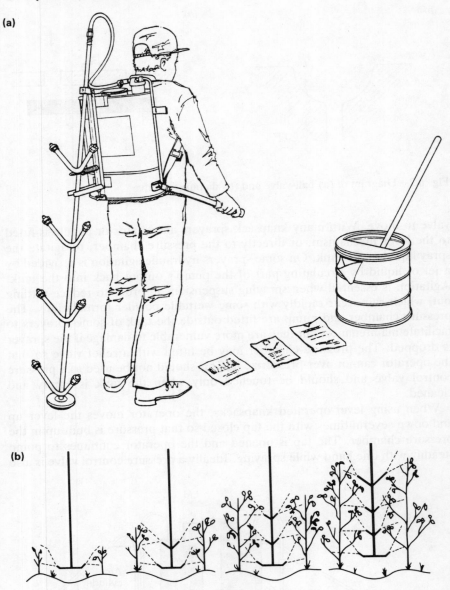

(b)

Fig. 6.4 (a) Knapsack sprayer with tailboom for spraying cotton (b) Variation in the number of nozzles with plant height

fitted adjacent to the tap. Most lever-operated knapsack sprayers are fitted with a simple lance with usually one or two nozzles at the end. Continuous operation of the lever makes it difficult to direct the lance precisely at a target, so in certain circumstances the compression sprayer is preferred.

Another problem is that the operator tends to walk towards where he is directing his spray and walk past foliage which has been treated, thus becoming contaminated with pesticide, particularly on the legs (Tunstall and Matthews 1965; Sutherland et al 1990). On various occasions, therefore, adaptations on the knapsack sprayer have been developed either to improve safety, obtain a better distribution of spray droplets or to increase the speed of spraying.

An example of this is the fitting of wide-angle nozzles on to the back of the spray tank for treating rice crops so that the operator walks away from the spray (Fig. 6.3) (Fernando 1956). An adjustable tail-boom (Fig. 6.4) has been shown to achieve better distribution of spray on cotton plants, thus improving control of bollworms (Tunstall et al 1961, 1965). A tailboom has also been used to spray coffee. To improve the speed of spraying, a horizontal boom has been developed for spraying more than one row of cotton at a time (Cadou 1959), and for applying fungicides to groundnuts (Johnstone et al 1975).

In addition to the conventional lance, some sprayers have a short boom with two or three nozzles fitted to the end of the lance. Extendable lances made of bamboo or aluminium may be used to spray trees up to about 6 m in height. A goose-neck at the end of a lance is useful for spraying some inaccessible sites; similarly, other specialized nozzle arrangements have been used to spray special targets such as pods resting on stems of cacao trees. The nozzles may be shielded so that herbicide sprays can be applied close to a susceptible plant or tree (Fig. 6.5).

The design and efficiency of operation of trigger valves on lances vary considerably. The handle should fit comfortably in the operator's hand, so that the valve is easy to operate. A clip mechanism to hold the valve open for prolonged spraying is useful, provided it can be released easily. Ideally there should be a clip to hold the valve closed when not in use. Unfortunately, many valves leak, particularly after abrasive particles have been sprayed, so that regular maintenance of the valve seating is needed with replacement springs. A test procedure for trigger valves has been described by L.B. Hall (1955).

Most commercially available lever-operated sprayers are strong enough only if used for short periods. When a farmer requires to spray a crop such as cotton several times during the season over 1 or 2 ha, then the more robust types of equipment are required. Mechanized durability tests can be carried out on different sprayers (Matthews et al 1969; Thornhill 1982).

The main faults have been poor linkage, inadequate strength of the lever, poor design or strength of certain components such as strap hangers, and the poor capacity of the pumps.

The performance of lever-operated sprayers has been recorded in the field by using a small portable recording pressure gauge. Comparison of number of strokes required to maintain various outputs and pressures can be a useful guide to the efficiency of the different sprayers commercially available (Matthews 1969).

Stretcher and other sprayers

Stretcher sprayers have rather heavy piston pumps which are operated by hand by a rocking motion of a long handle. Some pumps are operated by a foot pedal. The pump incorporates a very strong pressure chamber and pressures up to 10 bar or more can be obtained. They are designed for operation in one position with one or more hoses up to about 14 m in length, and lances to which various hydraulic nozzles are fitted. A separate container is needed and has to be carried separately from the pump. Two people are needed to carry the pump and operate it, while two or more others are needed to handle the lances. Stretcher sprayers are rather awkward to carry, but have been most useful for spraying large or tall trees or widely spaced bush crops. Similar sprayers are sometimes mounted on a wheelbarrow frame, together with a spray tank.

Compression sprayers

These sprayers, also referred to as pneumatic sprayers, have an air pump to pressurize the spray tank. The tank is never completely filled with liquid. A

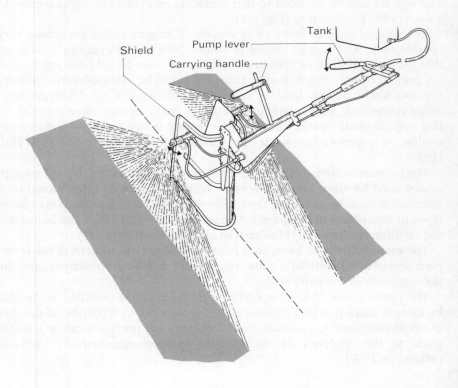

Fig. 6.5 Shielded nozzle for application of herbicide (Allman & Co. Ltd)

space is needed above the liquid so that air can be pumped in to create pressure to maintain the flow of liquid to the nozzle. Usually, a mark on the side of the tank indicates the maximum capacity of liquid at about two-thirds of total capacity. These sprayers vary in size from the small hand sprayers, suitable for limited use by gardeners, to large knapsack sprayers usually of 10 litre capacity and are used for spraying a wide range of pesticides. As no agitation is provided, these sprayers need to be shaken occasionally if using wettable powder formulations to prevent the suspension settling out.

Hand sprayers

A small tank usually made of plastic and 0.5–3 litres capacity is pressurized

Fig. 6.6 Hand-carried compression sprayer

by a plunger-type pump to a pressure of up to 1 bar (Fig. 6.6). Often a cone nozzle, the pattern of which can sometimes be adjusted, is fitted to a short delivery tube. The on/off valve is sometimes a trigger incorporated into the handle. They are useful for spraying very small areas, where it is inconvenient to pump continuously.

Knapsack or shoulder-slung compression sprayers

There are two types – the ordinary and pressure-retaining types. The pressure-retaining type used to be used extensively on tea estates and tsetse-control operations in groups of five or six sprayers operated from a central point, sometimes equipped with a motorized pump unit (Davies and Blasdale 1960). The management of support staff and vehicles in a large-scale tsetse-control programme with ground equipment was described by Wooff (1964). Because of the high pressure at which these sprayers are operated, the tank had to be very soundly constructed and subjected to a hydraulic pressure test at frequent and regular intervals. As these sprayers are very expensive, their use has declined except in Colombia where they are used to treat coffee. On the other hand, there are a large number of compression sprayers described below in which the air pressure is released before refilling the tank with liquid.

Non-pressure-retaining type

The compression sprayers (Fig. 6.7) are pressurized by pumping before

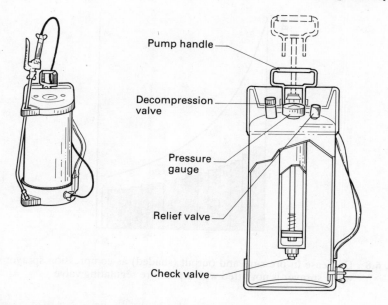

Pump handle

Decompression valve

Pressure gauge

Relief valve

Check valve

Fig. 6.7 Compression sprayers (courtesy British Crop Protection Council)

spraying commences, in contrast to the continuous pumping needed with lever-operated sprayers. This allows the operator to give more attention to directing the nozzle at the correct target. The pump is screwed in as part of the lid of the tank on the simpler and cheaper compression sprayers. When the pump has to be screwed in to the tank for each pressurization, the threads are liable to wear and limit the life of the sprayer, so the tank lid and pump should be separate. Ideally this type of sprayer should be fitted with a pressure gauge so that the operator knows what pressure is in the tank. A pressure gauge may not be provided, in which case the operator is instructed to pump a given number of strokes to achieve the working pressure. The pressure decreases very rapidly as soon as the operator starts spraying. Operators may have to stop and repressurize the tank before they can discharge the total contents from the tank. With the decrease in pressure at the nozzle while spraying (Fig. 6.8) droplet size increases. More uniform application is obtained by fitting a pressure-regulating valve to the tank outlet or lance. These valves must be cleaned and adjusted frequently, especially when wettable powders are used, to avoid sedimented material affecting the valve's performance.

On some occasions, some pressure may still be inside the tank when the operator has discharged the spray liquid and needs to refill the tank. This is released on the first quarter turn of the lid or pump, when a hissing sound indicates the escape of air. On some sprayers, the lid cannot be moved until

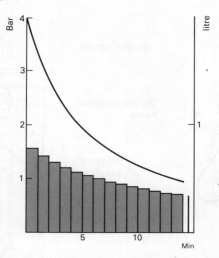

Fig. 6.8 Decrease in pressure and output (shaded) as compression sprayer empties when sprayer is not fitted with a pressure regulating valve

the pressure is released by a special valve, either in the lid or on the tank. The tank slightly expands and contracts during normal operation. To assess the durability of the tank, this is simulated by pressurizing the tank to 4.5 bar to 11 s, releasing the pressure and then repeating this for 12 000 cycles. The sprayer must be completely filled with water during this test. Further tests at 7 bar are usually carried out after dropping the sprayer in set positions to detect any weakness caused by the drop tests (L.B. Hall, 1955). Metal tanks should not have parts riveted in place as these are focal points for weakness. Some people may feel that plastic tanks will not be as strong as metal tanks but, in general, blow-moulded plastic tanks so far tested can stand pressures in excess of 7 bar, which is usually far above that obtained with the hand pumps provided with the equipment. Degradation of the plastic in sunlight (or UV light) has occurred, possibly by interaction with pesticides impregnated on the tank wall. The strength of these tanks is thus impaired. The base of the tank is usually provided with a skirt for protection against wear and also to enable the sprayer to stand firmly on the ground during pumping. On some sprayers, a footrest is attached to the skirt. The skirt serves as a backrest on some sprayers and is the lower fixing point for straps.

As with the lever-operated sprayers, the sprayer should have as large a tank opening as possible to facilitate filling and also to allow operators to put their gloved hands through the opening and clean the inside of the tank. Unfortunately, very few of the sprayers of this type have an adequately wide tank opening. Manufacturers often prefer placing the hose outlet at the base of the tank to avoid leaving any liquid in the tank and to eliminate a dip tube inside the tank, but this hose nipple is often broken when the sprayer is accidentally dropped. The better types of compression sprayers have the

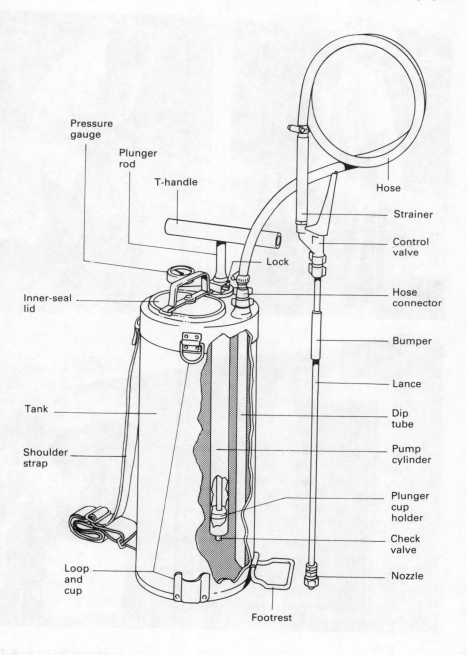

Pressure
gauge

Plunger
rod

T-handle

Hose

Strainer

Control
valve

Lock

Hose
connector

Inner-seal
lid

Bumper

Lance

Tank

Dip
tube

Shoulder
strap

Pump
cylinder

Plunger
cup
holder

Check
valve

Loop
and
cup

Nozzle

Footrest

Fig. 6.9 Compression sprayer to meet WHO specification (courtesy H. D. Hudson Manufacturing Co.)

Fig. 6.10   Residual spraying against mosquitoes in dwellings (photos WHO by R Da Silva)

hose opening at the top of the tank and a clamp is also provided to hold the lance when not in use. When the lance is left to trail in the mud while the sprayer is being refilled, the possibility of nozzle blockages is increased. Thornhill (1974a) has described the adaptation of a container used for dispensing soft drinks as a compression sprayer.

Compression sprayers of this type (Fig. 6.9) have been widely used on farms and also in vector-control programmes. WHO specification WHO/EQP/1.R4 was developed to ensure that reliable equipment was used to spray a residual deposit of insecticide on walls to control mosquitoes (Fig. 6.10).

The standard technique recommended by WHO for indoor residual spraying was to apply 757 ml/min through an 8002 fan nozzle when operated at 2.8 bar. Unless a pressure regulator is fitted, the working pressure decreases from 3.8 to 1.7 bar. Normally, the sprayer is charged initially to 3.8 bar with about fourteen pump strokes per bar (one stroke for each psi) and usually needs repressurization once during a 10 min period to discharge 7.5 litres. The lance is held 45 cm from the wall and moved at a steady speed of 0.64 m/s up and down the walls, covering a 75 cm swath each time. The same technique has been used for a number of different insecticides. The same type of sprayer has been used to apply larvicides, but the fan nozzle is replaced by a solid stream or cone nozzle. An experienced sprayman using a solid-stream nozzle with a 'swinging wand' pattern can treat an 8–10 m swath when walking at a steady 2 m/s. The nozzle is pointed above the horizontal so that the liquid trajectory reaches a maximum distance. If it is pointed down at the water surface there will be localized overdosing. The jet from the nozzle breaks up into a band of droplets which overlap with each swing of the lance. The solid-stream nozzle is also useful when directing spray into cracks and crevices in houses so that an insecticide is deposited in the resting sites of cockroaches and other household pests. A cone nozzle is used if a wider band of spray is needed on irregular-shaped objects in areas such as at the backs of sinks and boilers.

Pressure-retaining compression sprayer

Only a few manufacturers now offer this type of equipment as an extremely robust tank is needed to withstand very high pressures. One design had an internal rubber bag to retain the spray liquid, but the most widely used type had a double ball valve system (Fig. 6.11). Air was pumped into the container until a pressure of 3 bar was obtained. Spray liquid was then pumped into the container and, as the pressure could then reach 12 bar, a pressure-regulating valve on the outlet was essential to ensure a uniform emission rate. As the liquid was sprayed, the upper ball valve floated down and sealed on the bottom of the container to retain the air pressure.

Pressure-regulating valve

The type of valve suitable for use on knapsack compression sprayers differs

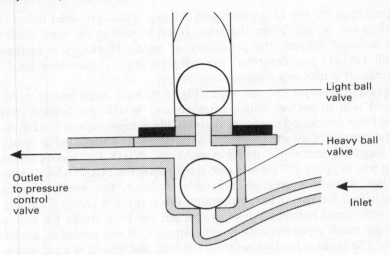

Light ball valve

Heavy ball valve

Outlet to pressure control valve

Inlet

Fig. 6.11 Ball valves from pressure-retaining compression sprayer

from that used on powered equipment in that there is no bypass return to the spray tank. The valve consists of a diaphragm, the movement of which is effected by a spring (Fig. 6.12). The tension of the spring can be adjusted by means of a screw so that output pressure can be varied from 0.71 to 4 bar, irrespective of the input pressure, which can be up to 18 bar.

Calibration of knapsack sprayers

The output of the sprayer should be checked by collecting and measuring the spray liquid emitted during 1 min. When using lever-operated sprayers, a pressure gauge should be fitted as close to the nozzle as possible and the lever operated evenly with a full stroke to maintain as uniform a pressure as possible. The operator will need to practise before achieving an even pumping rate. A pressure-regulating valve should be fitted on compression sprayers, otherwise the output will decrease as the tank empties. Having determined the output from the nozzle in litres/min, the rate per unit area treated can be calculated, knowing the swath width and walking speed.

$$\frac{\text{Output (l/min)}}{\text{swath (m)} \times \text{speed (m/min)}} = \text{volume application rate (l/m}^2)$$

Thus with a swath of 1 m and walking at 60 m/min and a flow rate of 0.6 litres/min, volume of spray per square metre is

$$\frac{0.6 \text{ litres/min}}{1 \text{ m} \times 60 \text{ m/min}} = 0.01 \text{ litres/m}^2 \quad \text{or} \quad \times 10\,000 = 100 \text{ litres/ha.}$$

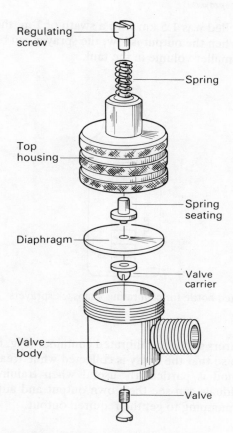

Regulating screw

Spring

Top housing

Spring seating

Diaphragm

Valve carrier

Valve body

Valve

Fig. 6.12 Pressure-regulating valve, exploded diagram

Alternatively, if you measure speed in km/hr, then

$$\frac{600 \times \text{output (l/min)}}{\text{swath (m)} \times \text{speed (km/hr)}} = \text{volume application rate (l/ha)}$$

thus if your flow rate is 2.2 l/min over a 1.7 m swath and your speed is 3.8 km/hr, then your application rate is 204 l/ha.

If the application rate is incorrect, other nozzles should be tried. When the most suitable nozzle has been selected, the volume applied can be rechecked by measuring the distance walked and time taken to spray a known quantity. For example if a full tank load of 15 litres is applied in 25 min the output is 0.6 litre/min which checks against the earlier calibration, the volume per hectare being given by

$$\frac{15 \times 10\ 000\ \text{m}^2\ (\text{ie l ha})}{\text{distance travelled (m)} \times \text{swath width (m)}} = \text{application rate (litres/ha).}$$

If the distance travelled was 1.5 km with a swath of 1 m, the application rate was 100 litres/ha. When the output is low, the sprayer can be calibrated more quickly by using a smaller volume in the tank.

Fig. 6.13   A calibrated bottle for calibrating knapsack sprayers

Some manufacturers supply a calibrated container (Fig. 6.13) which can be fitted to the nozzle so that the spray is collected while treating a known area (25m²). This method is particularly useful when training teams of spray operators as individuals can see their own output and adjust their speed of walking or rate of pumping to get the required output.

Handymist sprayer

In China, a lever-operated sprayer with an air pump has been designed by Tu Y.Q. (1990). The sprayer (Fig. 6.14) is essentially a compression sprayer but, in addition to pressurizing the liquid, some of the air is fed to a single twin fluid nozzle. The manually operated sprayer provides an inexpensive means of applying a very low volume mist sprayer and has been used to spray insecticides on rice and cotton.

Disposable container dispenser

Recently a disposable container dispenser (DCD) has been designed to fit manually operated sprayers, so that only water is put in the lever-operated knapsack or compression sprayer container. When the water passes through a specially designed trigger valve that incorporates a spray management valve (see p. 105) pesticide is metered into it at a set dilution rate. As there is no longer a need to measure out small quantities of pesticide product to put in the sprayer, operator contamination will be significantly reduced.

Fig. 6.14 Chinese 'Handymist' Sprayer with twin fluid nozzle; a lever-operated compression sprayer with air and spray liquid fed to the nozzle

Peristaltic pump

Liquid is forced through a piece of rubber of plastic tubing by progressive squeezing along the wall of the tube. A peristaltic pump, operated by rotating cams, attached to the wheel of a small sprayer can be used to deliver small volumes to individual nozzles.

7

Power-operated hydraulic sprayers

Various power-operated sprayers have been designed and range in size from small, hand-carried engine-driven pump units to large self-propelled sprayers (Fig. 7.1). Usually a series of nozzles are mounted along a boom (Fig. 7.2). Small units have a two- or four-stroke internal combustion engine with pump and are mounted on a knapsack, a wheelbarrow, or on a self-propelled or animal-drawn frame. These small sprayers are mostly used where treatment is not required over extensive areas. Otherwise the tractor-mounted boom sprayers with a pump driven from the power take-off (pto) is used, especially for field crops, to apply 50-500 litres/ha. Larger-capacity tanks may be mounted on trailers, or as saddle tanks alongside the tractor engine to spread the load more evenly. Some large sprayers are self-propelled instead of using the normal farm tractor, but these sprayers are used only on farms with sufficient flat land to allow the use of booms up to 27 m width and where the capital outlay is justified by their usage (see Chapter 17).

Tractor-mounted sprayers

Tank design

Most tractors have a standard three-point linkage on which the sprayer is mounted. The capacity of the tank is restricted by the maximum permitted weight specified for the tractor; half the sprayers in the UK have a tank of less than 750 litres capacity. Weights may be needed on the front end of the tractor, particularly the small tractors, to maintain stability. The farmer may prefer to use a smaller tank to reduce compaction of soil under the tractor paths, but if tank capacity is too low, frequent refilling may be required. The choice of spray tank size is also discussed in Chapter 17 in relation to other variables.

Fig. 7.1 Various types of power operated hydraulic sprayers (photos Allman & Co. Ltd)

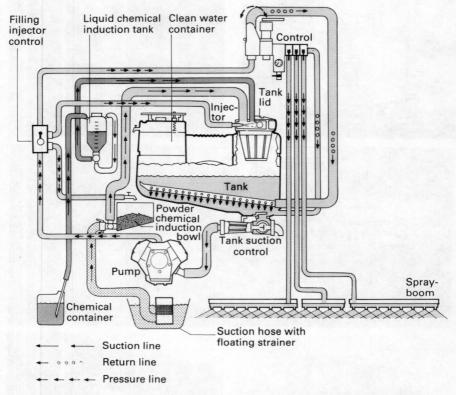

Fig. 7.2 Schematic diagram of the layout of a hydraulic sprayer (courtesy Allman & Co. Ltd)

Most modern sprayers have tanks constructed with a corrosion-resistant material such as multi-layer plastic. The tank should have a large opening (>300 mm) so that the inside can be scrubbed out if necessary. A large opening also facilitates filling the sprayer but to improve safety for the operator, new sprayers are fitted with a self-fill system and in some cases a separate low-level mixing chamber. A large basket-type filter should fit into the opening which is closed by a tight-fitting lid. The tank should have a drainage hole at its lowest point and a sight gauge visible to the tractor driver. The bottom of the tank should be fitted with a sparge-pipe agitator, that is a pipe with a line of holes along its length to give a series of jets of liquid to scour the tank bottom. Instead of a sparge pipe, a nozzle may be used to swirl the liquid over the tank bottom. Mechanical agitation is not recommended as, when the tank is nearly empty, the paddles may be only partly immersed and mix in air to cause foaming. A 200–1litre drum has been used as a cheap tank on a sprayer, but rust is liable to occur quite rapidly inside metal drums.

Pumps

A number of different types of pumps are used on tractor-mounted sprayers. Selection of a pump will depend on the total volume of liquid and pressure required to feed the nozzles and agitate liquid in the tank. The type of spray liquid will also influence the choice of pump, particularly the materials used in its construction. A comparison of pumps is given in Table 7.1.

Diaphragm pump

The basic part of the diaphragm pump is a chamber completely sealed at one end by a diaphragm (Fig. 7.3). The other end has an inlet and outlet valve.

Table 7.1 Summary of types of pumps

	Diaphragm	Piston	Centrifugal	Turbine	Roller
Materials handled	Most; some chemicals may damage diaphragm	Any liquid	Any liquid	Most; some may be damaged by abrasives	Emulsion and non-abrasive materials
Relative cost	Medium/ high	High	Medium	Medium	Low
Durability	Long life	Long life	Long life	Long life	Pressure decreases with wear
Pressure ranges (bar)	0–60	0–70	0–5	0–4	0–20
Operating speeds (rpm)	200–1 200	600–1 800	2 000–4 500	600–1 200	300–1 000
Flow rates (l/min)	1–15	1–15	0–30	2–20	1–15
Advantages	Wear resistant Medium pressure	High pressures Wear resistant Handles all materials Self-priming	Handles all materials High volume Long life	Can run directly from 1 000 rpm pto High volume	Low cost Easy to service Operates at pto speeds Medium volume Easy to prime
Disadvantages	Low volume Needs compression chamber	High cost Needs compression chamber	Low pressure Not self-priming Requires high-speed drive	Low pressures Not self-priming Requires faster drive for 540 rpm pto shafts	Short life if material is abrasive

(a)

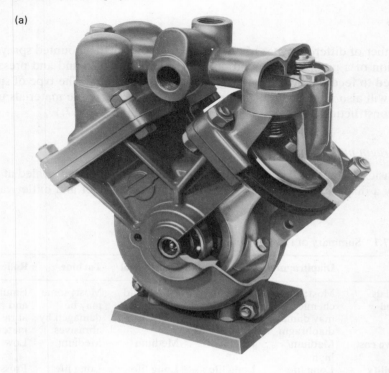

(b)

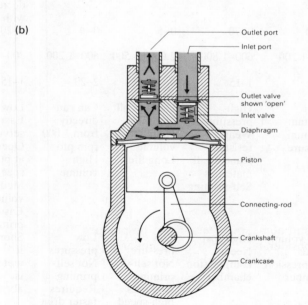

Outlet port
Inlet port
Outlet valve shown 'open'
Inlet valve
Diaphragm
Piston
Connecting-rod
Crankshaft
Crankcase

Fig. 7.3 (a) Diaphragm pump partly cutaway to show diaphragm and valves (photo Hardi UK Ltd (b) Diagram to show construction

Liquid is drawn through the inlet valve by movement of the diaphragm enlarging the chamber, and on the return of the diaphragm, it is forced out through the outlet valve. Some pumps have only one diaphragm, but usually two, three or more diaphragms are arranged radially around a rotating cam. This actuates the short movement of each diaphragm in turn to provide a more even flow of liquid instead of an intermittent flow or 'pulse' with an individual diaphragm. In any case a compression chamber, sometimes referred to as a surge tank, is required in the spray line if not incorporated in the pump to even out the pulses in pressure with each 'pulse' of the pump. These pumps are rather more complex as several inlet and outlet valves are required, but maintenance is minimal as there is less contact between the spray liquid and moving parts. Care must be taken to avoid using chemicals which may affect the diaphragms or valves. In general, diaphragm pumps are used to provide less than 10 bar pressure but maximum pressures of 15–25 bar are attainable.

Piston pump

Liquid is positively displaced by a piston moving up and down a cylinder, thus the output is proportional to the speed of pumping and is virtually independent of pressure (Fig. 7.4). Piston pumps require a positive seal between the piston and cylinder and efficient valves to control the flow of liquid. Owing to their high cost in relation to capacity, piston pumps are not used very much on tractor sprayers, but are particularly useful if high pressures up to 40 bar are required. A compression chamber is also required with these pumps. Piston pumps are less suitable for viscous liquids.

Centrifugal pump

An impeller with curved vanes is rotated at high speed inside a disc-shaped casing, and liquid drawn in at its centre is thrown centrifugally into a channel around the edge. This peripheral channel increases in volume to the outlet port on the circumference of the casing (Fig. 7.5). Centrifugal pumps are ideal for large volumes of liquid, up to 500 litres/min at low pressures. They can be used up to 5 bar, but the volume of liquid emitted by the pump decreases very rapidly when the pressure exceeds 2.5–3 bar. The pressure will increase slightly if the outlet is closed while the pump is running, and then slippage occurs without damage to the pump. Viscous liquids and suspensions of wettable powders and abrasive materials can be pumped. The seals on the shaft are liable to considerable wear as the pumps are operated at high speeds, but there is less wear on other parts as there are no close metal surface contacts. Instead of mounting a centrifugal pump directly on the pto, a belt or pulley drive is required to obtain sufficient rotational speed of the pump. The pump may also be driven hydraulically. Centrifugal pumps with a windmill drive are frequently used on aircraft spray gear. These pumps are not self-priming, and should be located below the level of liquid in the tank.

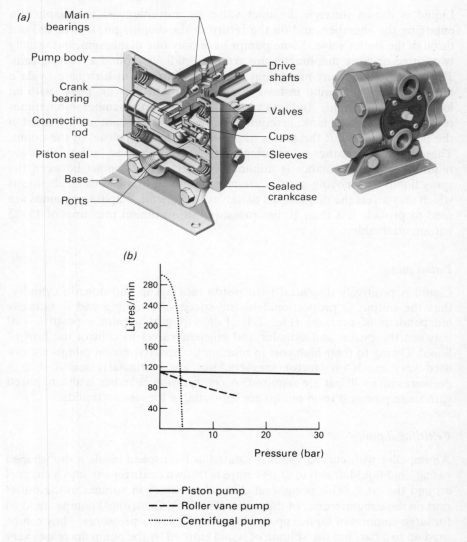

(a)

Main bearings

Pump body

Crank bearing

Connecting rod

Piston seal

Base

Ports

Drive shafts

Valves

Cups

Sleeves

Sealed crankcase

(b)

— Piston pump

‐ ‐ ‐ Roller vane pump

········· Centrifugal pump

Fig. 7.4 (a) Piston pump intact and cutaway diagram (photo Delavan Ltd) (b) Performance of piston pump related to other types. *Note*: A compression chamber or surge tank must be placed with either a piston or diaphragm pump to even out pulses of pressure

Pressure is increased in the turbine pump with a straight-bladed impeller in which liquid is circulated from vane to channel and back to the vane several times during its passage from the inlet to outlet port.

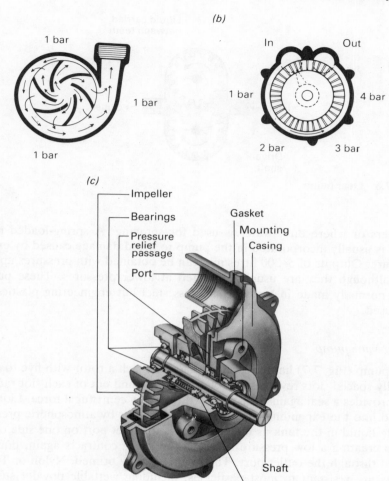

(a)

1 bar
1 bar
1 bar

(b)

In   Out
1 bar   4 bar
2 bar   3 bar

(c)

Impeller
Bearings
Pressure relief passage
Port

Gasket
Mounting
Casing

Shaft
Seals

Fig. 7.5 (a) Centrifugal pump (b) Turbine pump (c) Cutaway diagram to show construction of turbine pump (photo Delavan Ltd)

Gear pump

Gear pumps (Fig. 7.6) are seldom used and have been superseded by either the roller-vane or diaphragm pumps. The gear pump consists of two elongated meshed gears, one of which is connected to the tractor. The gears revolve in opposite directions in a closely fitting casing, the liquid being carried between the casing and the teeth to be discharged as the teeth enmesh once more. Any damage or wear to the gears or the casing results in a loss in efficiency, therefore these pumps should not be used to spray wettable

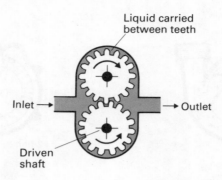

Fig. 7.6 Gear pump

powders or where dirty water is used for spraying. A spring-loaded relief valve is usually incorporated in the pump to avoid damage caused by excess pressure. Outputs of 5–200 litres/min can be obtained with pressures up to 6 bar, although they are usually operated at lower pressures. These pumps were normally made in brass or stainless steel but engineering plastics are also used.

Roller-vane pump

This pump (Fig. 7.7) has an eccentric case in which a rotor with five to eight equally spaced slots revolves. A roller moves in and out of each slot radially and provides a seal against the wall of the case by centrifugal force. Liquid is forced into the expanding space between the rotor by atmospheric pressure on the liquid in the tank as the rollers pass the inlet port on one side of the pump creating a low pressure area. As the space contracts again, liquid is forced through the outlet port. The pump is easily primed. Nylon or Teflon rollers are resistant to most pesticides, including wettable powder suspensions. Rubber rollers are recommended to pump water and wettable powders when the pressure does not exceed 7 bar. However, sand particles contaminating water supplies can abrade and damage the pump, so a filter between the spray tank and the pump inlet is essential to reduce the damage. The rollers can be replaced when necessary or the whole pump returned to the manufacturers for reconditioning. The case is usually made of cast iron or corrosion-resistant Ni-Resist, and has replaceable Viton, Teflon or leather shaft seals. The pumps are usually designed to operate at pto speeds of 540–1000 rpm with outputs from 20 to 140 litres/min, with pressures up to a maximum of 20 bar, although at higher pressures output and pump life are reduced. Output is approximately proportional to speed. The roller-vane pump is compact in relation to its capacity and is readily fitted to the pto and attached to a torque chain on the tractor. Before mounting, the pump shaft should be turned by hand, or with the aid of a wrench, to check that it turns easily in the proper direction.

(a)

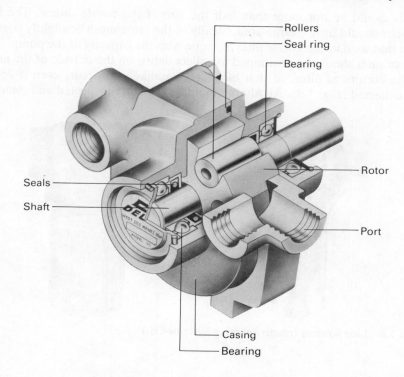

(b)

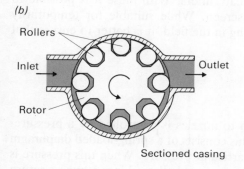

Fig. 7.7 Roller-vane pump (a) Cutaway diagram to show construction (photo Delavan Ltd) (b) Diagram showing action of pump

Filtration

Careful filtration of the spray liquid is essential to prevent nozzle blockages during spraying. Apart from a filter in the tank inlet, the pump must be protected by a filter, or line strainer, on its input side (Fig. 7.2), and each individual nozzle should have a filter. At the nozzle the apertures of the filter mesh should be not more than half the size of the nozzle orifice. The line strainer should have a large area, ideally of the same mesh or slightly coarser than that used in the nozzle filter, to cope with the capacity of the pump. The line strainer should be positioned to collect debris on the outside of the mesh at the bottom of filter, so that blockage is unlikely to occur, even if debris has collected (Fig. 7.8). All filters should be regularly inspected and cleaned.

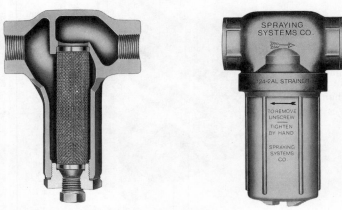

Fig. 7.8   Line strainer (photo Spraying Systems Co.)

Some manufacturers provide 'self clean' filters. With these it is possible to back flush debris collected on the screen. While suitable for temporarily cleaning the filter to complete spraying in the field, it is better to ensure that the screen is cleaned each day.

Pressure control

Flow of spray liquid from the pump to nozzles is controlled by a pressure-regulating valve (PRV) (Fig. 7.9). This consists of a spring-loaded diaphragm or ball valve which can be set at a particular pressure. When this pressure is exceeded, the valve opens and the excess liquid allowed into a bypass return to the spray tank. Liquid returned to the spray tank in this way provides hydraulic agitation of the spray liquid. The return flow should be through a suitable agitator at the bottom of the tank to ensure thorough circulation of the liquid. Some sprayers have a separate flow line to the agitator in addition to the bypass line from the pressure-regulating valve. When the pressure gauge is mounted next to the valve, readings have to be checked against pressures measured at the nozzles, so that account is taken of any drop in pressure between the valve and the nozzles. The drop in pressure to the end of a boom depends on the capacity of the boom, output of the nozzles and input pressure. It is important that the bore of the boom is adequate for the

(a)

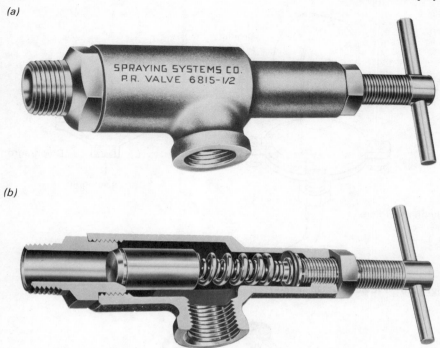

(b)

Fig. 7.9 (a) Pressure relief valve (photo Spraying Systems Co.) (b) Cutaway diagram to show construction

nozzles being used. Ideally, the output and pressure of liquid from the pump is in excess of total requirements of the nozzles, so that hydraulic agitation in the spray tank is continuous and sufficient to keep wettable powders in suspension, even when spraying at maximum output. Unfortunately, pressure gauges do not remain reliable under field conditions, and the gauge and sprayer calibration should be checked regularly. The life of a pressure gauge can be increased if it is protected by a diaphragm (Fig. 7.10). A gauge should have a large dial to facilitate reading.

Between the pressure-regulating valve and the nozzles, an on/off valve is positioned so that it can be easily operated by the tractor driver. Often there is a simple mechanical lever for the driver to operate, but for the totally enclosed safety cabs, electrically operated solenoid valves (Fig. 7.11) are required for remote control and to avoid pipes containing pesticides being in the cab. Some electronic devices are available to provide the tractor driver with a digital display of the area covered, output, speed and other variables. When the spray boom is divided into three sections, left, right and central, the main valve is often a seven-way valve, so that individual sections, pairs or the whole boom can be operated. This is particularly useful when the edges of fields are being treated and part of the boom is not required. On

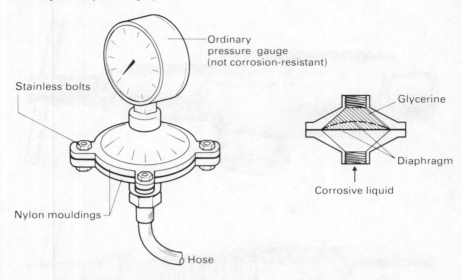

Fig. 7.10 Pressure gauge isolator

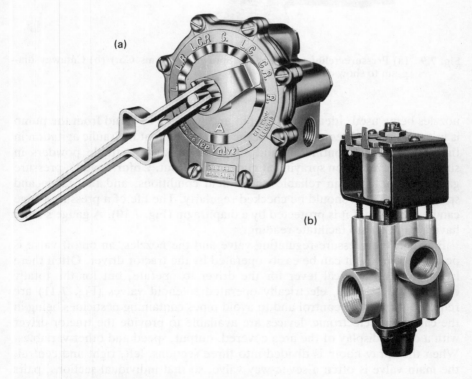

Fig. 7.11 (a) Solenoid valve (b) Seven-way tap for different boom sections (photos Spraying Systems Co.)

some sprayers liquid in the boom can be sucked back to the tank when the valve is closed. This may result in excess foaming and care must be taken to avoid damage to the pump if the sprayer is empty.

Spray booms

For most farmers the width of the boom is fixed. A suitable boom width for the fields can be calculated from

$$\text{boom width} = \frac{\text{area requiring treatment}}{\text{time available} \times \text{tractor speed}} = \frac{m^2}{h \times m/h}$$

thus if a farmer has a 100 ha field which needs treating in 3 days (6 h actual spraying per day) at a speed of 8 km/h, the minimum boom width required is 6.94 m (ie 7 m). Sprayer requirements should be based on completing the spray programme within 3 days in any 1 week to allow for rain, wind, equipment maintenance and other delays. On this basis, 1 m of boom is required for each 13.5 ha to be treated. A survey in the UK (Ministry of Agriculture, Fisheries and Food 1976) revealed that 59 per cent of farm sprayers had a boom between 10 and 12.5 m wide. However, the trend is to lengthen spray booms to give a wider swath, 18 m booms being popular, and reduce the number of wheelings across fields. Some booms are so wide that support wheels have to be fitted. Variation in spray deposit is liable to increase with wide booms due to greater movement of the end of the boom relative to the ground unless the land is very even. The pump output in litres/min is given by

$$\frac{\text{swath (m)} \times \text{application rate (litres/ha)} \times \text{velocity (km/h)}}{600}$$

for example with a 24 m boom travelling at 8 km/h, the pump capacity required to apply 200 litres/ha is 64 litres/min to allow for agitation. In practice, the cereal farmer also needs to choose a boom width related to the width of the seed-drill.

Boom design

Most booms are mounted at the rear of the spray tank, although some are in front of the tractor, particularly for band applications of herbicides so that the farmer can see the position of the nozzles in relation to the rows. The front boom position should not be used when spraying insecticides as the operator moves towards the spray. Booms are generally designed in three or more sections so that the outer sections can be folded for transport and storage. During spraying, the outer sections are often mounted so that they are moved out of the way by any obstruction which is hit. Manufacturers have used various methods to pivot and fix the boom sections for easy handling. Normally, the booms are unfolded by hand, but on some sprayers, positioning of the boom can be controlled through the hydraulic system without the operator leaving the tractor.

During field spraying, uneven distribution of pesticide is caused by movements of the boom, including vertical bounce, horizontal whip or both, which is accentuated as booms increase in width. Due to the yawing, the boom may be stationary at times in relation to ground speed, so causing an overdose of pesticide. The rolling movement varies the height of nozzles relative to the crop and thus the pattern of overlap is affected. Ideally, the boom should be as rigid as possible over its length and mounted centrally in such a way that as little as possible of the movement of the tractor is transmitted to the boom. Any breakaway mechanism should be strong and return the outer boom quickly and positively into its correct position. Booms constructed as stiff cantilevers have been shown to be better than other types (Nation 1982). An inclined-link boom suspension was developed to allow articulation between the boom and sprayer in both rolling and yawing planes (Nation 1985). Instead of a passive suspension, a boom can now be fitted with an active suspension in which a sensor detects the height of the boom relative to the crop and controls its position (Frost 1984; Marchant 1987; O'Sullivan 1988; Frost and O'Sullivan 1988; Marchant and Frost 1989).

Nozzles on spray boom

A wide range of hydraulic nozzles (see pp. 102–13) can be used on a boom. The nozzle body may be screwed into openings along the boom, but often the boom incorporates special nozzle bodies clamped to the horizontal pipe (Fig. 7.12). Sometimes the liquid is carried to the nozzles in a plastic tube so that spacing between nozzles can be adjusted by sliding the nozzle body along the boom (vari-spacing). Choice of nozzle tip depends very much on the material being sprayed, the volume of liquid needed and the ultimate target, so that the output (litres/min), spray pattern quality and angle and droplet size are appropriate. Cone nozzles are preferred for application of insecticides and fungicides to foliage, while fan nozzles are mostly used when treating the soil with any pesticide. To reduce drift, herbicides are sprayed at low pressure, usually 0.7–1 bar, and a guard may be needed to protect the

(a)

(b)

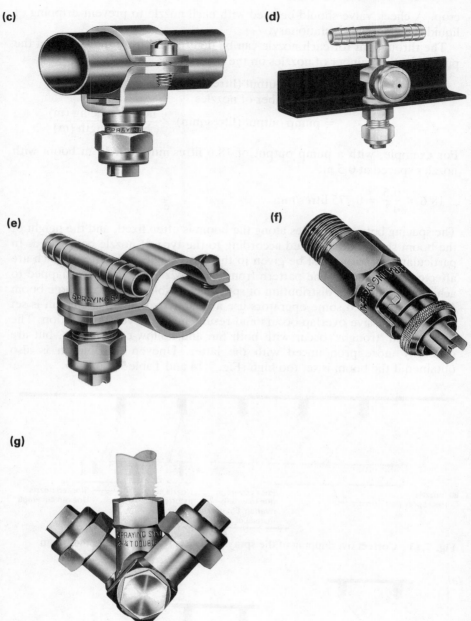

(c)

(d)

(e)

(f)

(g)

Fig. 7.12 Different systems of fitting nozzles to spraybooms (a) Conventional nozzle body screws into boom (b) Nozzle tip screws directly into boom (c) Nozzle body clamps to pipe boom (d) Nozzle fixed to L section boom with hose between nozzles (e) Vari-spacing (f) Bayonet fitting of nozzle tip and (g) double swivel nozzle on downpipe (photos Spraying Systems Co.)

crop. A check valve should be used with each nozzle to prevent dripping of liquid if the sprayer is stationary.

The throughput for each nozzle can be determined from the output of the pump and the number of nozzles on the boom, thus

$$\text{nozzle throughout (litres/min)} = \frac{\text{pump output (litres/min)}}{\text{number of nozzles}}$$

$$= \text{pump output (litres/min)} \times \frac{\text{nozzle spacing (m)}}{\text{boom length (m)}}$$

For example, with a pump output of 18.6 litres/min on a 12 m boom with nozzles spaced at 0.5 m,

$$18.6 \times \frac{0.5}{12} = 0.775 \text{ litres/min}$$

The spacing between nozzles along the boom is often fixed, and the height of the boom should be adjusted according to the type of nozzle being used. In particular, attention must be given to the spray angle and pattern which are affected by pressure. The pattern from each nozzle has to be overlapped to achieve as uniform a distribution of spray as possible across the whole boom (Fig. 7.13); indeed some operators use a double overlap. If the boom is set too low, excessive overlap occurs and results in an uneven distribution. The 'peaks' and 'troughs' occur with both fan and hollow-cone nozzle, but are generally more pronounced with the latter. Uneven distribution is also obtained if the boom is set too high (Fig. 7.14 and Table 7.2).

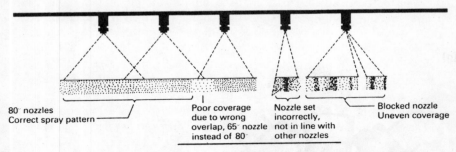

80° nozzles
Correct spray pattern

Poor coverage due to wrong overlap, 65° nozzle instead of 80°

Nozzle set incorrectly, not in line with other nozzles

Blocked nozzle
Uneven coverage

Fig. 7.13 Correct overlapping of the spray pattern is required across the boom

(a)

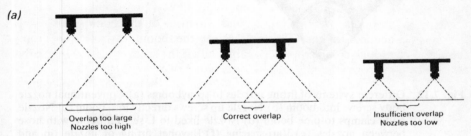

Overlap too large
Nozzles too high

Correct overlap

Insufficient overlap
Nozzles too low

Fig. 7.14 Correct height above the crop is essential

Table 7.2 Variation in boom height (cm) above crop or ground with different nozzle spacing along the boom and spray angles

Nozzle angle (°)	Nozzle spacing along boom (cm)		
	46	50	60
65	51	56	66
80	38	46	50
110	24	27	29

Nordby and Skuterud (1975) recommend that the boom should be about 40 cm high, the working pressure of fan nozzles should not exceed 2.5 bar, and that to reduce the risk of spray drift herbicide sprays should be applied only when wind speeds are less than 3 m/s. Downwind drift is increased if the boom height is increased (P.C.H. Miller 1988). Lower boom heights are possible if a fan spray is directed back at an angle instead of pointing vertically down on the crop (some nozzles are specially made to do this) or if wide-angle nozzles are used, but the wider the spray angle, the greater the risk of producing more very fine droplets. The distribution can be checked by spraying water on to a dry surface or placing strips of water-sensitive paper across the swath, or by adding a dye such as lissamine scarlet to the water, a record of the distribution being obtained by spraying across a band of white paper. If the spray pattern is uneven, the throughput of individual nozzles can be checked (see p. 126). A computer model showed that for a boom set at the optimum height, the coefficient of variation increases continuously with increases in boom roll angle, due to the changes in nozzle height, rather than a change in the angle of the spray (Mawer and Miller 1989). Electronic instruments to measure flow rate can be used to check the evenness of the output across a boom, but the actual output should be checked by collecting liquid in a calibrated container.

Most tractor-mounted booms have nozzles arranged in a horizontal line and directed downwards, but for spraying some crops to get underleaf cover, 'drop legs' are fitted so that nozzles can be directed sideways or upwards into the foliage (Fig. 7.15). When tail-booms were used on tractor equipment, they were pivoted on the horizontal boom and held by a strong spring. Also, the bottom section of the boom was mounted on a flexible coupling to avoid damage if the boom touches the ground. In front of the booms, a curved guard was needed to ease the passage of the boom through the crop. Movement through some crops is possible only if the sprayer is used regularly along the same rows and in the same direction.

Felber (1988) described the mounting of a bar below the boom which deflects plants, such as wheat, to provide a gap that allows the nozzles to be set at crop height and spray to penetrate inside the crop canopy. The main advantage is for fungicide application where deposition lower in the canopy is required.

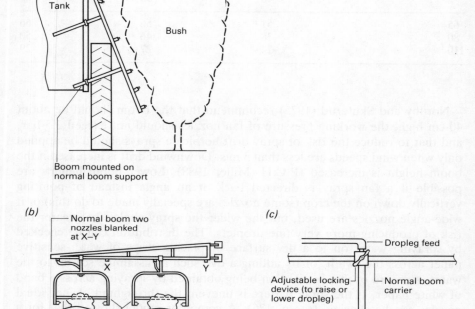

Fig. 7.15 Specialist booms for certain crops (a) Soft fruit (b) Strawberries (c) Brassicas

Some chemicals are applied in a band, usually 18 cm wide, along the crop row to reduce the cost of chemical per hectare. Band spraying requires a higher standard of accuracy in the selection and positioning of the nozzles which are often mounted on the seed-drill (Fig. 7.16).

Calibration of a tractor sprayer

The importance of careful calibration cannot be overstressed. One method is to select the gear to a pto speed of 540 revs/min and forward speed which gives an acceptable level of boom movement. Next mark out a 100 m and with the tractor moving at the required speed as it passes the first mark, time how long it takes to cover the 100 m to the next marker. The forward speed

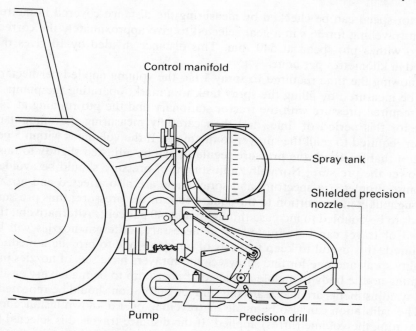

Control manifold

Spray tank

Shielded precision nozzle

Pump

Precision drill

Fig. 7.16 Tractor-mounted band sprayer

km/h = 360 ÷ the measured time (sec). Measure the nozzle spacing (m) and calculate the output required per nozzle as follows:

$$\frac{\text{Volume application rate (l/ha)} \times \text{speed (km/h)} \times \text{nozzle spacing (m)}}{600}$$
$$= \text{nozzle output (l/min)}$$

For example $\dfrac{200 \text{ l/ha} \times 6 \text{ km/h} \times 0.5\text{m}}{600} = 1 \text{ l/min}$

The nozzle is then selected from the information in the manufacturer's charts to emit the correct volume at the appropriate pressure and achieve the spray quality required. With the spray boom set up, the output of the nozzles is checked. Another method which can be used to check the calibration of the sprayer is to calculate the time required to spray 1 ha, thus:

$$\frac{600}{\text{swath (m)} \times \text{speed (km/h)}} = \text{time required (min)}$$

Note the effective swath is the distance between each nozzle along the boom multiplied by the number of nozzles, for example if 30 nozzles are spaced at 50 cm intervals the swath is

$$\frac{30 \times 50 \text{ cm}}{100 \text{ cm}} = 15 \text{ m}$$

Tractor speed can be checked by measuring the distance covered in metres when travelling for 36 s in a gear selected to give approximately the correct speed with a pto speed at 540 rpm. This distance divided by 10 gives the speed in kilometres per hour.

Knowing the time required to spray 1 ha, the volume applied per hectare can be measured by filling the spray tank to a mark, operating the pump at the required pressure with the tractor stationary and the pto running at 540 rpm for this period of time, and then carefully measuring the amount of water required to refill the sprayer to the mark. If the volume is within 5 per cent of that required, the pressure regulator can be adjusted slightly to raise or lower the pressure. Normally, adjustment of pressure should be avoided because droplet size spectrum and spray angle are also affected and nozzle throughput is in proportion to the square root of the pressure; thus pressure needs to be doubled to increase throughput by 40 per cent. Alternatively, the speed of travel can be adjusted or, if necessary, different nozzles will be required. It is useful to keep two sets of nozzles – one to provide a medium quality spray and one for fine sprays. Some sprayers have sets of nozzles in a rotating nozzle body to enable a change of nozzle tip to be made very easily. If any adjustments are necessary the sprayer calibration should be repeated.

The calibration can also be made by travelling over a known distance and measuring the volume (litres) applied. If the distance travelled is selected by dividing the boom width (m) into 1 000, the volume measured multiplied by 10 is in litres per hectare. The pump pressure and speed of travel must be constant.

With a band spray, the application rate can be calibrated as described above, but as only a proportion of the area is actually sprayed, the rate per treated area will be higher in proportion to the ratio between the width of the treated plus untreated band and the treated band, thus:

$$\text{volume applied to superficial area (litres/ha)} \times \frac{\text{width of treated + untreated band}}{\text{width of treated band}}$$
$$= \text{volume applied to band}$$

for example 20 litres/ha is applied but confined to band 20 cm bands along rows 100 cm apart. Thus:

$$20 \times \frac{100}{20} = \text{litres/ha on the treated bands}$$

Details of any calibration of the sprayer should be recorded for future reference (Table 7.3).

A sprayer should be cleaned and checked regularly. A stock of spare parts should be readily available. In particular, it is wise to keep spare nozzle-tips and take some to the field during spraying. If a nozzle is blocked, a replacement can be quickly fitted to avoid the need to clean a blocked nozzle in the field. The output of each nozzle should be checked periodically to ensure that it has not increased (see p. 129). The cost of a replacement nozzle is negligible in comparison with costs of the pesticides sprayed. The interval between checks will depend on the volume and type of liquid sprayed.

Table 7.3   Record of calibration

| Tractor – make | | | – registration | | | | litres |
| Sprayer – make | | | – tank capacity | | | | |

Calibration	Tractor gear	Throttle setting (rpm)	Ground speed (km/h)	Nozzle-tip size	Pressure (bar)	Output (l/h)	Area per load* (ha/tank)
1							
2							
3							

* $\dfrac{\text{Tank capacity (litres)}}{\text{output (litres/ha)}}$

Swath matching

Matching the end of one swath with the next is not easy, especially in a closely spaced cereal crop. Wheelings in the crop for fertilizer and pesticide application reduce yield, especially at advanced stages of crop growth. Nearly 4 per cent of losses can occur on a 10 m swath. The problem is overcome by not drilling or burning off gaps for the wheels of the tractor. These are referred to as 'tramlines' (Fig. 7.17). Tillering, and more grains per ear on the plants adjacent to the gaps, almost compensates for the reduced plant

Fig. 7.17   Tramlines through a cereal crop (photo J. W. Chafer Ltd)

population. With tramlines there is a small saving in seed, operations subsequent to drilling are quicker, and late applications, if needed, are more likely to be applied at the correct time.

The tramline system requires the width of the seed drill, fertilizer spreader and sprayer to match (Fig. 7.18). Tramlines are established by blocking appropriate drill coulters at the required intervals across the field. The seed cut-off mechanism can be operated automatically on certain drills. Tractor

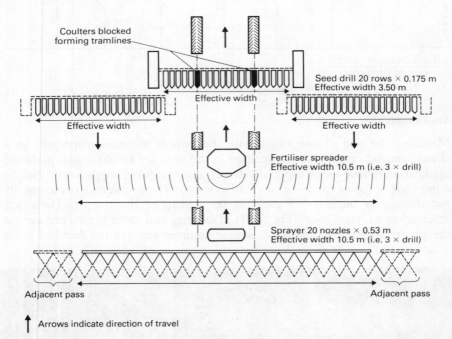

Fig. 7.18 Diagram to show formation of tramlines by matching seed drill, fertilizer spreader and sprayer. *Note:* Three passes of seed drill is equivalent to one swath with sprayer

tyre widths may also necessitate a slight displacement of the coulters on either side of the tramlines. Other systems are available, for example when the seed drill cannot be altered, a herbicide such as paraquat is sprayed behind the tractor wheels. The disadvantages of the technique are that it is more expensive, accurate marking is essential and great care is needed to avoid spread of the herbicide to other areas. The headland operations and weed control must be carefully planned to ensure continuity of clean tramlines. Increased attention to rabbit and hare control may be required, since these vertebrate pests may use tramlines as 'runs' into the fields. It is well worth spending some time measuring out each swath and having fixed marks to indicate the centre of each swath, even on row crops. Damage to bushy plants, like cotton, caused by the passage of the tractor is less than expected owing to plant compensation – even when a tractor with only a 48 cm clearance at the front axle is driven over two rows of cotton – and is more profitable than growing an alternative low crop, such as groundnuts, along the 'pathways' (Tunstall et al 1965).

On some sprayers a large blob of foam or dye is released at the end of the boom to indicate the edge of the swath, but the tractor driver has great difficulty in aligning the end of his boom with the blobs of foam unless he has a system of mirrors in a horizontal periscope. One mirror is mounted on the tractor and another large one at the end of the boom. The mirrors must be mounted to avoid vibration. Another method is to tie a long length of twine to each end of the spray boom. As the tractor is turned, the end of the twine attached to the boom is brought back over the other end trailing in the crop. A weight may be attached to the string. The use of automatic headland markers has also been tried. A small battery-powered wheeled marker moves across the headlands while remotely controlled by the tractor driver. Headlands cleared of all obstructions and uneven ground are needed for the passage of the marker. Often the crop is sown right up to the edge of the field and no headland is available for turning. When the turn is made inside the crop, the crop will be overdosed if spray is applied during the turns. It is preferable to spray two swath widths around the field, and then treat the remainder of the field by spraying swaths parallel to the longest side of the field. The pto is kept running during the turn to keep the spray liquid agitated, but the valve to the boom is closed throughout the turn (Fig. 7.19).

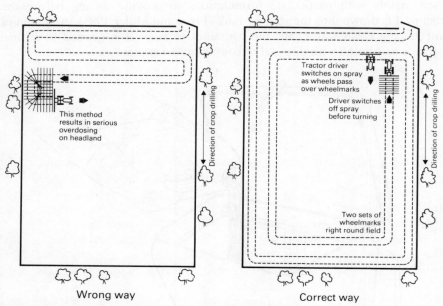

Fig. 7.19   The correct and wrong way to spray a field (courtesy Schering Agrochemicals Ltd)

However, some farmers now have an untreated headland which is managed separately to conserve wildlife in the hedgerows.

Refilling the sprayer

If the spray tank is not fitted with a sight gauge, a flickering of the pressure-gauge needle indicates that the sprayer is empty. The pto should be disengaged immediately to avoid the pump running while 'dry'. Most operators prefer to empty the tank before refilling to avoid errors in calculating and measuring the amount of chemical for a different volume of liquid. Nevertheless, if the sprayer empties while in the middle or far side of a field, time is wasted in returning to the refilling point, and a smaller quantity may be required for the last load to avoid having any spray left over. If possible the farmer should have detailed measurements of his field, so that with accurate calibration the appropriate amounts of chemical can be calculated beforehand for each load, thus reducing the time for ferrying to refill the sprayer.

To increase safety to the operator, the trend is to use the sprayer pump to fill the tank partially with water and then transfer and mix a measured quantity of pesticide using a closed system to reduce direct contact with the chemical (Brazelton and Akesson 1987). Such systems include the use of a suction probe to use the sprayer pump to draw chemical from its container (Fig. 7.2). Sprayers may also have an induction bowl (Fig. 7.20) which is used mainly with particulate formulations to provide mixing before the chemical is drawn into the sprayer tank (Frost and Miller 1988). A system of introducing clean water into the container to rinse it is needed to reduce hazards associated with disposal of contaminated containers.

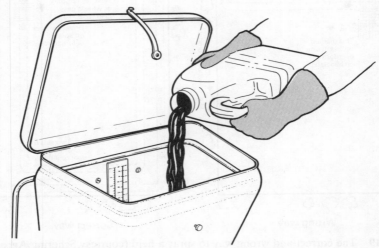

Fig. 7.20   Low-level mixing bowl

Some systems have now been developed in which the unopened pesticide container is totally enclosed inside a chamber before it is punctured to release the contents (Fig. 16.3, p. 346). The chamber acts as a measuring and storage unit, and where possible the size of the pesticide container is selected to match the requirements for the day's spraying operation. The chamber must be fitted with a system to rinse the container and in some the empty container can then be crushed to prevent re-use.

Alternatively, the farmer may have a separate water tank or mixing facility with its own pump to refill the sprayer. Great care must be taken to avoid contamination of the water source.

Metered spraying

Uniform application with the equipment described so far depends on a constant tractor speed and constant pressure. Forward speed may vary, so systems are needed to regulate the flow of liquid to the nozzles. A variation in speed from 0 to 80 km/h must be considered when herbicides are applied to railway tracks (Amsden, 1970). Some systems incorporate a metering pump which is linked to the pto or sprayer wheel, and a proportion of spray may or may not be returned to the tank. Pump output must be proportional to the forward speed, so a diaphragm or piston positive displacement pump is needed; gear or roller-vane pumps are unsuitable. When the pump – usually a piston pump with an adjustable stroke – is driven by the sprayer wheel, a second pto pump is needed for agitation and refilling the tank (Fig. 7.21). The main disadvantage is that the power required to drive the metering

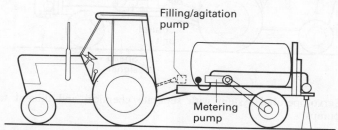

Fig. 7.21 Metering pump system

pump is high, 10 hp being needed to supply 500 litres/ha through a 12 m boom. This can be overcome by using the ground-wheel pump at low pressure and a separate pto pump to boost pressure to the nozzles. These systems are relatively simple to operate, but droplet size is also affected when flow rate is adjusted by pressure. The operator should try to keep within ± 25 per cent of the selected speed so that the pressure is not greatly affected. Other systems include a centrifugal regulator linked to the sprayer wheel and

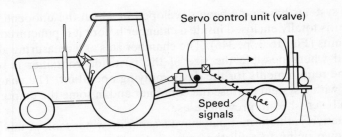

Fig. 7.22 Output controlled by electronic sensing of forward speed

metering pumps or valves operated electronically by the forward speed of the sprayer (Fig. 7.22). The more complex electronic systems are expensive, and their use is limited unless specialized maintenance facilities are available. All systems linked to the rotation of the pto or wheel may be affected by wheel slip causing underdosing or overdosing, so the metering device must be operated by a trailed wheel rather than a driving wheel (Amsden 1970). The spray is already mixed in the sprayer tank with these automatic regulating systems. Ultimately, the chemical and diluent may be kept in separate tanks (Fig. 7.23) and using an in-line mixing system with the concentrate of spray

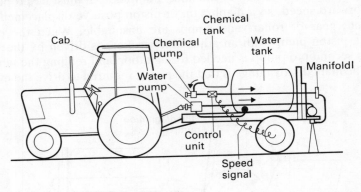

Fig. 7.23 Servo-operated system with separate chemical and diluent tanks and pumps

affected by forward speed (Hughes and Frost 1985). Unused chemical can then be readily returned to the store. Frost (1990) has described a novel metering system in which the flow of water is used to control the flow rate of the chemical, making the system independent of the characteristics of the chemical (Fig. 7.24).

In another closed system, a piston pump with a ceramic piston to withstand the effects of the pesticide concentrate is used to meter the chemical into a mixing chamber. An electric stepper motor, controlled from the tractor cab, is used to adjust the length of pump stroke and thus the input of chemical

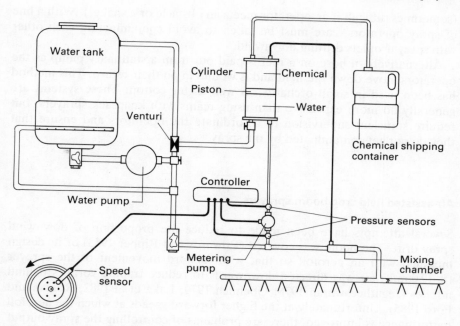

Fig. 7.24 Pesticide injection metering system (from Frost 1990)

into the water that is pumped separately into a mixing chamber and thence to the nozzles (Landers 1988). Humphries and West (1984) describe a similar system which uses compressed air to force the pesticide to the mixing chamber.

Regardless of which system is used, the sprayer must be properly calibrated, and worn parts, especially nozzles, replaced regularly.

Portable-line sprayers

A flexible boom or hose can be used when a horizontal boom on a tractor cannot be used in orchards or forests, or because the land is undulating. Sufficient labour is needed to carry the hose. Various types of portable line have been used. In cotton fields operators spaced at 4 m intervals carry an interconnecting hose on a short mast, supported in a waist strap. The interconnecting hose is liable to stretch unless strengthened by Terylene or a similar fibre in the hose wall. The portable line is connected to a spray tank and pump, mounted on a trailer driven or pushed along pathways across the field. The main difficulties with the system are the need for a pressure regulator at each operator to compensate for the pressure drop along the line, and the need for each operator to walk at the same speed as the tractor (sometimes difficult if an operator meets an obstacle or a snake!). With a line of spray operators, care must be taken to avoid contaminating each other with spray droplets drifting downwind.

Alternatively, a hose on a reel is paid out from a stationary pump as the operators move down the field and is wound in on their return. This method has been used in small orchards, as well as for cotton. These systems are generally no more expensive than using teams with knapsack sprayers, but require sufficient supervision to co-ordinate the operators and ensure that they do not get contaminated by the spray.

Air-assisted field crop boom sprayers

Several attempts have been made to reduce the proportion of downwind spray drift. The boom was covered (Edwards and Ripper 1953) or its design modified with an aerofoil so that with forward movement of the sprayer (8–12 km/h) air was directed downwards to reduce trailing vortices behind nozzles (Jegatheeswaran 1978; Goehlich 1979; Lake et al 1982; Rogers and Ford 1985). Unfortunately at the higher forward speeds at which the aerofoil performance is improved, there are problems of controlling the spray output while accelerating and slowing down at the field edges.

More recently penetration into crop canopies and reduced risk of spray drift has been achieved by using a centrifugal or axial fan to deliver air through a sleeve mounted above the boom and nozzles (Fig. 7.25). Air from the sleeve produces a curtain which directs the spray cloud into the crop canopy, and reduces lateral displacement of the spray in a crosswind. The reduction in spray drift permits the use of nozzles with a finer spray or allows a faster forward speed (Taylor et al 1989). Conversely the air curtain can increase drift in the absence of crop foliage on which the droplets can impact due to the deflection of air by the ground. Other details of equipment used in air-assisted sprayers are given in Chapter 10.

Some farmers have also used twin-fluid nozzles (see p. 107), especially as the reduction in drift increases the number of days on which a spray may be applied for optimal timing of a pesticide (M.J. May 1991; Nettleton 1991).

Incorporating herbicides

Some volatile herbicides, such as trifluralin and dinitramine, must be incorporated into the soil to prevent loss by volatilization or photodecomposition by sunlight. Incorporation also reduces the rainfall requirement and places the chemical in close contact with the weed seeds or roots for best control. The implements used for incorporation will vary with different herbicides,

(a)

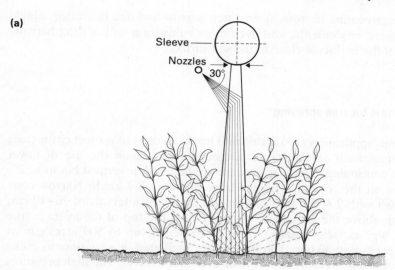

Sleeve

Nozzles

O   30°

(b)

Fig. 7.25 Air-sleeve sprayer (a) Diagram to show relative position of airflow and spray from nozzle (b) Sprayer

depending on the distribution of chemical required. Power-driven devices are required to incorporate herbicides evenly to a precise depth. A. Walker et al (1976) reported that a single pass with a rotovator gave an even distribution of trifluralin in the top 5 cm of soil, ie its working depth. Incorporation to half its working depth was obtained by cross-cultivations with a rotary power harrow, reciprocating harrow, springtined harrow and disc cultivator. Single passes of these implements, and even cross-cultivation with a drag harrow, left much of the herbicide close to the soil surface.

Special booms for tree spraying

High-volume application (100 litres/tree) has been used to protect citrus trees in some areas such as California, although research on the use of lower volumes is continuing. These sprays are applied with a vertical boom which is mounted on the rear of a tractor travelling at 2–2.4 km/h. Narrow-cone nozzles (16°) with 2.4–3.5 mm orifices are placed at intervals of 30–40 cm, from 45 cm above the ground to no lower than the top of the average tree height. A high-capacity pump is needed to deliver up to 500 litres/min at pressures exceeding 30 bar. The nozzles are oscillated in a continuous cyclic rotation pattern once every 0.5 – 0.6 m of forward travel. The high pressures and volume with this system are essential to penetrate the peripheral 'shell' of foliage and achieve up to 88 per cent coverage, including the upper central parts of the trees (Carmen 1975).

To treat small trees it is possible to mount the boom within a tractor-

Fig. 7.26 Animal-drawn sprayer, with engine-driven pump

mounted shield to reduce the effect of wind on spray dispersal (Cooke et al 1977). With the need to reduce spray drift in orchards there is likely to be an increase in the use of 'tunnel' sprayers (Fig. 10.12, p. 239).

High-volume fungicide sprays (>800 litres/ha) for the control of coffee berry disease in East Africa have been applied with twin horizontal overhead booms covering eight rows (2.7 m apart) on a single pass (Pereira and Mapother 1972). Nozzles were operated at 54 bar. Deposits were predominantly on upper surfaces and decreased from the top to lower parts of the coffee, so that effective control of the disease depended on redistribution of the fungus. Overhead spraying is less effective against coffee leaf rust as these spores germinate on the undersurface of leaves. Pereira (1972) has also described an inverted 'hockey-stick' boom at the rear of a tractor-mounted sprayer operated at 7 bar for both overhead and lateral spray application of coffee trees. Variable-cone nozzles, 250 mm apart, on the vertical section of boom were angled to spray up through the branches to improve coverage on the berries. Small fruit trees can be sprayed with an overhead boom.

Some experiments have been made with a cross-flow fan driven by a hydraulic motor to provide an air flow to improve penetration of a tree canopy. The main advantage of the system is that a fan unit can be mounted on a hydraulically manoeuvred boom so that the nozzles are placed relatively close to the foliage.

Animal-drawn sprayers

Animal-drawn sprayers have been used where farmers have draught animals such as oxen, as in Central Africa. The tank, boom and pump are usually mounted on a suitable wheeled frame. A high-clearance frame is needed for some crops (Fig. 7.26). These sprayers can be operated even when conditions are too wet to allow the passage of a tractor, and the crop is not damaged by the animals. The pump can be driven by a small engine or by means of a chain drive from one of the wheels on the frame. When the latter is used, the pump has to be operated for a few metres to build up sufficient pressure at the nozzles before spraying starts. If wheel slip occurs, spray pressure will decrease.

8

Controlled droplet application

Controlled droplet application (CDA) emphasizes not only the importance of applying the *correct size of droplets* for a given target, but also the *uniformity of droplet size* to allow use of minimum volumes and dosages to achieve effective control. A narrow droplet spectrum is generally shown by a vmd/nmd ratio less than 2 when a sample of spray is measured with a laser light diffraction system. The size of the droplet is chosen in relation to its intended target (see p. 47) (Bals 1975b; Matthews 1977).

Most use of CDA has occurred in insecticide/fungicide application where it is a logical extension of the ULV concept, the objective of which is to apply the *minimum volume* compatible with achieving economic control. Water-based formulations, such as wettable powders and emulsifiable concentrates may be applied by CDA techniques at very low volumes (VLV). However, when droplets smaller than 100 μm are applied the risk of evaporation shrinking the size of droplets while airborne has to be counteracted by the addition of an adjuvant of low volatility, such as an oil or molasses. An adjuvant may also be needed to enhance redistribution, cuticle penetration or rainfastness. When ULV/CDA sprays are applied using small droplets it is usual to have a low-volatility formulation. Specific ULV formulations for use with CDA machines have been made available in many countries , particularly in Africa where the technique is widely used in semi-arid areas. Total volumes are generally in the 0.5 – 3.0 l/ha range.

Using larger droplets for herbicide application generally means water-based formulations can be applied without a risk of drift due to evaporation from droplets. Such use of a more uniformly sized large droplets will reduce drift compared to sprayers producing a wide spectrum of droplet sizes (Lloyd et al 1986).

Controlled droplet application came into prominence in the UK when efforts were made to improve herbicide application by reducing *drift* and improving the timeliness of applications, where weather – rain and wind – and soil conditions can limit the number of days suitable for applying sprays. Fraser (1958) and others had previously stressed the need for a nozzle which produces a *narrow* spectrum of droplet sizes to increase the proportion of spray reaching its intended target. The very wide range of droplet sizes produced by most nozzles results in off-target losses due either to drift or run-off or both, so efficiency of conventional nozzles is poor. Drift can be reduced but not eliminated by changes in formulation, thus average droplet size is increased when viscosity additives are added to sprays (Butler et al 1969), or by operating nozzles at low pressure. Alternatively, granules may be applied instead of sprays, but poor retention of these on foliage has limited their use against many pests.

One method of controlling size of droplets within fairly narrow limits is by using centrifugal-energy nozzles (eg spinning discs or cages), with which droplet size can be adjusted by varying their rotational speed. As pointed out in Chapter 5, these nozzles operate efficiently only when the volume of spray applied is restricted to prevent flooding resulting in sheet formation. Ideally, a suitable formulation and flow rate are selected so that at a given rotational speed droplet formation is from ligaments with a minimal number of satellite droplets. As indicated in Chapter 5, very uniform droplets can be produced if the flow rate is low enough to avoid ligaments being produced. A narrow droplet spectrum is also achieved with an electro-dynamic nozzle described in Chapters 5 & 9.

The spray volume required depends not only on the selected droplet size but also on the number of droplets required on a target surface. When a spray is evenly distributed over a flat surface, the same number of droplets per unit area ($100/cm^2$) is achieved with as little as 500 ml/ha when 46 μm droplets are applied as 1.8 l/ha with 70 μm droplets or 200 l/ha with 340 μm droplets (see pp. 47–9). If fewer droplets are needed to control a pest, then less liquid is needed per unit area. In practice as little as 18 litres per hectare of certain herbicides has given good weed control when 300 μm droplets provided an average of $14/cm^2$. In some cases in the UK as little as 2.5 l/ha has been used in upland pastures and 8 l/ha for bracken control using smaller droplets.

Hand-carried, battery-operated spinning-disc sprayers

These lightweight sprayers have a plastic spray head with small dc motor which drives a rotating disc, a liquid reservoir (a screw-on bottle), a handle and a power supply. Various designs are available to provide particular droplet spectra and to accommodate different types of battery (Fig. 8.1). The main use of these sprayers has been on tropical crops, especially cotton, where farmers have great difficulty or find it impossible to collect sufficient water for conventional spraying techniques (Matthews 1989a, 1990). Cauquil

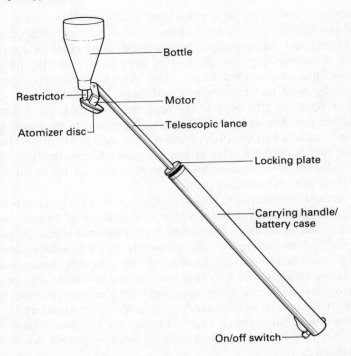

Fig. 8.1 Hand-carried spinning disc sprayer. Diagram to show component parts

(1987) reported on the extensive use of these sprayers to treat approximately 1 million hectares of cotton in West Africa. This equipment has also been recommended for locust and armyworm control, in forestry and by local authorities for pest control in urban situations.

Disc design

Initially two discs were used so that any liquid splashed from the rear of the front disc was collected and thrown off the rear disc. In practice, most liquid which reached the rear disc did so where the two discs were joined by pegs (Boize and Dombrowski 1976). Movement of air, caused by the fan action created by the two discs, increased power consumption. Most of these sprayers now have a single disc to minimize power consumption.

A number of different disc designs have been used (Fig. 8.2) including smooth edged or toothed, 'saucer' or 'cup' shaped and with or without grooves leading to the peripheral teeth. The size of the disc is also dependent on the particular sprayer. In practice discs with uniform distribution of liquid via grooves to the teeth improve ligament formation even at higher flow rates to produce a narrow droplet spectrum.

Different disc designs

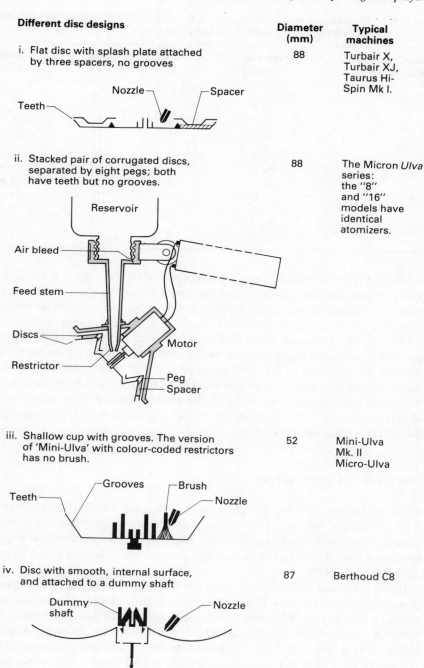

	Diameter (mm)	Typical machines
i. Flat disc with splash plate attached by three spacers, no grooves	88	Turbair X, Turbair XJ, Taurus Hi-Spin Mk I.
ii. Stacked pair of corrugated discs, separated by eight pegs; both have teeth but no grooves.	88	The Micron *Ulva* series: the "8" and "16" models have identical atomizers.
iii. Shallow cup with grooves. The version of 'Mini-Ulva' with colour-coded restrictors has no brush.	52	Mini-Ulva Mk. II Micro-Ulva
iv. Disc with smooth, internal surface, and attached to a dummy shaft	87	Berthoud C8

Fig. 8.2 Diagrams to show variations in the design of discs (from A. C. Arnold 1983) except the 'Ulva' adapted from Boize and Dombrowski (1976)

Table 8.1 Droplet size with spinning cup of the 'Mini-Ulva' spraying a low volatility oil (HLP40) (from Johnstone and Johnstone 1976)

Rotational speed of disc (rpm)	Flow rate (ml/min)	vmd	nmd	Ratio vmd/nmd
9 000	4.0	71	36	2.0
	8.5	64	40	1.6
	26.0	71	50	1.4
	60.0	81	55	1.5
12 000	4.0	56	27	2.1
	8.5	54	33	1.6
	26.0	52	39	1.3
	60.0	60	38	1.6
15 000	4.0	40	29	1.4
	8.5	41	32	1.3
	26.0	45	33	1.4
	60.00	64	37	1.7

Droplet spectra from a Micro-ULVA disc (Table 8.1) spraying an oil shows an optimum flow rate for a given disc speed when the ratio of vmd/nmd is minimal. When a water-based formulation is used, it must contain sufficient wetting agent (0.5–5.0 per cent v/v) to ensure good distribution on the disc.

One of the problems when liquid is fed near the centre of a disc is to prevent liquid entering the motor along its shaft. A separate baffle plate or spinner may be fixed to the shaft between disc and motor. Alternatively, the centre of the disc can incorporate a cylindrical baffle which interleaves with corresponding channels moulded in the motor housing. The motor should always be run for a few seconds after stopping the flow of liquid, so that all the liquid is spun off the disc.

Disc speed and power supply

Most of the sprayers are designed to operate with zinc-carbon (Leclanche-type batteries) but the performance of these varies depending on the manufacturer and their storage life. High power transistor-type batteries provide a longer service life than single-power cells (Matthews and Mowlam 1974; Beeden 1975).

Battery voltage decreases as power is used, so unless the motor speed is governed, droplet size will increase as voltage and thus the motor speed declines. Resting the batteries allows repolarization to occur and the voltage partially recovers. Normally, a period of 15–20 min spraying would be followed by a rest of 5–7 min to change the bottle and spraying would be for no longer than 2 h per day. Spraying for longer than 2 hours per day is possible with the herbicide sprayers and the newer sprayers which have a lower power requirement. On some of the older types of sprayers different sets of batteries were used to extend the spraying period. The different sets should be numbered, as they are used in sequence. Care must be taken when changing batteries that they are all inserted correctly and that wires and connections are not damaged. Some operators prefer to carry the batteries in a separate shoulder-slung container, while others find that the connecting cable tends to catch on plants. Ideally, the weight of the batteries in the handle is used to counterbalance the weight of spray liquid at the nozzle.

Alternative power services are available (A.C. Arnold 1985). These include the use of alkaline manganese, zinc-air (E.V. Rogers 1975; Matthews and Thornhill 1974) or rechargeable batteries. Both lead acid and nickel-cadmium type rechargeable batteries have been used. An ac mains or solar-powered recharger may be used, the latter being particularly important in rural areas without electricity supplies.

Rechargeable 12. V 4 Ah, 'D'-type nickel-cadmium batteries provide a more even voltage supply for long periods of spraying. These batteries can be used for one full day's spraying if recharged overnight at 400 milliamps. The initial cost of batteries and charging unit is high, and approximately 80 spray days have been calculated as needed to break even in cost compared with alkaline manganese batteries (E.V. Rogers 1975).

Rechargeable batteries are more important when spinning discs are used for VLV sprays as the increase in flow rate resulting from using higher volumes decreases the life of zinc carbon batteries (Fig. 8.3) and reduces rotational speed of the disc (Fig. 8.4) (Beeden 1975). The considerable

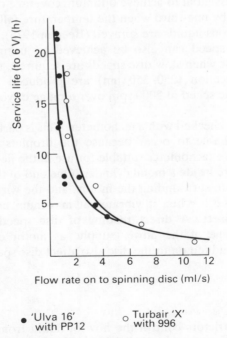

● 'Ulva 16' with PP12   ○ Turbair 'X' with 996

Fig. 8.3 Life of battery with different flow rates (after Beeden 1975)

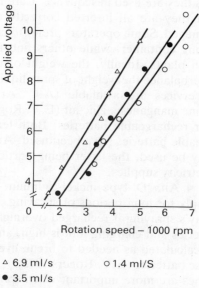

△ 6.9 ml/s   ○ 1.4 ml/S
● 3.5 ml/s

Fig. 8.4 Effect of voltage on disc speed at different flow rates (after Beeden 1975)

increase in battery life at very low flow rates (0.5 ml/s) is important where narrow swaths are essential to achieve optimum coverage of a target. Battery life can be reduced by one-third when the temperature falls from 27 to 15°C, and when more viscous liquids are sprayed (Beeden 1975).

A constant disc speed can also be achieved by using a motor with a mechanical governor when slow disc speeds are required, thus large droplets for herbicide application (250–350 μm) are produced by direct droplet formation with a disc speed of 2000 rpm over a range of voltages (Bals 1975a 1976).

Disc speed can be checked with a tachometer. This is particularly important if phytotoxicity is liable to occur because the droplets are too large. A relatively inexpensive tachometer suitable for use in the field, the 'Vibratak', consists of a thin wire inside a metal cylinder. One end of the cylinder is held against the back plate surrounding the motor, and the wire is pushed out of the cylinder (Fig. 8.5). When it vibrates at maximum amplitude the rpm reading is taken direct. A direct reading of disc speed is preferable to measuring the voltage of the power supply, as motor efficiency and the amount of spray liquid fed on to the disc also affect disc speed.

Control of flow rate

Interchangeable restrictors control the flow of liquid from the reservoir by gravity to the disc. A vacuum inside the reservoir is avoided by incorporating

Fig. 8.5 Measuring disc speed of rotation with a 'Vibratak' (Photo BP Co. Ltd)

an air bleed into the design so that constant pressure is obtained. The air bleed may be along a small channel at the base of the thread in the socket or through a very fine bore tube. Apart from the size of the restrictor, flow rate is also affected by viscosity of the formulation, which may change with the temperature (Cowell and Lavers 1988) (Fig. 8.6).

Flow rate should be checked, prior to spraying and during spraying if there is a marked change in temperature, by timing the period to spray a known quantity of liquid, preferably with the discs rotating. Comparison of the flow rate between different formulations can be made by using the restrictor separately from the sprayer. In general, the lowest effective flow rate is chosen to reduce the load on the motor and thus avoid increased power consumption and droplet size. Some restrictors or feed stems are colour coded (Table 8.2).

On some sprayers a filter is inserted between the reservoir and the restrictor to prevent blockages.

A plastic bottle used as a reservoir for the spray liquid is screwed directly into the restrictor. Such bottles have either a 1- or 0.5-litre nominal capacity, but the latter is preferred to reduce the weight at the end of the handle. If necessary the operator can carry a reserve supply of spray liquid in a plastic bottle mounted on the shoulder or on a knapsack frame (Fig. 8.7). On one sprayer designed for herbicide application, a 2.5–1litre bottle is fitted to the battery case at the end opposite to the spinning disc and acts as a counterbal-

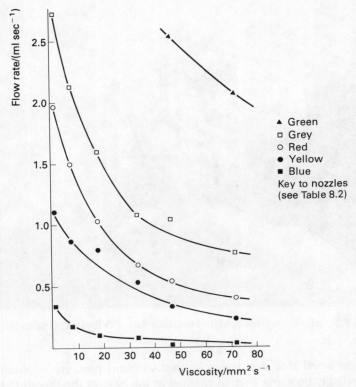

Fig. 8.6 Effect of viscosity on flow rate with different restrictors (from Cowell and Lavers 1988)

Table 8.2 Orifice diameter for one range of colour-coded restrictors

Colour	Diameter of orifice
White	0.50 mm
Brown	0.65 mm
Blue	0.78 mm
Yellow	1.00 mm
Orange	1.30 mm
Red	1.56 mm
Black	1.60 mm
Grey	2.00 mm
Green	2.90 mm

ance to the rest of the sprayer. The bottle must be screwed in firmly to avoid leakage (Fig. 8.8). Contamination of the outside of a bottle must be avoided, particularly when sprays contain oil, as some bottles are difficult to hold

Fig. 8.7 Spinning disc sprayer with knapsack spray tank

when wearing rubber gloves. Some agrochemical manufacturers provide pre-packed containers to fit directly on the sprayer.

The use of very or ultra low volumes makes the distribution and sale of pre-formulated, pre-packed chemicals economically feasible. This has advantages in eliminating mixing and filling operations which is where there is the greatest potential risk of operator contamination. Indeed, when this type of sprayer was introduced, this was seen as one the potential advantages over conventional knapsack sprayers.

Swath width and track spacing

Movement of droplets after release from a nozzle depends on their size, wind velocity and direction and height of release above the crop (or ground). As discussed earlier in Chapter 4, large droplets deposit quickly by gravity with

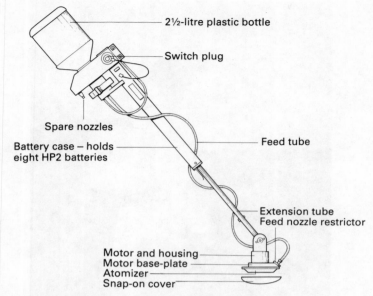

2½-litre plastic bottle

Switch plug

Spare nozzles

Battery case – holds
eight HP2 batteries

Feed tube

Extension tube
Feed nozzle restrictor

Motor and housing
Motor base-plate
Atomizer
Snap-on cover

Fig. 8.8 Diagram to show Spinning disc herbicide applicator

minimal displacement by the prevailing wind. An unshielded 80 mm disc producing 250 μm droplets has a swath of approximately 1.2 m. In contrast 70 μm droplets may be blown more than 10 m downwind if released 1 m above the crop, even when wind velocity is less than 7 km/h. Convective air turbulence could carry such droplets much further. Swaths up to 20 m downwind of the operator can be treated effectively with droplets less than 100 μm diameter under certain circumstances but there is a risk of thermal air movement taking such small droplets upwards away from a crop.

The overall distance downwind over which sufficient droplets are deposited is referred to as the swath, whereas the distance between successive passes across a field is more appropriately referred to as the track spacing. Often the term 'swath' is used synonymously with track spacing.

Choice of track spacing used in incremental spraying will depend on the behaviour of the pest and the type of foliage of a host crop, as well as wind velocity affecting the amount of downwind displacement across each swath. Insects exposed on the tops of plants, for example the leafworm *Alabama argillacea*, can be controlled by spraying across wide swaths. Similarly, a 15 m track spacing has been used to spray aphids on wheat in the UK but, when an insect is feeding on the lower part of plants and penetration of the foliage is needed, narrow track spacings are essential so that more droplets are carried by turbulence between the rows. A 3 m track spacing has been used to spray tall bracken. Too wide a track spacing should be avoided as variations in wind velocity may result in uneven distribution of spray. Incremental spraying by overlapping swaths generally improves coverage

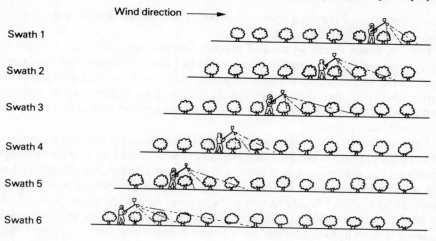

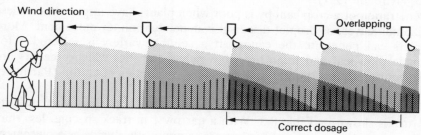

Fig. 8.9   (a) Overlapping of swaths from successive passes across the field as the
operator moves upwind. *Note*: Swath width will vary with wind speed
(b) Diagram showing area of correct dosage

(Fig. 8.9). Narrow swaths (0.9) gave better control of *Helicoverpa* on cotton
than when a wide track spacing (4.5 m) was used, even though a lower
concentration of spray was used on the narrow track spacing (Matthews
1973). Raheja (1976) obtained no difference in the yield of cow-peas when
spraying at 2.5 litres/ha over 1.8 or 3.6 m wide track spacings, mainly against
Maruca.

Sometimes adjacent swaths can be displaced not only in space but also in
time (sequential spraying) (Joyce 1975), thus twice-weekly sprays over a
double-width swath may be preferred with a less persistent chemical or if
rain reduces the effectiveness of a weekly spray. An increase in the frequency
of application without increasing the total volume of spray per unit time may
improve deposition, as the chance of sprays being applied **under different
wind conditions** is increased, thus a change in wind direction and an amount
of turbulence exposes other leaf surfaces. A spray repeated when the wind is
from the opposite direction is ideal for improved spray coverage. When more
frequent sprays are logistically possible, it is feasible for the farmer to use a

lower dosage and repeat a spray if necessary to compensate for the effect of rain or vigorous plant growth.

For any given swath width and droplet size an increase in the volume of spray may not improve control, as the greater number of droplets produced are carried in the same volume of air to more or less the same positions within the crop; thus Matthews (1973) obtained no differences in yield of cotton when 0.5 and 1.0 ml/s flow rates were examined with one-, two- and five-row track spacings.

On some crops the track spacings should be changed in relation to the size of the plants, thus, as the area of foliage increases, reduction in track spacings increases the volume per unit area. On cotton in Central Africa, sprayed at 0.5 ml/s with a special solution formulation, the track spacing was reduced from six to four and finally to two rows as plant height increased from 25 cm to 25–50 cm and >50 cm, respectively, (Matthews 1971). A two-row track spacing is used until plants are knee height (approximately 50 cm), and then a single row when wettable powder formulations are applied (Mowlam et al 1975; Nyirenda 1991).

Penetration of a crop canopy is poor when plants have large, more or less horizontal leaves, as droplet dispersal is dependent on air movement. More droplets can penetrate the canopy if a suitable variety is selected – for example okra leaf and frego bract have characteristics which breeders are endeavouring to incorporate into commercial varieties of cotton (Parrott et al 1973).

The time required to spray 1 ha with different track spacings when walking at 1 m/s is shown in Table 8.3. With a narrow 1 m track spacing, less than half a person day is required per hectare to apply a herbicide as a placement spray at 10 litres/ha, in contrast to over 30 person days needed to hand-hoe weeds. Even less time is needed using incremental spraying with wider track spacings.

Table 8.3 Time (min) required to treat a square field (one ha) at a walking speed of one m/s. Extra time is required to mix the spray, replace the bottles and carry the materials to the field

Track spacings (m)	Time (m)
1	168
2	85
5	35
10	18
15	13

$$\text{Time to treat area (min)} = \frac{\left(\dfrac{\text{area to be treated (m}^2\text{)}}{\text{area covered per second (m}^2\text{)}} \right) + \text{field width across rows (m)}}{60}$$

Spraying procedures

Incremental drift spraying

Wind direction is noted so that the spray operator can walk progressively upwind across the field through untreated crops. Smoke pellets can be used to assess wind direction (Fig. 8.10). Alternatively a piece of thread can be attached to a wire fixed to the spray head. Spraying commences 1 or 2 m inside the downwind edge of the field. The disc speed is checked before the spray liquid is prepared. The bottle is filled and then screwed on to the sprayer. When the operator is ready to spray, the motor is switched on and the disc allowed to reach full speed. The sprayer is held either with the handle across the front of the operator's body, or over the operator's shoulder with the disc above the crop, pointing downwind, so that droplets are carried away from the operator while walking through the crop. The bottle is then inverted as liquid is gravity fed to the disc, but if the operator stops for any reason or reaches the end of the row, the sprayer should be turned over

Fig. 8.10　Measuring wind direction with smoke pellet (photo: BP Co. Ltd)

again to stop the flow of liquid and avoid overdosing. The inversion of the bottle at the end of each row ensures that the spray remains well mixed. The motor is not switched off while the operator walks along the edge of the field to the start of the next swath. If there is more than one operator in a field, great care must be taken to avoid walking in each other's spray cloud. An extra swath outside the upwind edge of the field may be necessary. At the end of spraying, the motor is left running for a short period to remove any liquid from the discs. Ten litres/ha is applied when the operator walks at 1 m/s (÷) spraying a track spacing of 1 m wide (÷) with a flow rate of 1 ml/s (×). If any of these variables is changed the volume (10 litres/ha) is divided or multiplied as indicated by the sign in the brackets. Thus, with a 5 m swath: $10 \div 5 = 2$ litres/ha.

The spinning disc is normally held about 1 m above the crop. It may be necessary to hold it lower while spraying the first swath along the leeward side of a field to reduce the amount of chemical which may drift outside the treated area. Similarly, the nozzle may be held lower during the final swath on the windward side of a field, to cover the edge of the field. Nozzle height can be lowered if necessary when the wind velocity increases but, if the area being treated is sufficiently large, a wider track spacing can be used to take advantage of the wind. Simple anemometers are available to check wind velocity which should be 2–15 km/h One small simple anemometer has a pith ball which moves up a vertical tube according to the strength of the wind (Fig. 8.11). Extreme conditions, such as a dead calm or a strong, gusty wind, should be avoided.

Fig. 8.11 Measuring wind velocity with a simple anemometer. (Photo BP Co. Ltd)

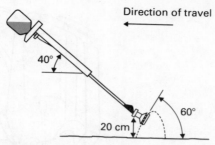

Fig. 8.12 Position of 'Herbi' sprayer behind operator

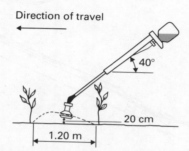

Fig. 8.13 Holding 'Herbi' sprayer in front of operator

Placement spraying

When spraying herbicides, the disc is held only a few centimetres above the weeds so that downwind displacement of the spray is negligible. The disc is held behind the operator at 60° from the ground (Fig. 8.12) to avoid the hollow-cone pattern from a horizontal disc. The operator does not walk over treated surfaces with this method but, if greater control of the position of the swath is needed, the less poisonous chemicals can be applied with the disc tilted slightly away and in front of or to the side of the operator (Fig. 8.13). Also, a wider swath can be achieved by mounting two or more discs on a hand-held boom, a practice used in plantation agriculture. Sprayers with a shrouded disc have been developed to allow adjustment of the swath width as well as apply a narrow droplet spectrum.

To avoid the use of batteries, a pneumatically driven disc has been developed on a knapsack-type sprayer fitted with a manually operated air-pump (Fig. 8.14).

Motorised fan-assisted spinning-disc sprayers

Discs can be mounted in front of a fan which provides a directional airstream

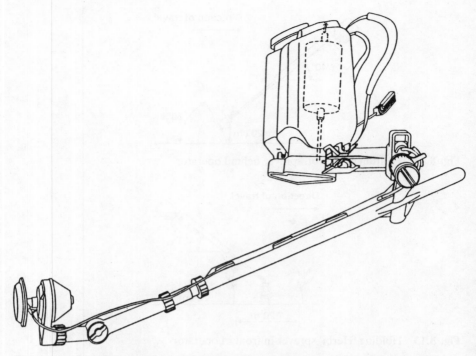

Fig. 8.14 'Birky' sprayer in which a manually operated air-pump provides power to rotate disc

so that insecticide and fungicide sprays can be applied in warehouses, glasshouses and other enclosed areas where natural air movement is insufficient to disperse the spray droplets. The power required to move air is much greater than that required to produce the droplets. Small battery-operated sprayers with a fan have been used in glasshouses but the period of operation is limited, even when rechargeable batteries are used. An ac electric motor (Fig. 8.15) can be used if a mains electricity or a portable generator power supply is available, but a trailing cable is a disadvantage. For greater mobility several sprayers have been designed with two-stroke engines (Fig. 8.16). Discs mounted directly on the drive shaft of the engine operate at a relatively slow speed of about 6 000 rpm, and some larger droplets may collect on operators, particularly on their hands, if sprayers are too close to their body. Higher disc speeds can be achieved by mounting the disc on a separate shaft from the fan, but fitted with a small propeller driven by the airstream. Hand-carried sprayers of this type are not very comfortable to use because of the vibrations from the engine: their weight is also a disadvantage. Machines with gravity feed of liquid are rotated to stop and start spraying, thus a modified carburettor is needed to allow the engine to run in different positions. Slight pressure in the spray tank to force liquid to

Fig. 8.15 Electrically driven spinning-disc sprayer with fan (photo Turbair Ltd.)

Fig. 8.16 Disc with fan mounted on shaft of two-stroke engine – Turbair 'Tot' (photo Turbair Ltd)

the nozzle is achieved, if necessary, by positioning an air line from in front of the fan or engine to the tank. Changes in the direction of the airstream cause leaves to flutter and collect droplets more efficiently.

Vehicle-mounted sprayers with centrifugal-energy nozzles

Boom sprayers

Spinning-disc nozzles have also been developed for tractor-mounted boom sprayers. Much of the early experimental work was on herbicide application at volumes less than 50 1/ha. Initially stacked 'Herbi' discs were mounted on a vertical shaft and shrouded to attempt to provide spray distribution equivalent to a fan-jet nozzle and with adjustable swath (Taylor et al 1976; Cussans and Taylor 1976). The spray pattern from a horizontal disc is a hollow cone, so to obtain the fan-type distribution, the top four of the five discs were shrouded. Half the spray was collected on shrouds in two 90° segments and drained to the disc immediately below it. Additional liquid was fed to the discs except the unshrouded bottom disc from which comparatively little of the spray was emitted (Fig. 8.17). As the initial trajectory of droplets depends on their size, the distance between the five spray units was adjustable in the horizontal plane, so that the total swath width varied from 4.5 m with

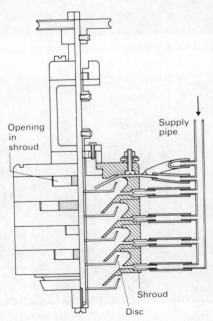

Fig. 8.17 Cross-section of five – disc unit used on WRO experimental sprayer

Table 8.4 Theoretical mean number of droplets (N/cm^2) at various volume application rates and droplet sizes

Droplet diameter (μm)	Volume of droplet (μl)	$N/cm2$		
		5 (litres/ha)	15	45
150	1.77×10^{-3}	28.3	84.8	254.0
250	8.19×10^{-3}	6.1	18.3	54.9
350	2.25×10^{-2}	2.2	6.6	19.8

150 μm droplets to 6.0 m with 350 μm droplets. Spray distribution obtained with 150, 250 and 350 μm droplets applied at 5, 15 and 45 litres/ha was comparable with conventional spraying (Taylor et al 1976). The theoretical droplet density achieved with these treatments is shown in Table 8.4. Spray recovery on the ground 30 cm under the boom when 15 litres/ha was applied with 150, 250 and 350 μm droplets was equivalent to 8.7, 9.7 and 12.7 litres/ha, respectively, when the swath was displaced by a 1.8 m/s wind. With winds up to 5.4 m/s the swath was displaced by up to 3 m downwind.

Studies with reduced dosages to highlight differences in control were encouraging, but later when full dosages were applied by others biological results were variable. In many trials CDA gave no better results or were slightly inferior to conventional hydraulic nozzles, particularly with herbicides applied at only 10 – 20 l/ha. Merritt and Taylor (1977) reported less effective control with contact herbicides such as ioxynil, bromoxynil and bentazone with 250 um droplets, presumably because of inadequate coverage. More consistent results were achieved at 30–60 l/ha. (Mayes and Blanchard 1978) Some of this variability was apparently due to poor penetration and increased variability of deposits in dense crop canopies. However, other studies have shown the effects of formulation and concentration can be highly significant (Merritt 1976, 1980; Robinson 1978), so that for optimal use of CDA, changes in formulation are required for the very low volumes applied.

As an alternative to multiple shrouded discs on each unit, Heijne (1978) studied a single large spinning cup (Micron 'Micromax') with individual grooves to 180 teeth (Fig. 8.18), which allowed herbicide, insecticide and fungicide application by alteration of the disc speed (See also Bode et al 1983). Several versions of this nozzle have been supplied commercially with electrical or hydraulic drive systems. However, a lack of registration of products suitable for very low volume application, some problems with the maintenance required, higher initial cost and perceived poor penetration of dense crop canopies when sprays are emitted in a horizontal plane limited acceptance of CDA sprayers despite the advantages of low volume and drift reduction.

A shrouded vertical large (14 cm) spinning disc was developed to improve penetration (Morel 1985) with good results achieved with 25 l/ha. More development on shrouded spinning discs is in progress (Bals, personal communication). Air-assisted spinning discs have also been used, and others have been adapted by using an electrostatic charge. The charged sprays of insecticides and fungicides gave increased deposits which gave better control at equivalent dosages, and enabled a decrease in rates of application without loss of biological effect. However, penetration of herbicides to weeds hidden beneath the crop plants is often less than with hydraulic sprayers unless

(a)

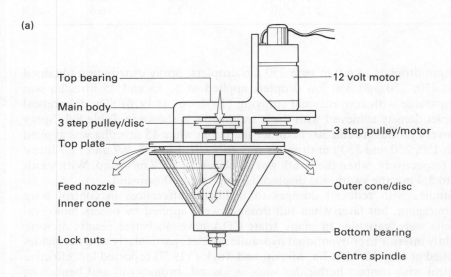

Top bearing — 12 volt motor

Main body

3 step pulley/disc

Top plate — 3 step pulley/motor

Feed nozzle — Outer cone/disc

Inner cone

Lock nuts — Bottom bearing

— Centre spindle

(b)

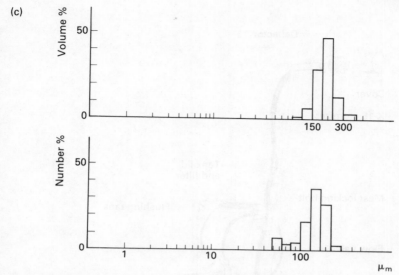

Fig. 8.18 (a) Diagram to show large spinning-cup ('Micromax') (b) Tractor sprayer fitted with 'Micromax' units (c) Example of droplet spectrum

formulations and the charging system are adjusted to give the least charge on the droplets (Griffiths et al 1984). Results also depended to some extent on the type of pesticide (systemic *v.* contact) and behaviour of the pest so variable biological results were obtained in some crops despite increasing total deposition (Cayley et al 1984 1987).

Vehicle-mounted 'drift' sprayer

This type of equipment has been used to apply insecticides to control locust hoppers (Symmons et al 1989), as ULV spraying is accepted as standard. Similarly droplets can be drifted through the nesting sites of the Red-billed Quelea *Quelea quelea* (Dorow 1976). Low dosages of fungicides have also been applied to cereals with similar equipment.

A series of spinning discs are mounted on a shaft through which spray liquid is pumped from a reservoir mounted on the vehicle. The discs are driven by an electric motor by means of a pulley system (Fig. 8.19).

Air-assisted sprayers

Sprayers with various types of rotating cage or discs mounted in an airstream are considered in Chapter 10.

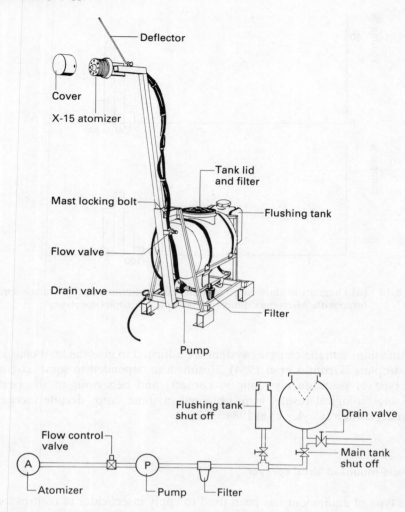

Fig. 8.19 Vehicle-mounted 'Ulvamast' sprayer (Micron Sprayers Ltd)

9

Electrostatically charged sprays

The possibility of improving spray deposition by electrostatically charging droplets has received much attention (Bailey 1986; Law 1987; Marchant 1987; Matthews 1989b). Three methods of charging agricultural sprays have been used:

1 Induction charging of conductive liquids
2 Ionized field charging of either conductive or non-conductive liquids.
3 Direct charging of semi-conductive liquids.

In each case, the normal balance of positively charged protons and negatively charged electrons is disturbed by movement of electrons, so that additional electrons provide a negative spray droplet, while a deficit of electrons makes the spray positive.

Induction charging

When a high-voltage electrode, positioned close to where spray liquid is emitted from a nozzle, is positively charged, a conductive liquid, such as a water-based pesticide spray at earth potential has a negative charge induced on its surface by the attraction of electrons. If the electrode was negative, the reverse occurs and electrons repelled from the liquid to earth provides a positively charged liquid. As the droplets are formed, the charge is retained on them. A conductive liquid is needed so that the charge transfers from earth to the liquid jet in the very short time while it passes the electrode. The level of charge induced per unit area of surface will be proportional to the voltage applied to the electrode.

The charge on the spray droplets is the opposite of that on the electrode, so some spray is liable to be attracted on to the electrode, which if wetted, is liable to short circuit the power supply. An air stream is used on some nozzles to blow droplets away from the electrode and keep it dry (Law 1978).

Ionized field charging

A high voltage applied to a needle point can create an intense electric field around it that is sufficient to ionize molecules of the surrounding air. A positively charged conductor will repel the positive ions created, while the electrons which are released in the ionization process will be attracted to the conductor and neutralise some of its charge. With a negatively charged conductor, the reverse is true and positive ions are attracted back to the conductor. Great care is needed to protect the fragile needle and avoid reverse ionization. The level of charge is dependent upon the dielectric constant of the spray, its surface area, the electrical characteristics of the corona discharge and the time within the ionized field.

When a stream of liquid passes near to the ionizing tip of the needle, the charged ions produced are attracted to the liquid and carried away by it. The needle is usually negatively charged as higher voltages are required to create an equivalent positive corona. Liquids with a wide range of conductivities can be charged with this method (A.J. Arnold and Pye 1980).

Direct Charging

When a semi-conductive spray liquid with an electrical resistivity in the range 10^4–10^6 ohm m is exposed to a high voltage (15–40 kV) as the liquid emerges through a narrow slit, mutual repulsion between different portions of the liquid overcomes surface tension so ligaments are formed and these break up into droplets due to axisymmetric instabilities. The level of charge on the droplets represents the maximum that can be attained and is called the Rayleigh limit (Rayleigh 1882). The initial droplet size distribution is initially bimodal, but the very small satellite droplets are attracted to an earthed electrode or 'field adjusting electrode' positioned close to the nozzle, so that essentially a monodisperse spray is produced. The size of droplets is reduced for a particular flow rate by increasing the applied voltage. Increasing the flow rate without changing the voltage will increase droplet size. There is therefore a complex interaction of electrical, viscous and surface tension forces, affecting droplet size with the resistivity and viscosity of the spray liquid being particularly important factors.

(a)

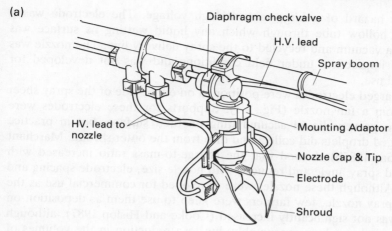

Diaphragm check valve

H.V. lead

Spray boom

HV. lead to nozzle

Mounting Adaptor

Nozzle Cap & Tip

Electrode

Shroud

(b)

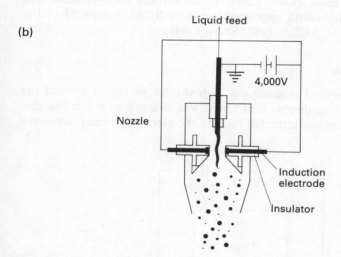

Liquid feed

4,000V

Nozzle

Induction electrode

Insulator

Fig. 9.1 (a) Hydraulic nozzle with induction charging (b) Diagram to show position of electrodes relative to the spray

Electrostatically charged nozzles

Induction charging nozzles

Hydraulic nozzles

In one system, spray to hydraulic spray nozzles was charged at a potential of up to 10 kV and the electrode was earthed (Marchant and Green 1982). The distance between the spray tank and individual nozzles was lengthened by using a long coil of narrow bore tube to increase the electrical resistance and reduce the hazard of a large tank at high voltage. The electrode was a perforated hollow tube through which any liquid wetting its surface was sucked by a vacuum and recycled to the spray delivery line. The nozzle was considered impractical under field conditions and was not developed for commercial use.

Later charged electrodes were positioned on either side of the spray sheet emitted from a fan nozzle (Fig. 9.1). Supports for these electrodes were designed to prevent liquid accumulating on their surface, but in practice small charged droplets did collect and drip from the outer shroud. Marchant et al (1985a, 1985b) showed that the charge-to-mass ratio increased with voltage and spray angle and reduced with nozzle size, electrode spacing and pressure. Although these nozzles were developed for commercial use as the EXACT spray nozzle, few farmers were keen to use them as deposition on the crop was not significantly increased (Cooke and Hislop 1987), although drift downwind was less, presumably due to a reduction in the volumes of small droplets in the spray cloud. Later studies showed no significant drift reduction and at higher wind speeds airborne drift 5m downwind of the nozzle increased (P.C.H. Miller, 1988; Sharp 1984).

Spinning disc atomizers

Marchant (1985) described mounting a high-voltage electrode around the periphery of an atomizer disc so that liquid was charged as it left the disc (Fig. 9.2). Charged droplets attracted back to the electrode were reatomized

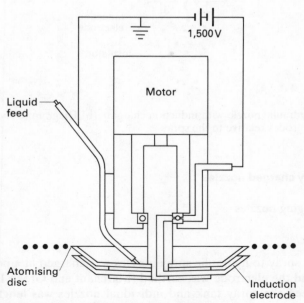

Fig. 9.2 Spinning disc nozzle with induction charging

as the electrode rotated at the same speed as the atomizer. Carlton and Bouse (1980) also designed an induction charged spinner to study the use of charged sprays from aircraft.

Air-shear nozzles

The advantage of an air-shear nozzle (Fig. 9.3) for induction charging is that the air stream carries the charged spray away from the electrodes. Law (1978) embedded an electrode at the outlet of an air-shear nozzle and has done extensive studies of charged sprays with his equipment. The nozzle was

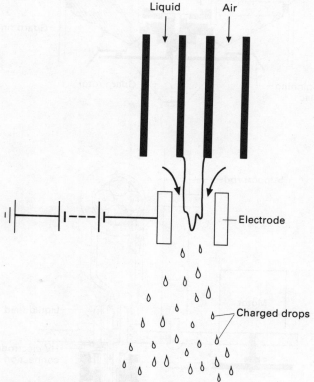

Fig. 9.3 Twin-fluid nozzle with induction charging

optimized to produce 30-50 μm vmd (volume median diameter) droplets at flow rates of 75-100 ml/min with a-10mC/kg charge to mass ratio by charging with voltages less than 1 kV and a power consumption less than 25-50 mW (Law 1987). However, overall power consumption is at least one hundred times greater as a large compressor is needed to provide the air needed to atomize the spray. This nozzle is more suited for use on a tractor mounted sprayer, but commercial use has been limited. Flow rate can be increased to 500 ml/min. with suitable modification of the nozzle (Frost and Law 1981).

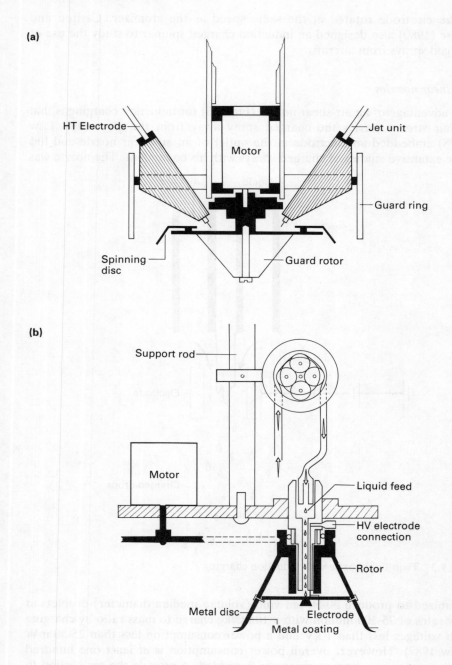

(a)

HT Electrode

Jet unit

Motor

Guard ring

Spinning disc

Guard rotor

(b)

Support rod

Motor

Liquid feed

HV electrode connection

Rotor

Metal disc

Electrode

Metal coating

Fig. 9.4 (a) Spinning disc nozzle with ionic charging (b) Arrangement with inverted 'Micromax' spinning cup and peristaltic pump – the 'Jumbo' rotary atomizer

Another air-shear nozzle for use on an orchard sprayer was designed with a high voltage petal electrode mounted opposite the liquid outlet (Inculet et al 1981).

Piezo-electric nozzle

A piezo-electric nozzle designed to produce monosized droplets for experimental work was improved by incorporating an induction charging system to ensure that the droplets did not coalesce (Stent et al 1981). The apertures are extremely small, so only well filtered solutions can be applied with this nozzle.

Reichard et al (1987) also used an electrostatic charge on a Bergland and Liu (1973) droplet generator to provide uniform sized droplets for laboratory studies.

Ionized field charging nozzles

Most studies have been with a rotary atomizer with a needle mounted so that it does not quite touch the liquid moving over the surface of a disc, which is coated with a thin metallic layer (Fig. 9.4). When a water-based spray, a good conductor is used, the liquid is charged to a potential close to that of the needle, so the spray liquid must be isolated if a high voltage is used. A.J. Arnold (1983, 1984) used a peristaltic pump to maintain a constant flow from separate small spray containers to each individual nozzle, an inverted spinning grooved cup with toothed edge, the 'Micromax', which was referred to as the 'Jumbo' atomizer. In tests with a smaller disc, droplet size was significantly reduced as the applied voltage increased, thus with an oil spray fed at 20 ml/min and disc speed of 3 500 revs/min, droplet size decreased to about 50 μm at 30 kV from 200 μm with an uncharged disc (A.J. Arnold and Pye 1980).

Electrodynamic nozzle

A hand-carried battery-operated sprayer has been developed in which the electrodynamic nozzle is manufactured as an integral part of the pesticide container, and known as a 'Bozzle' (Fig. 9.5). The nozzle is made of a special plastic material to conduct the electric charge to the oil-based formulation as it is fed by gravity through the very narrow annulus. When the 'Bozzle' is screwed into the 'stick', a flange on the nozzle makes contact with a cable from a high-voltage generator in the handle of the 'stick'. Four batteries in the handle provide six volts that is converted by the generator to 24 kV fed to the nozzle. The flow of liquid to the nozzle is determined by the restrictor and air-bleed in the 'Bozzle' and is set during manufacture in relation to each

(a)

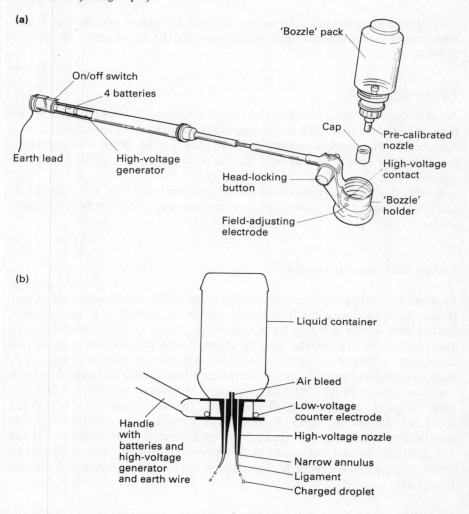

(b)

Fig. 9.5 (a) Electrodynamic sprayer (ICI Agrochemicals) (b) Diagram showing 'Bozzle'

product. The protective cap over the 'Bozzle' is colour coded to indicate the flow rate. At the far end of the handle away from the 'Bozzle' is a switch that has be kept depressed during spraying. As soon as the switch is released the voltage to the 'Bozzle' is discontinued, although there will be a small residual charge on the nozzle until it leaks away or the nozzle is deliberately earthed.

Spray liquid emerging from the nozzle annulus immediately forms a series of ligaments around its periphery. The diameter of these ligaments is remarkably uniform as the flow of liquid is evenly spread around the annulus. These ligaments break up into droplets and smaller satellite droplets, the trajectories of which are affected by the locally intense electrical field close to the nozzle so they are captured on the counter electrode (Coffee 1980). Movement of the charged cloud can also be influenced by electrodes of the same polarity as the nozzle, and known as 'deflectrodes' (Coffee 1981; Chadd 1990). The charged spray cloud may also be discharged by reverse ion bombardment (Coffee 1980; Morton 1982).

A linear version of the electrodynamic nozzle (Coffee 1980, 1981; J.M. Wilson 1982) has also been designed and some preliminary evaluation has been done with these mounted on aircraft.

Direct charging of a hydraulic nozzle at voltages up to 70 kV has been tried but is difficult to do safely with conductive liquids (Moser and Schmidt 1983).

Deposition studies

Uncharged spray droplets may sediment on to mainly horizontal surfaces due to gravity or be impacted on vertical surfaces by their velocity by movement in air currents (see also Chapter 4). The latter applies to the smallest droplets with a low terminal velocity, but these droplets can be readily carried by the air movement around targets such as stems and leaves, or carried upwards on thermals due to convection. Improvement of the collection efficiency of these small droplets (<100 μm) can be achieved by the addition of an electrostatic charge on the spray. Deposition of charged droplets is influenced in several ways. There is the electrical field between the nozzle and the nearest earthed object, the space cloud effect, an induced field effect (Fig. 9.6) and there is often a naturally occurring electro-potential gradient near the surface of the ground as positive polarity in the upper atmosphere induces a negative charge on the earth's surface. Modelling of deposition of charged aqueous spray in barley has been reported by Lake and Marchant (1984).

Nozzle effect

When a charged nozzle is relatively close to a crop, the electrical forces exerted on the droplets are much greater than the gravitational force, thus Coffee (1979) calculated that for a 100 μm droplet, with a charge at about 75 per cent of the Rayleigh limit , the initial electrical force acting on it would be about 50 times the gravitational force. Computer simulation suggests that the terminal velocity of a 100 μm droplet would be increased about 16 times to approximately 5 m/s. and that with an air velocity of 4 m/s, the electrical force would be about 20 per cent greater than the air drag force (Marchant 1980). Thus droplets of the same charge as the nozzle would be repelled from the nozzle towards the nearest earthed object, and their trajectories would be less affected by the air movement above or within the crop canopy. In

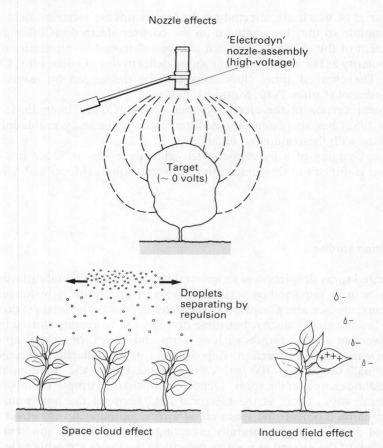

Fig. 9.6 Diagram to show nozzle, space cloud and induced charge effects on deposition

some cases droplets would travel upwards against gravity when the nearest earthed surface is the undersurface of a leaf.

Space cloud effect

The spray cloud containing a large number of droplets of the same polarity expands rapidly as each droplet repels its nearest neighbour. The spray cloud thus creates its own electrical field which influences the trajectory of the individual droplets. While the effect of the electrical field from the nozzle is relatively short-lived and occurs only while the nozzle is above the crop, the space cloud effect continues after the passage of the sprayer as long as there is still a cloud of droplets. Some of the droplets, repelled outwards away from the centre of spray cloud move upwards and could be carried by convection away from the sprayed area, thus P.C.H. Miller (1989) and Western and Hislop (1991) have reported no reduction in spray drift with charged hydraulic spray nozzles, in contrast to data reported by Sharp (1984). However, much depends on droplet size and volatility of the formulation and, provided the droplets do not become too small, there is generally a downward movement of the spray cloud and downwind drift is less than with uncharged sprays (Johnstone et al 1982). Penetration into a crop canopy will depend on the openness of the canopy structure so that the space cloud can force droplets into air spaces between branches and leaves.

Induced field

One effect of a charged space cloud is to induce an opposite charge on an earthed target surface, thus when the droplets are positively charged, electrons are attracted to the crop surface from earth. Due to the inverse square relationship of force to distance, the opposite induced or image charge will attract droplets to a surface only when they are very close to the surface (ie less than 1 cm). Thus deposition of the smallest droplets is enhanced as, if uncharged, these would be the most likely to be carried by air currents around some target surfaces. If, however, there is an excessively resistive pathway to earth, a charge due to deposition of droplets could raise the electrical potential on a target and consequently diminish further deposition (Law 1989).

Factors affecting deposition

Pointed leaves

Laboratory experiments readily confirm increased deposition on artificial spherical targets (Fig. 9.7), thus Law and Lane (1981) reported a seven-fold increase compared with an uncharged spray. Similarly A.J. Arnold and Pye (1980) obtained up to eight-fold increases in deposition of oil sprays at 30 kV. However, when a needle point was present on a target, deposition was significantly reduced (Law and Lane 1982) due to a gaseous discharge which flows from the point to the charged spray cloud (Coffee 1971b). This is caused by the electric field intensification and ionization at the point caused by the spray cloud's space charge. A single point can account for up to 80 per cent of the total charge exchange between a target and the incoming spray cloud (S.C. Cooper and Law 1985). The counterflow of positive ions drawn from an earthed target point by a negatively charged spray reduces the charge/mass of the adjacent spray cloud so that the effect of the space-charge (approx 25 uC/m3) is so decreased that droplets are no longer forced into air

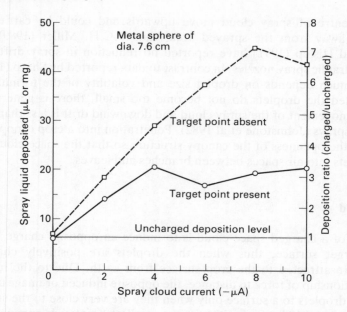

Fig. 9.7 Comparison of charged and uncharged spray deposition on a sphere, with and without a needle point (from Law 1980)

spaces between foliage. Laser Doppler studies showed that the induced corona from a point affected the momentum and charge of approaching charged droplets and in some locations repelled droplets from targets (Law and Bailey 1984). Nevertheless, droplet deposition was 20 per cent better if the spray cloud was negatively charged compared with a positively charged spray (Cooper and Law 1987a). Attempts to overcome this problem by using a bipolar instead of unipolar charged spray, failed to improve deposition (S.C. Cooper and Law 1987b).

The effect of the ionization from pointed leaves will also vary between different types of foliage, and the way in which the plants are spaced apart in the field. Initial studies were with artificial targets. Giles and Law (1985) using cylinders of different diameters and spacings achieved better deposition (a) closer to the top of the cylinders, (b) the wider the spacing between cylinders and (c) the larger the diameter of the cylinders. Later Law et al (1985) used fluorometric analysis to examine deposition on different segments of cereal leaves, broadleaved weeds under the cereals and the soil. Charging droplets in the 1.5–4.5 mC/kg range increased deposition on all plant surfaces and reduced residues on the soil, but deposition was not uniform and was not improved by increasing air velocity from 2 to 4 m/s. When an external voltage was applied to a cylinder mounted just behind the spray cloud in an attempt to repel charged droplets further into the crop canopy, the electric field of 37 kV/m did not increase deposition significantly , but the gaseous charge exchange between the spray cloud and leaf surfaces was exacerbated via undesirable leaf-tip coronas (Fig. 9.8).

Lane and Law (1982) examined the level of deposit on cotton plants affected by drought stress, but plant moisture levels did not significantly affect the transient charge-transfer ability of the plants.

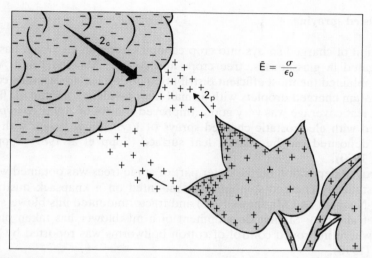

Fig. 9.8 Partial neutralization of charged spray droplets by leaf-tip ionization (from Law 1980)

Evaporation

As Law (1978) was using aqueous sprays in small droplets, he was concerned that evaporation would affect the charging of the sprays. By study of a 3mm diameter droplet held within a closed cabinet in which humidity was controlled, Law (1986) concluded that surface charge did not affect vapour movement, and the evaporating liquid did not dissipate the charge. As non-conductive vegetable oils are sometimes added to sprays to reduce the effect of evaporation on droplet size, and enhance rainfastness of deposits, Law and Cooper (1987) investigated the use of oil-based sprays through an induction charged nozzle. A combination of formulated vegetable oil with surfactant was suitable for induction charging and a charge-to-mass ratio of 4.1 mC/kg was achieved.

A hand-carried unit of the charged rotary atomizer was used in field trials to investigate spray coverage of cotton and soybeans (A.J. Arnold and Pye 1980). Spray deposits on these crops was increased with increasing voltage (0–30 kV), but canopy penetration was not improved by the higher voltage. These trials indicated that with aqueous sprays the more rapid trajectory of charged droplets overcame to some extent the effect of evaporation in high ambient temperatures. However, studies by Lake et al (1980) had calculated that droplets of a given size produced by a hydraulic nozzle are airborne for less time than one in free fall from a horizontal spinning disc, even when the spray is charged.

Air-assisted spraying

Projection of charged sprays into crop canopies using an airstream has been investigated in glasshouses, tree crops and cotton. Abdelbagi and Adams (1987) obtained the most efficient distribution of droplets for whitefly control using 18 μm charged droplets with a small fan providing 2 m³/min air flow as abaxial leaf coverage was very good. Improved control of *Aphis gossypii* was achieved with electrostatic charged sprays of *Verticillium lecanii* with more spores deposited on the abaxial leaf surface (Sopp el al 1989; Sopp and Palmer 1990).

Improved deposition on the outer part of apple trees was obtained with an electrostatically charged spinning disc mounted on a knapsack mistblower (Afreh-Nuamah and Matthews 1987) and tractor mounted mistblowers (J.G. Allen et al 1991). Similar development of a mistblower has taken place in China where improved control of cotton bollworms was reported by Shang and Li (1990).

Commercial development of electrostatic spraying

Attempts to market electrostatically charged hydraulic nozzles on tractor-

Fig. 9.9. Farmer using an 'Electrodyn' sprayer

mounted boom sprayers or air-assisted orchard sprayers have met with little success due to higher cost of equipment and the relatively small improvement in spray deposition obtained under field conditions. In the USA, a motorized unit with lance incorporating an air-atomizing nozzle based on work by Law (1978) has been promoted mainly for use in glasshouses (Lehtinen et al 1989). In the charged mode, bifenthrin gave better control of *Trialeurodes vaporariorum* than an uncharged spray or a high volume (1 200 l/ha) treatment (Adams et al 1991). Improved deposition achieved on strawberries indicated that the dosage of captan could be reduced by 50 per cent (to 1.12 hg ai/ha) in a charged spray (80 l/ha) and achieve similar persistence as the full rate of uncharged high volume spray (1870 l/ha) (Giles and Blewett 1991). The main commercial development has been with the hand-carried 'Electrodyn' sprayer, used principally on cotton in Africa and South America (R. Smith, 1988; Matthews 1990).

The 'Electrodyn' sprayer

The sprayer is used in a similar manner to the hand-held spinning disc sprayers with the nozzle held downwind of the operator, usually 0.4 m above the crop (Fig. 9.9). The nozzle may be held closer and directly above small plants to reduce loss of spray on the soil, whereas once there is sufficient foliage, the nozzle can be held between 2 and 3 rows with a swath of 1.0–2 m depending on the distance between rows. The 'Bozzle' is inverted and as soon as liquid starts to drip, the switch is held in the 'on' position while walking through the crop. Immediately before reaching the end of the row, the 'Bozzle' is turned over to stop the spray before the operator leaves the row. Without the intervening plants outside the crop, continued spraying is liable to contaminate the operator. The operator turns, keeping the stick downwind and re-enters the crop as the 'Bozzle' is inverted to spray again. The Bozzle has an air-bleed to control the flow rate in relation to the viscosity of the formulation. The flow-rate is indicated by the colour-coded cap (Table 9.1), with the average droplet size between 40 and 80 μm. The nozzle has no moving part, so power consumption is low and the four batteries to power the 26 kV generator should be effective for treating 30 hectares. The sprayer

Table 9.1 'Electrodyn' sprayer flow rates

Colour-coded nozzle cap	Flow rates (ml/min)	Vmd at 25 kV
White	1.5	28
Yellow	3.0	43
Blue	6.0	55

was developed primarily for spraying cotton and cowpeas for which there is a suitable range of insecticides to suit an integrated pest-management programme. However, as new more active pesticides are registered, it should be possible to extend the range of products suitable for electrodynamic spraying. Future development of tractor-mounted equipment with a linear nozzle will depend on the extent to which future legislation imposes greater restrictions on existing application techniques.

10

Air-assisted sprayers

Air-assisted sprayers provide an airstream in which droplets are projected towards the target. A number of different types of sprayers have been manufactured with an artificially induced airstream and have been referred to as 'air-blast' 'concentrate' (Potts 1958), 'mistblower' and 'air-carrier' sprayers. The subject is reviewed by Hislop (1991). The term 'mistblower' should be restricted to those sprayers which produce droplets in the range 50–100 μm. Air-assisted sprayers are particularly useful when spraying large targets, such as trees, and have replaced the expensive and time-consuming use of hand-held lances in orchards. In Brazilian cocoa, however, Pereira (1985) preferred the use of lances due to the relatively short life of two-stroke engines and constant maintenance required. Air-assisted spraying has also been used for field crops using horizontal projection of the spray over several rows, but with increasing concern about spray drift, downwind projection of spray using an air-curtain to entrain the droplets in an airstream into the crop canopy has received increasing attention (see Chapter 7).

The airstream may be used to break up liquid into droplets, so twin-fluid nozzles (see pp. 114–16) often form an essential part of air-assisted sprayers. Alternatively, droplets may be produced by a hydraulic- or centrifugal-energy nozzle mounted in the airstream. When droplet formation is independent of airshear, emphasis can be given to the volume of air required and a lower velocity airstream may be adequate to carry droplets to the target. Droplet size is affected by the position of the orifice of hydraulic nozzles in relation to the direction of the airstream, which varies the amount of shear as the spray liquid reaches the airstream. Both cone- and fan-type nozzles have been used on air-carrier sprayers. Wide-angle cone nozzles permit very efficient break-up of the spray, but when larger droplets are needed, a narrow cone is used. The position of the nozzles in relation to the air outlet is important for achieving proper mixing and projection of droplets in the airstream.

Irrespective of which method of droplet formation is used, droplets less than 100 μm are more easily conveyed in the airstream, thus Potts (1946) found that in a particular airstream, droplets of 60–80 μm diameter were carried 46 m, while the larger droplets (200–400 μm) travelled only 6–12 m. This factor is particularly important when projecting spray vertically upwards into a crop canopy; gravity affects the trajectory of the large droplets, increasing fallout which results in considerable wastage of pesticide on the ground as well as increasing the risk of operator contamination. However, registration authorities are concerned that it is the small droplets that contribute principally to spray drift.

Air-assisted sprayers are often used to apply the same quantity of pesticide in one-tenth of the volume normally applied with hydraulic sprayers, but can be adapted for ULV application by restricting the flow of liquid to the nozzle.

Fans

The central feature of an air-assisted sprayer is the fan unit although a few sprayers rely on a compressor to provide air to twin-fluid nozzles. Four main types of fan described below are used, namely propeller fans, centrifugal fans, cross-flow and axial fans (Fig. 10.1). The propeller fan is simplest, but is suitable only for very low pressure systems and is generally used in conjunction with a centrifugal-energy nozzle (Fig. 8.15, p. 201).

An axial fan has blades of 'aerofoil' shape similar to an aeroplane wing, with a blunt leading edge and a thin trailing edge. In an axial fan, air is accelerated in the same direction as it was moving, whereas air is drawn in at the centre of a centrifugal fan and discharged at 90° to its entry. Axial fans are used to move larger volumes of air at low pressure, and the air velocity is usually insufficient for the use of shear nozzles. The performance of the fan depends on the shape and angle or 'pitch' of the blades in relation to the direction of rotation. Air pressure can be increased, within limits, by increasing the blade pitch or hub diameter, but this reduces the airflow. The clearance between the tip of the blade and the casing is also critical for optimum efficiency.

The centrifugal fan is similar to a centrifugal pump and consists of a wheel with blades rotating in a 'volute' or scroll casing. There are three types of fans:

1 those with the tip of the blade curved forwards (ie in the direction of rotation to provide a 'scoop' effect)
2 straight radial blade fans
3 those with tip of the blade curved backwards to provide a smoother flow of air.

The forward curved fan is run at a slower speed (rpm), and the backward curved fan faster, than a radial blade, when moving the same volume of air

(a)

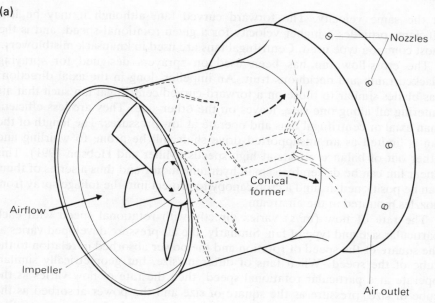

Nozzles

Airflow

Impeller

Conical
former

Air outlet

(b)

Air in

'L'-flow

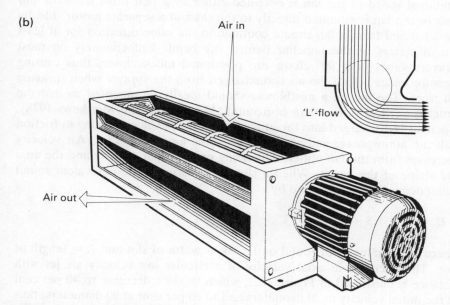

Air out

Fig. 10.1 (a) Axial flow fan (b) Cross-flow fan

at the same velocity. The forward curved fan, although it may be less efficient, provides a higher velocity for a given rotational speed, and is the most common type used. Centrifugal fans are used in knapsack mistblowers.

The cross-flow fan has been used on sprayers designed for spraying blackcurrants and deciduous fruit. An impeller, long in the axial direction, has blades similar to those on a forward-curved centrifugal fan such that air entering all along one side, leaves on the other side. They are less efficient than axial or centrifugal fans and operate at low pressures. The length of the fan is limited as an unsupported drive shaft will be prone to whirling and other out of balance effects at high speeds (Miller and Hobson 1991). This linear fan can be driven by a small hydraulic motor and thus a series of them can be positioned around a crop canopy to project into the foliage spray from nozzles mounted in the airstream.

The rate of flow (m³/s) varies directly with rotational speed with each particular size and type of fan. Similarly, the air pressure developed varies as the square of the speed of rotation and the power absorbed in relation to the cube of the speed. When fans of different size, but geometrically similar, operate at a particular rotational speed, then the rate of flow varies as the cube of size, pressure as the square of size and the power absorbed as the fifth power of the size. Thus, generally, an increase in fan diameter rather than fan speed is a more efficient way of increasing rate of flow. The rotational speed of the fan is obtained either by a belt drive from the pto shaft or the fan is mounted directly to the shaft of a separate motor. Ideally the airstream from a fan should continue in the same direction for at least two diameters of the impeller before any bend. Unfortunately on most sprayers vanes or a 90° elbow are positioned much closer, thus causing pressure losses before the air is discharged from the sprayer when spraying tall trees. The fan on a mistblower should ideally be mounted so that the outlet is vertical rather than horizontal (MacFarlane and Matthews 1978). When air is discharged into the atmosphere it loses velocity owing to friction with the atmosphere, and also entrains some air with the jet. Air velocity decreases from the fan outlet, depending on its initial velocity and the area and shape of the outlet. When a slot outlet is used, the equivalent round outlet diameter is determined by

$$D = W \times 1.3 + \frac{\sqrt{L}}{4}$$

where D = diameter of round outlet, W = width of slot and L = length of slot. The decrease in axial velocity of a circular low-velocity air jet with distance is illustrated in Figure 10.2, which shows a decrease to 40 per cent of the initial velocity at 20 diameters and to 10 per cent at 90 diameters, thus if the initial velocity at the 5 cm diameter nozzle is 50 m/s, then at 200 cm the velocity has decreased to 10 m/s. In practice, lower velocities are usually recorded under field conditions (Potts and Garman 1950). The discharge tube should have the largest circular opening to achieve maximum throw of droplets, but there is an optimum diameter for a given air capacity and air

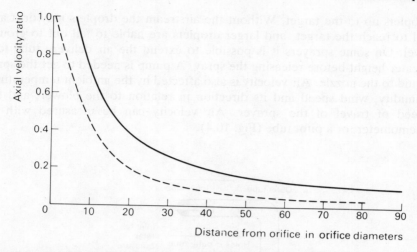

Fig. 10.2 Axial velocity of an air jet with distance. Theoretical velocity (solid line) – data from Potts and Garman (1950) – obtained with sprayers of different diameters and air velocities (broken line) (after Fraser 1958)

pressure. The velocity field with contours of equal velocity from an air jet is illustrated in Figure 10.3. At the mouth of the discharge tube, a turbulent mixing region surrounds an air core at the initial velocity, but at about five diameters this air core disappears.

When spraying tall trees, it is better to establish a column of air moving up into the canopy then spray briefly and continue the flow of air to carry the

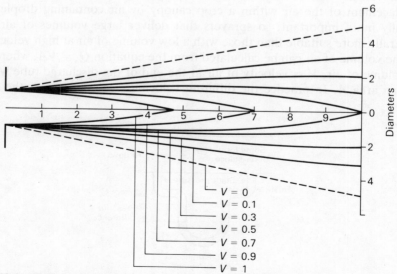

Fig. 10.3 Velocity field of a symmetrical air jet (after Fraser 1958)

droplets up to the target. Without the airstream the droplets may drift and fail to reach the target, and larger droplets are liable to fall out to ground level. On some sprayers it is possible to extend the air delivery tube to a greater height before releasing the spray. A pump is needed to get the spray liquid to the nozzle. Air velocity is also affected by the ambient temperature, humidity, wind speed and its direction in relation to the blower, and the speed of travel of the sprayer. Air velocity can be measured with an anemometer, or a pitot tube (Fig. 10.4).

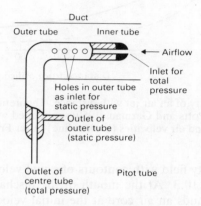

Fig. 10.4 Measuring air velocity from a mistblower

Air velocity can be important when projecting spray up into trees but displacement of the air within a crop canopy by air containing droplets is usually more important, so sprayers that deliver large volumes of air are generally more suitable than those with a low volume of air at high velocity.

The volume of air can be calculated from the equation $Q_a = VA$, where Q_a = volume of air, V = velocity of air at the end of the discharge tube which has an area A. In practice the different velocities recorded across the area

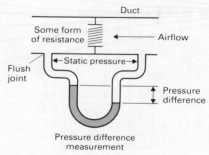

Fig. 10.5 Measuring air volume from a mistblower

have to be integrated. In the laboratory, the volume of air moving through a duct can be calculated more accurately by measuring the differential pressure across two orifices partially separated by a sharp-edged plate mounted in a smooth-bore pipe so that the upstream pipe is 20 × pipe diameter and downstream 5 × pipe diameter (Fig. 10.5). Another method is to deliver the volume of air into an enclosed space and, while maintaining no pressure change, measure the volume of air expelled using a previously calibrated standard fan.

Pumps

Low liquid pressures are usually sufficient to feed spray to the nozzles, so any of the pumps described in Chapter 7 can also be used as well as peristaltic pumps. As a high-speed drive is available, simple centrifugal pumps are suitable, but diaphragm and piston pumps are frequently used. On some sprayers spray is fed into the airstream by gravity; on others air pressure from the fan is used to pressurize the spray tank and force liquid to the nozzle.

Fig. 10.6   Knapsack mistblower (note: engine exhaust should be guarded)

Motorised knapsack mistblowers

The development of portable air-carrier sprayers which are invariably referred to as motorized knapsack mistblowers (Fig. 10.6) has followed development of light-weight, two-stroke engines. Efficient use and maintenance of these sprayers depends on understanding the operation of a two-stroke engine (Fig. 10.7). Sutherland (1980) gives information on the operation and maintenance of mistblowers. They are used in vector control as well as crop protection.

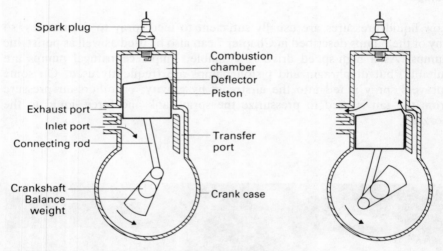

Fig. 10.7   Diagram showing the operation of a two-stroke engine

Two-stroke engine

When the piston is moving up the cylinder to compress the fuel/air mixture, the inlet and outlet ports are covered initially but then, as the piston continues to travel upwards, it creates a partial vacuum and uncovers the inlet port. This vacuum causes a depression in the carburettor inlet and air passing over the petrol jet collects a metered quantity of fuel. This fuel/air mixture is mixed and drawn through the inlet port into the crankcase. Meanwhile the previous charge of fuel/air mixture, compressed in the combustion chamber, is ignited while the piston is still ascending the cylinder. Momentum of the piston carries it over top dead centre and the expansion of the burning gases provides the power stroke, the downward movement of the piston. After a short distance, the exhaust port is uncovered and the burnt gases escape. As the piston moves down, the fuel/air mixture in the crank-case is compressed and, when the transfer port is opened, it is forced into the combustion chamber ready to be compressed by the next upward stroke of the piston.

The fuel for two-stroke engines is a mixture of oil and petrol, usually in the ratio of 1:24. The correct mixture is marked on the fuel tank or the cap of the fuel tank. The most suitable oil is 30 SAE oil. Multigrade oil should never be used, because the additives it contains may cause engine failure. Similarly, only regular low-grade petrol should be used. If a two-stroke oil is used, check whether it has the same viscosity as SAE 30, as a different mixture may be required. A siliconized oil is more suitable for certain small two-stroke engines. The fuel is fed from a tank by gravity to the float chamber of the carburettor through a needle valve (Fig. 10.8). The float is designed to

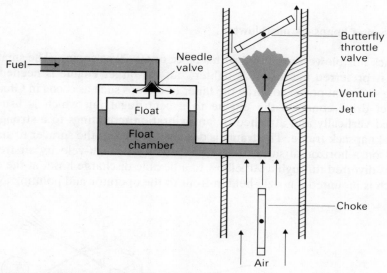

Fig. 10.8 Principles of a simple carburettor

maintain the required level of fuel in the float chamber but, when starting the engine, a tickler knob can be used to 'flood' the carburettor to provide a richer fuel/air mixture. Air is drawn through a filter which should be cleaned regularly to prevent dust and grit entering the engine. The flow of air is speeded up by a narrowing of the tube known as the venturi. The increase in speed causes a decrease in air pressure which draws in fuel through a jet. A throttle valve controls the volume of fuel/air mix entering the combustion chamber, hence the speed and power of the engine. The throttle is operated by means of a flexible cable (Bowden cable) connected to a lever which needs to be conveniently placed for the operator. It may be fixed to the bottom of the left-hand side of the knapsack frame, or close to the on/off tap on the spray line. A choke or 'strangler' restricts the flow of air through the venturi. The choke is used to enrich the fuel mixture when starting the engine.

Ideally, fuel should be drained from the tank and carburettor when the sprayer is being stored, especially in hot climates, otherwise petrol may evaporate, affecting the petrol/oil ratio. Oil deposits in the carburettor may make it difficult to start the engine. If it is necessary to stop the engine in the field, even for short periods, this should be done by closing the fuel tap rather than by shorting the electrical circuit. The latter is useful in an emergency, but restarting is easier if the carburettor has been run dry.

The engine is usually provided with a recoil starter but, when a pulley wheel is provided as part of the starter, the engine can be started by using a rope or strap. The starter mechanism should be fully covered by a cap while the engine is running to prevent the operator touching a moving part. Electronic ignition is now provided on some engines.

The design of knapsack mistblowers

Knapsack mistblowers are fitted with a 35 cc or 60–70 cc engine. The smaller engine is preferred because it is lighter, but the larger engine is needed to provide sufficient power to spray the taller tree crops such as cocoa in Ghana. A direct drive connects the engine to a centrifugal fan which is usually mounted vertically and attached by anti-vibration mountings to a strong L-shaped knapsack frame. The frame is designed to permit the sprayer to stand upright on a horizontal surface. The fan produces a high-velocity airstream which is diverted through a 90° elbow to a flexible discharge hose, at the end of which is mounted a nozzle held in front of the operator and pointing away

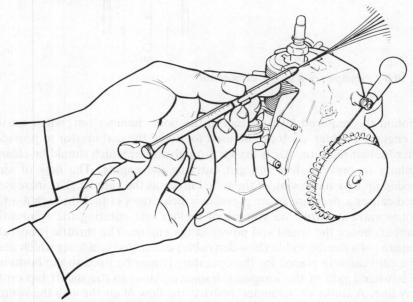

Fig. 10.9 Vibrating wire tachometer ('Vibratak')

from him. Lavabre (1971) suggested that contamination of the operator would be less if the nozzle was mounted at the rear of the sprayer, pointing in the opposite direction to the operator's movement. The 90° elbow is eliminated on one model by having a horizontally mounted fan. Engine speed is often 6 000 rpm, but some engines achieve higher speeds. The correct engine speed is needed to provide the correct air velocity and volume at the nozzle; so it should be checked regularly with a suitable tachometer (Fig. 10.9). A faulty engine should be returned to the workshop for maintenance.

The most common nozzle fitted to these sprayers is a shear nozzle, but various adaptations have been tried to improve droplet formation. An even flow of air over the liquid orifice assists droplet formation, so on some sprayers there is an orifice on either side of an aerofoil-shaped section mounted across the outlet of the discharge tube (Fig. 5.18, p. 116). On some sprayers liquid is spread over a fixed disc but spinning discs and cups (Fig. 10.10) have also been used on mistblowers to spread the liquid thinly before reaching the air jet (Clayphon and Thornhill 1974b; Takenaga 1972; A.E.H. Higgins 1964, Hewitt 1991).

The spray tank is mounted above the engine/fan unit and normally holds at least 10 litres. A large opening facilitates filling, and this should be protected by a basket-type filter with a fine mesh to prevent nozzle blockages. The lid normally must seal the top of the tank completely when the tank is slightly pressurised (0.2 bar) to force the spray liquid to the nozzle. Liquid feed to the nozzle is improved on some sprayers by using a small pump driven on the engine shaft. This is particularly useful when the nozzle is held high for spraying trees, but such pumps have not been very durable in the field. The flow of viscous liquids can be improved by deliberately increasing the temperature of the liquid by passing the spray through a heat exchanger, consisting of a copper tube wound spirally around the engine silencer. When the sprayer has a small pump, the heated spray liquid can be recirculated to the tank (Thornhill 1974b). The base of the tank may be sloping to the outlet to facilitate the flow of dusts or granules when the sprayer is converted for dry applications. The tank outlet is provided with a fine-mesh filter (100 mesh) when adapted for very low flow rates.

The spray liquid is metered to the nozzle by an adjustable barrel-type valve, a sleeve with four different orifices, or a set of restrictors. The latter is preferable as the appropriate restrictor can be fixed in the workshop, whereas other types may be set incorrectly or changed by inexperienced operators during spraying (Jollands 1991). Most sprayers have a separate ON/OFF valve. This can be replaced with a trigger valve to facilitate intermittent spraying of individual targets.

The basic weight of a motorized knapsack sprayer may be as much as 14 kg when empty, so they are much heavier to carry than other types of knapsack sprayer. The straps are usually provided with a non-absorbent pad over the shoulder and a backrest to improve operator comfort and reduce the effect of engine vibration.

The performance of different knapsack mistblowers can vary considerably

(a)

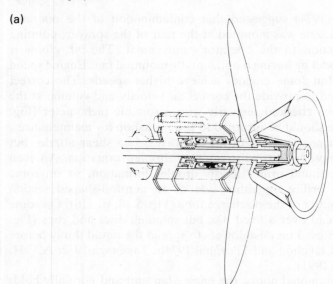

(b)

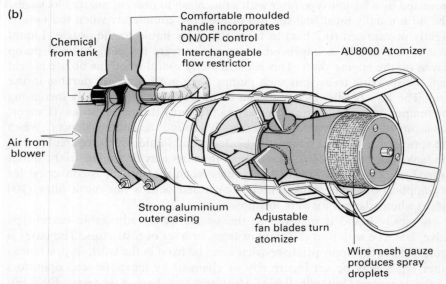

Chemical
from tank

Comfortable moulded
handle incorporates
ON/OFF control

Interchangeable
flow restrictor

AU8000 Atomizer

Air from
blower

Strong aluminium
outer casing

Adjustable
fan blades turn
atomizer

Wire mesh gauze
produces spray
droplets

Fig. 10.10 (a) Spinning-disc (Micron 'X–1') (b) Micronair AU8000 mounted on Knapsack mistblower

despite the use of the same basic design (Table 10.1). Clayphon (1971) examined the suitability of a range of mistblowers for cocoa spraying. The main criteria in choosing these sprayers depend on whether it is intended to spray trees or to drift spray over field crops. For the former, the vertical throw, that is the height to which droplets can be carried by the airstream, is

Table 10.1 Comparison of the performance of two knapsack mistblowers (adapted from Clayphon 1971)

	Mistblower A	Mistblower B
Engine capacity (cc)	35	70
Fuel tank capacity (litres)	1.25	1.5
Fuel consumption (litres/ha)	0.9	1.6
Air velocity at nozzle (m/s)	66	74.6
Air volume		
at fan outlet (m³/min)	7.9	14.7
at nozzle (m³/min)	5.2	8.2
Flow rate (litres/min)	0.7–1.8	0.04–2.8
Horizontal throw* (m)	13.7	16.8
Vertical throw* (m)	6.1	9.75

Note: * Measured at maximum flow rate

the most important factor; thus, for some cocoa and citrus, spray is needed up to 10 m, whereas the horizontal throw is more important when field crops are sprayed.

Horizontal throw can be determined by measuring droplet density on Kromekote cards fixed to stakes vertically 1 m above ground level. A typical layout has ten rows, each with seven stakes which are spaced at 1.5 m between rows and 0.75 m within the rows. The first row is 3 m from the spray nozzle, which is directed down the centre line of the target layout when spraying a dyed liquid for about 5 s. The width of the airstream with entrained droplets is also measured by using a series of stakes.

Vertical throw is similarly measured with Kromekote cards fixed horizontally at 30 cm intervals to a rope which is raised over a pulley to a tower, so that the highest target is at least 12 m and the lowest 4 m above the ground. The nozzle is held at an angle 1.5 m above ground level and 3 m from the rope and directed upwards to spray the targets with the minimum interference from natural air movements. Both upper and lower surfaces of the targets are examined to assess droplet density.

Deposition from a mistblower can be increased by adding an electrostatic charge to the spray (Afreh-Nuamah and Matthews, 1987). Improved deposition has also been reported with similar equipment in China (Shang and Li 1990) (see also Chapter 9).

Method of using a knapsack mistblower

The correct petrol/oil mixture is poured through a fine-mesh filter into the fuel tank. The spray tank is then filled through the filter and the tank lid replaced tightly. Any on/off switch on the engine is turned on, the petrol tap opened and the carburettor allowed to fill with fuel. The choke lever is moved to the 'closed' position and, with the throttle closed, the engine is started by pulling sharply but evenly on the recoil starter. The starter rope should be allowed to rewind slowly and **not** released to snap back. When the engine starts, the choke can be moved to the 'open' position and the throttle opened up to allow the engine speed to increase to maximum revs. While spraying, the engine should be run at full throttle and not allowed to idle for long periods. When the engine is running smoothly the sprayer is put on the operator's back. This can be accomplished by the operator holding the harness straps with crossed hands and swinging the sprayer up on to his or her back. When someone is helping the operator, this person should avoid touching the hot exhaust which should be surrounded by a guard. If the machine is used for a long period, the operator should wear ear muffs.

The nozzle should be directed downwind so that natural air movements assist dispersal of the droplets away from the operator. If the nozzle is pointed upwind, droplets are liable to be blown back on to the operator. The discharge tube should be held at least 2 m from the target to allow dispersal of the droplets, as the air velocity close to the nozzle may exceed 80 m/s. Operators should walk at an even pace through the crop and close the spray liquid tap whenever they stop, to avoid overdosing part of the crop. When spraying cocoa to control mirids, spray is directed principally up the trunk and into the tree canopy.

Calibration of knapsack mistblower

The sprayer is operated with a small quantity of spray liquid to check that the engine and nozzle are functioning correctly, and until all the liquid has been emitted from the nozzle. The on/off valve is closed and a known quantity of liquid sufficient for a minimum of 1 min spraying, is poured into the tank. The spray liquid should be the pesticide formulation to be used or a liquid with the same viscosity. With the engine operated at maximum revs, the time taken to spray the measured volume is measured with a stop-watch. This calibration procedure should be repeated at least twice.

Tractor-mounted or trailed air-assisted sprayers

These are of four basic types

1 The airstream from an axial fan is deflected through 90°, and nozzles are mounted close to the outlet (Fig. 10.11a).
2 The airstream is provided by one or more centrifugal fans.
3 The airstream is by a cross-flow fan, particularly suitable for low, trellis or spindle pruned trees.

(a)

(b)

Fig. 10.11 Tractor-mounted or trailed orchard sprayers with axial fan (a) Commandair (b) Berthoud sprayer

4 Sprayer with centrifugal fan or axial fan, but air-ducted along horizontal boom with nozzles at intervals along the boom (Fig. 7.25, p. 181).

The first three types are designed primarily for spraying trees, although they are also used to spray field crops. Special attachments are available to suit application of spray to different crops such as grapevines and deciduous fruit trees (Fig. 10.11b). When the same volume (400 l/ha) was applied, similar spray coverage was reported for both axial fan and cross-flow fan sprayers in one series of trials in which the air volume was 8.3 m^3/s and 11.1 m^3/s and air velocity was calculated as 21.5 m/s or 29.6 m/s respectively (Raisigl et al 1991).

Few boom-type air-assisted sprayers were developed, mainly because of a loss in efficiency along the boom (Zucker and Zamir 1964). Taft et al (1969) and Takenaga (1976) described air-assisted boom sprayers, but recently an airsleeve type boom has been introduced to project the spray downwards from hydraulic nozzles (Taylor et al 1989; Cooke et al 1990; Hadar 1991; Taylor and Andersen 1991). Renewed interest in these sprayers has been due to the need for increased penetration of crop canopies (for example to control whiteflies on cotton and replace aerial sprays with ground equipment), to improve spray distribution on both leaf surfaces, reduce spray drift and continue to use hydraulic nozzles, including those in the 'fine' spray quality (see also Chapter 7).

Tractor equipment in orchards

The choice of fan design for spraying orchards depends on the performance of different air jets in terms of penetration into tree canopies. In general, distribution of spray is improved by increasing the volume of air applied to replace most of the air within the crop canopy with air containing spray droplets. Thus, when the amount of energy available is fixed, the ratio of volume/velocity should be as large as possible. Sufficient velocity is needed to cause movement of leaves to assist deposition and allow droplets to penetrate to the inner part of the canopy. Velocity is also important to direct spray towards the top of trees, or if the horizontal distance between the target and air outlet is large. In bush apple trees J.M. Randall (1971) concluded that the optimum performance required a volume of 13.4 m^3/s at an outlet velocity of 31 m/s. Uniformity of deposit was improved if the forward speed of the tractor was as slow as economically possible (ie 2.75 km/h was better than 6.5 km/h) but actual speed will depend on wind conditions and the type of plantation.

Much of the energy from an axial-flow fan is lost, however, when air is deflected through 90° to aim spray at trees. When the air has to be deflected, a series of guide vanes should be mounted in the passage to the outlet. Deflector plates can also be used to broaden the air jet. The type and position of nozzles in the airstream is chosen so that the volume of liquid is distributed as evenly as possible into the canopy.

Fig. 10.12 Tunnel sprayer in which the spray is protected from the wind and contamination of the soil is reduced

Studies on orchard spraying (Hale 1975) led to the development of a sprayer (Fig. 10.11a) which displaced 7.8 m³/s of air to each side of the machine at a velocity of 20.7 m/s. However, the trend is to reduce the height and prune crop canopies to facilitate harvesting so the 'volume' to be sprayed is less in modern intensive orchards, for which most axial fan air-assisted sprayers are poorly suited (Cross 1991). A large proportion of the spray may be carried above the tops of the trees and lost as spray drift, especially when low volumes (100 l/m) are applied (Planas and Pons 1991).

Most growers now use much less spray liquid (<600 l/ha) than was customarily used on tree crops (>2 000 l/ha) (Cooke et al 1975a); Morgan (1964), Sutton and Unrath (1984) and Ras (1986) have all advocated that growers assess the volume of their tree canopies and adjust the spray volume and spray concentration accordingly.

Care must be taken when mounting air-driven centrifugal nozzles on a tractor-mounted mistblowers to ensure that the correct rotational speed is achieved (Cross 1991; Sander 1991). In New Zealand, a belt-drive system was adopted and the spinning discs mounted in a slower airstream so that droplet size was controlled by rotational speed rather than by high-speed airshear which affected ligament production. On this system a flushing tank was incorporated into the design so that particulate deposits were removed from the discs as soon as spraying ceased.

Good penetration of citrus trees was achieved with a sprayer having contra-rotating twin centrifugal fans with the fan outlets modified to cause the direction of the air blast (4.7 m³/s at 60 m/s) to fluctuate through approximately 90° to the hinge axis of the outlet, at a frequency of 20–25 cycles/min (Johnstone 1970). Oscillation of the airstream caused sufficient leaf movement to permit droplet penetration into the canopy.

Concern about environmental pollution, especially when spray is projected up and beyond the crop canopy, demands efforts should be made to reduce tree size and adjustments made to the sprayer to direct spray sideways into the canopy. Recent developments include the use of a mobile canopy over the trees (Fig. 10.12) to protect the spray from the wind and to recycle spray liquid not deposited on the crop and in the Netherlands an unmanned sprayer has been developed (van de Werken 1991). This system is ideal for relatively small closely spaced trees (Morgan 1981) and as less spray is wasted downwind (Fig. 10.13) their use will increase because authorities will impose more restrictions to reduce pollution. Many of these sprayers were fitted with hydraulic nozzles but reduced volumes are considered more suitable to decrease the volume of spray that is recycled.

Other sprayers with gaseous-energy nozzles

Exhaust nozzle sprayer

This sprayer (Fig. 10.14), which uses vehicle exhaust gases to pressurize the spray tanks and also to shear the spray liquid into droplets, was designed specifically for the control of locust hoppers as mentioned in Chapter 2 (Sayer 1959; Sayer and Rainey 1958). The latest unit has twin 50-litre tanks and sits on the back of a suitable vehicle (Land Rover, Jeep, Unimog, etc.) so that the controls can be operated through the rear of the cab (Watts et al 1976). Exhaust gases are directed through a flexible hose to the spray tank and the nozzle, the orifice size of which is selected to suit the particular vehicle and engine. A steady pressure of between 0.2 and 0.4 bar should be used during spraying but, to protect the engine from higher back pressures, a safety valve prevents pressures exceeding 0.5 bar. The exhaust system of the vehicle has to be checked for leaks and modified so that the end pipe is pointing to and level with the rear of the vehicle to avoid the connecting hose being swept off when driving through scrub.

The sprayer is operated so that a wind across the line of travel takes the plume of 70–90 μm vmd droplets projected upwards downwind away from the vehicle. Swath width will vary with wind speed, but can be up to 100 m with 10–15 km/h winds. Vehicles are normally operated in low gear and care must be taken to avoid overheating of the engine at high ambient temperatures. At 1.2 litres/min output and vehicle speed of 10 km/h, the application rate is 0.3 litre/ha over a 240 m swath. Twin tanks are provided so that two different insecticides can be applied, if necessary, thus a persistent chemical can be applied in remote areas of bush, but a less persistent and safe insecticide is needed if hoppers are feeding on food crops, or near villages.

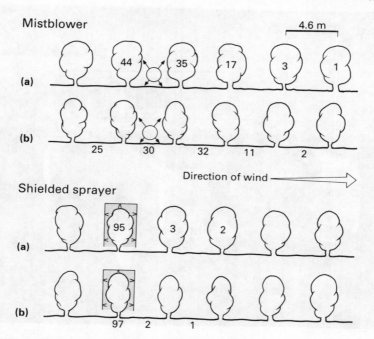

Fig. 10.13 Percentage of spray deposited on treated rows and adjacent rows downwind by conventional orchard sprayer and tunnel sprayer (a) on foliage and (b) on the ground.(from Morgan, 1981)

Due to problems with vehicle exhausts, alternative equipment is now being used to control locusts (Symmons et al 1989) (Fig. 8.19, p. 206).

Sprayers with vortical nozzles

Small sprayers with an electric motor to drive a blower unit have incorporated a simple vortical nozzle to produce a mist (Fig. 10.15). These sprayers have been used mainly inside buildings, especially warehouses, to control pests of stored products (Fig. 10.16). Output is regulated by interchangeable nozzle orifices or by a variable restrictor/tap unit. The latter is less efficient, as resetting it to the same position for subsequent treatments is often difficult. Larger droplets which may fall out close to the machine are produced when flow rate is increased as the proportion of air volume for each unit volume of liquid is decreased. This effect also applies when aerosol sprays are obtained with a two-stage vortical nozzle. A range of machines with different capacities is available (Table 10.2). On these, a power-driven high-performance blower provides air to the nozzle. On some machines the spray tank is also pressurized to approximately 0.25 bar, but flow rate depends on viscosity

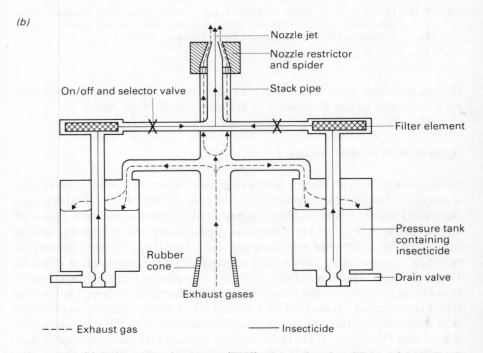

Fig. 10.14 (a) Exhaust nozzle sprayer (ENS) mounted on Land Rover (photo Evers & Wall Ltd) (b) Diagram of the layout of an ENS

Fig. 10.15 Hand-carried aerosol sprayer with electric motor (Microgen Corp.)

Fig. 10.16 Propane-powered machine for use in warehouses. (photo Wellcome Environmental Health)

Table 10.2   Performance of two sprayers with vortical nozzles

Droplet size (μm vmd)

	Sprayer A	Sprayer B
kW	1.86	12
Insecticide tank capacity (litres)	1.0	56
Weight (kg)	6.1	146
Air velocity (m/s)	109	196
Air volume (m³/s)	0.0095	0.1
Droplet size (μm vmd)	17	17 (at 140 ml/min)
Max. flow rate of light oil (ml/min)	30	590

which changes with temperature, so a thermometer and flow meter are fitted to the control panel of these sprayers. A pointer on the flow meter is useful to indicate the flow rate required. The flow meter should be protected by filter. The flow rate is more effectively controlled by fitting a positive displacement pump which can be operated in relation to the forward speed of the vehicle. A digital display of the pump output can be fitted in the vehicle cab.

This equipment eliminates some shortcomings with thermal fog equipment, and the larger versions, such as the 'Leco HD' and 'Microgen ED2–20' (Fig. 10.17), mounted on a truck are widely used to apply ULV insecticides in vector-control programmes. A typical truck-mounted unit has a 12 kW four-stroke engine with direct drive to a Roots-type blower. Older models have a belt drive which needs to be checked regularly to ensure that the blower is operated at the right speed. Some machines have a single vortical nozzle which is usually directed upwards at an angle of 45°. Other types have two or more nozzles which can be adjusted to point in various directions. Units with several nozzles on a vertical boom have been used experimentally to apply microbial insecticides to tree crops (Falcon and Sorensen 1976). The sprayer can be operated from a remote-control panel mounted in the cab of the vehicle. On some machines, the engine can be started and stopped as well as adjusting the flow rate and speed of the engine. The cab vehicle must be well sealed to prevent aerosol droplets contaminating the driver. Inside the cab the operators need not wear ear protectors, which are needed when working close to the machine.

Dosage recommendations are based on flow rate and vehicle speed, and intended swath width. Spray droplets emitted while the truck is driven, in particular around urban areas, drift downwind through and between dwellings, but variations in air velocity and direction, especially around buildings of different shapes and sizes, will affect the actual distribution of a pesticide. Penetration into houses is poor so that vector species such as *Acdes aegypti* resting inside dwellings may not be controlled (Perich et al 1990, Goose 1991). Wind speed should not exceed 1.7 m/s for maximum efficiency.

Fig. 10.17   Aerosol generator with four-stroke engine suitable for mounting on a vehicle showing in the foreground, the control unit that is mounted in the vehicle cab (Microgen Corp.)

Optimum results against mosquitoes are obtained when they are actively flying in the evening and inversion conditions occur. The initial droplet size of these aerosols is generally below 30 μm (Mount et al 1975), so correct formulation of pesticides is essential to prevent their size decreasing further. Undiluted technical malathion (95 per cent ai) has been very successfully used with the equipment because of its relative involatility, but the range of flow rates is affected by its viscosity. Specifications for vehicle-mounted aerosol machines are now published by WHO. (Anon 1990).

These sprayers, particularly the smaller versions (Figs. 10.15 and 10.16), have been used successfully in warehouses and glasshouses; for example, aerosol sprays have been very effective against whiteflies (Mboob 1975). On the small machines, droplet size is affected by flow rate more than on the larger machines, presumably because the air volume emitted through the nozzle is insufficient to shear liquid effectively at high flow rates. When used in enclosed spaces, for example in the food industry, the motor can be converted for propane gas fuel (Fig. 10.16) so that environmental contamination is reduced.

Warehouses can also be treated by space sprays using an insecticide formulated in a compressed gas (CO_2) supplied in cylinders that are fitted to a spray gun (Fig. 10.18). Slatter et al (1981) reported on the use of non-residual synthetic pyrethroids applied through a cone nozzle 0.5 mm orifice at a nominal output of 6 g/s at 5 000 kPa. Immediately after discharge the droplets of insecticide plus solvent had a VMD of approximately 9 μm. Efficient control of flying insects was achieved.

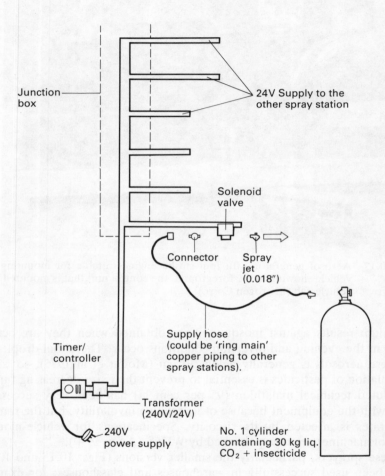

Fig. 10.18 'Pestigas' system

11

Fogging

Fog is produced when aerosol droplets, having a diameter less than 15 μm, fill a volume of air to such an extent that visibility is reduced. The obscuring power of a fog is greatest when droplets are 1 μm in diameter. Insecticides and some fungicides are applied as fogs.

In thermal fogging machines, pesticide, usually dissolved in an oil of a suitable flash-point, is injected into a hot gas and vaporized. A dense fog is formed by condensation of the oil when discharged into the atmosphere. Using a laser system, the volume median diameter of a sample of fog is usually less than 3 μm, but most fogging machines also produce droplets larger than 15 μm diameter (Table 11.1), especially if too large a volume is applied and partial vaporization occurs. When wettable powder formulations are applied using fogging machines (see also Chapter 3), it is better to have

Table 11.1 Droplet sizes measured at different distances from a pulse-jet fogging machine used at four flow rates (from Munthali 1976)

| Distance from nozzle (m) | Flow rate (ml/min) | | | | | | | |
| | 109 | | 193 | | 270 | | 370 | |
	nmd	vmd	nmd	vmd	nmd	vmd	nmd	vmd
A Collected on cascade impactors 1 m above ground								
1	11	26	14	29	14	36	19	27
2	12	26	16	31	15	24	14	26
10	12	24	15	23	18	26	20	30
B Sedimented on slides placed 1 m above ground								
0.15	35	44	36	110	35	105	48	80
2.5	23	30	33	41	30	40	40	85
10	54	50	40	58	34	47	33	42

Fig. 11.1 Fogging inside a glasshouse (photo ICI Agrochemicals)

an electrically powered agitation in the tank to keep the formulations in suspension.

Fogging is particularly useful for the control of flying insects, not only through contact with droplets, but also by the fumigant effect of a volatile pesticide. It is mainly used to treat unoccupied enclosed spaces, such as warehouses, glasshouses (Fig. 11.1), ships' holds and farm sheds, where the fog will penetrate in accessible cracks and crevices. Fogging has also been used to treat sewers. All naked flames must be extinguished and electrical appliances disconnected, preferably at the mains, when fogging in confined spaces. In the case of pilot lights, sufficient time must be allowed for gas in the pipes to be used up. Thus in glasshouses, automatic ventilation, irrigation and CO_2 systems should be switched off and the glasshouse kept closed as long as possible after fogging. Care must also be taken in buildings in which there may be a high concentration of fine dust particles in the air. A single spark can set off an explosion when more than 1 litre of a formulation containing kerosene if fogged per 15 m³. Overdosing may be confined to localised pockets of fog which exceed the explosive limit. Fogging rates are usually less than 1 litre/400 m³, but this lower rate can be ignited by a naked flame. The volume of the space to be treated should be calculated and the machine's output calibrated carefully so that the correct dosage is applied.

Foliage should be dry, with temperatures between 18° and 29°C, and fogging should be avoided in high humidity conditions, and in direct sunlight to minimize risk of phytotoxic damage. Application is often better if made in the evening. Plants needing water should not be fogged. The small droplets (<15 μm) eventually sediment on horizontal surfaces. Experiments with fogging on microbial insecticide *Bacillus thuringiensis* confirmed that 95 per cent of the spray was deposited on the upper surfaces of leaves (Burges and Jarrett 1979). Such deposits have little or no residual effect unless a persistent chemical has been fogged, so reinfestation can take place readily from neighbouring areas. Also, not all stages in the life cycle of a pest may be affected by a pesticidal fog; for example, whitefly adults are readily killed, but egg and pupal stages on the undersurface of leaves are less affected. When fogging indoors the lowest flow rate possible should be used to reduce the proportion of large droplets in the fog. The machine is moved gradually through the space to be treated by an operator wearing appropriate protective clothing. Alternatively the fogger can be mounted on a trolley and pulled through the building by a rope so that the operator can stay outside. Operators who attempt to fog large spaces from a doorway will not achieve an even distribution of pesticide unless there is sufficient air circulation to spread the fog away from the nozzle. Some buildings have shuttered openings at intervals along the exterior walls so that treatments can be carried out from outside. Fogging at higher volume rates and with a greater proportion of larger droplets, sometimes referred to as a 'wet' fog will leave a heavier deposit on foliage. This may provide a longer residual effect, but at high flow rates foliage close to the nozzle is liable to be damaged by an overdose of large droplets.

Fogs can be used outdoors, when advantage can be taken of temperature inversion conditions, usually either early morning or early evening, so that the fog remains close to the ground. The fog is released as close to the ground as possible, or directed towards ground level and drifted across the area to be treated. Wind velocity should not exceed 6 km/h or the fog will disperse too quickly. Fogs have been used extensively for the control of adult mosquitoes and other vector or nuisance insects, but in urban areas preference has been given in some countries to ULV aerosols, to avoid traffic hazards associated with fogs and also to avoid the use of large volumes of petroleum products as diluents. In forests, the retention of fog has been utilized to control cocoa mirids and to treat tall trees, such as rubber, but the chimney effect caused by the spaces between individual trees lifts the fog (Khoo et al 1983). Unless there are inversion conditions the fog is likely to be sucked rapidly out of the canopy by air movement above it. Water is injected into the hot gases closer to the combustion chamber on some machines deliberately to cool the fogging temperature and help keep the fog closer to the ground.

Great care must be taken to avoid inhalation of fog, as the smallest droplets are not trapped in the nasal area and may be carried into the lungs. Research has shown that the particle sizes most likely to reach the lungs are 10 μm (Swift and Proctor 1982). Proper protective clothing must be worn; this includes a full-face respirator for many pesticides and special fogging carriers. After application, all doors to enclosed spaces should be locked until after the required ventilation period, usually not less than 5–6 h. Treated areas should be marked with suitable warning notices.

Fogging machines

Several types of thermal fogging equipment are available. Ideally, all types should be started outdoors and never in an area partially fogged.

Pulsejet

These machines are usually hand- or shoulder-carried machines (Fig. 11.2); a larger more powerful model is trolley mounted. These machines consist essentially of a fuel tank and pesticide tank, a hand-operated piston or bellows pump, spark plug, carburettor and long exhaust pipe. A few

Fig. 11.2 Pulse-jet hand-carried thermal fogger (photo Motan gmbh)

machines have an electrically operated pump. Some machines have a translucent tank so that the quantity of pesticide remaining can be easily seen. Where the tank is detachable it is easy to change to a different chemical when necessary, if a spare tank is used. The provision of a large tank opening facilitates not only refilling but also cleaning. To start the machine both tanks are pressurized, using the pump. Instead of a hand-operated piston or bellows pump, some machines now have an electrically powered pump that can be connected to a battery for example by using the connection to a car battery. An initial mixture of fuel and air is supplied through non-return valves into a combustion chamber, and is ignited by a high-tension spark obtained from a battery-powered vibrator or mechanically operated magneto connected to the plug for a few seconds. The fuel is regular-grade petrol and about 1 litre/h is used on the smaller machines. Once the machine has started, the high-tension spark is no longer required and can be stopped. The exhaust gases from the combustion chamber escape as a pressure wave at high velocity through a long pipe of smaller diameter than the combustion chamber, and draw in a fresh charge of fuel and air through a non-return valve. If operating with the correct mixture, there are about eighty pulsations per second, slightly irregular with maximum noise. By means of a non-return valve the pesticide tank is also pressurized, and when the machine has warmed up, after about 2 min running, a valve is opened to permit the flow of pesticide solution through an interchangeable or variable restrictor into the end of the exhaust pipe. On some machines suitable for oil-based formulations only, the inlet is nearer to combustion chamber to give a more complete vaporization of the liquid. Some machines have a variable restrictor, but these are generally difficult to set repeatedly at the appropriate position. Droplet size can be increased by increasing the flow of solution; the larger droplets sedimenting usually within 2–3 m from the nozzle (Table 11.1). On some machines the liquid is injected into the hot gases through two openings on opposite sides of the exhaust pipe. This gives a better distribution and break-up of the liquid. The temperature to which the liquid is exposed is lowered by an increase in flow rate, but even at low flow rates there is a very rapid decrease in temperature as soon as the fog is formed (Table 11.2). At

Table 11.2 Temperatures measured at different distances from a small pulse-jet fogging machine (from Munthali 1976)

Flow rate (ml/min)	Temperature °C at			
	Injection of insecticide	End of pipe	0.8 m from nozzle	2.5 m*
0	524	379	78	28.5
109	420	318	72	28.5
193	350	290	65	28
270	318	185	52	32
370	196	178	40	32

Note: * Essentially ambient temperature

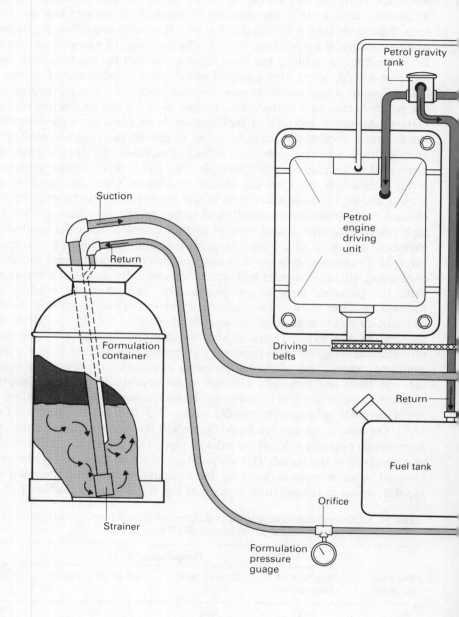

Fig. 11.4 Diagram to show layout of 'TIFA' 100E fogging machine. (courtesy Tifa (CI) Ltd)

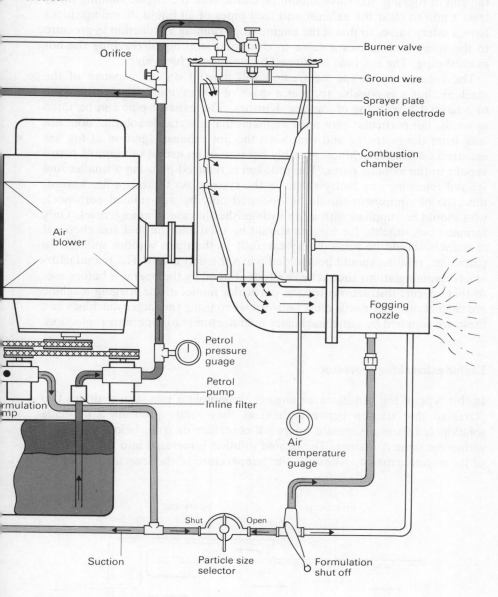

air inlet screen

Orifice

Burner valve

Ground wire

Sprayer plate

Ignition electrode

Combustion chamber

Air blower

Fogging nozzle

Petrol pressure guage

Petrol pump

Inline filter

Formulation mp

Air temperature guage

Suction

Shut   Open

Particle size selector

Formulation shut off

the end of fogging, the valve should be closed with the engine running for at least 1 min to clear the exhaust and feed pipes of all liquid. Some machines have a safety valve, so that if the engine stops running a reduction in pressure to the spray tank causes a valve to close and stop liquid reaching the hot exhaust pipe. The machine is stopped by closing the fuel valve.

The volume which can be treated will depend on the capacity of the machine, but it is possible to treat a space of 200 m³/min and cover an area of 3 ha in 1 h with fog of varying densities. The exhaust pipe can be tilted upwards, but particular care must be taken that insecticide solution does not leak from the restrictor and run down the hot exhaust. Ignition of fog has occurred on some machines, possibly because of an excess of unburnt petrol vapour in the exhaust gases. The problem is reduced by using a smaller fuel jet and renewing any faulty valves in the system. As there is a fire hazard, this type of equipment should be operated only by well-trained personnel, who should be supplied with a fire extinguisher in case of emergencies. Only formulations suitable for fogging should be used and the fuel and chemical containers should be refilled very carefully in the open without spillage. In particular, refilling should be avoided when the unit is hot. The manufacturer's recommendations should be carefully studied by the operator before use, so that specific instructions for the particular model of the fogging machine are carried out. Ear muffs must be worn when using the larger machines and these are supplied by some machinery manufacturers for operator protection.

Engine exhaust fog generator

In this type of fog generator an engine, sometimes a two-stroke, drives two plates so that friction between them as they rotate preheats a pesticide solution fed from a separate knapsack container or from below the plates within the same container. The heated solution is metered into the hot gases of the engine exhaust. Although the temperature of the insecticide solution

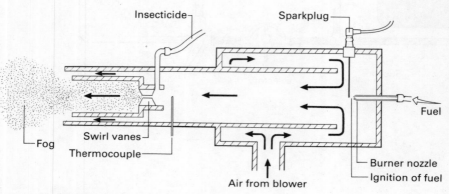

Fig. 11.3 Thermal fogging nozzle

is lower compared with other fogging machines, breakdown of *Bacillus thuringiensis* formulations occur, as the duration of exposure to a high temperature is longer. Since the insecticide solution is separated from the spark plug, the fire hazard is considerably reduced.

Large fog generators

Larger fog generators, developed originally for military use, such as the Todd Insecticide Fog Applicator of 'TIFA' (Figs 11.3 and 11.4) have a petrol engine to operate an air blower and two pumps. The blower supplies a large volume of air at low pressure to a combustion chamber, in which petrol pumped by a gear pump from a second fuel tank is ignited by a spark plug to heat the air. Some models do not have a pump and the fuel tank is pressurized from the blower. The hot gases at 500°–600°C pass through a flame trap to a distributor head to which the insecticide solution is delivered by a centrifugal pump. A small proportion of the hot gases partly vaporizes the liquid in a stainless steel cup in the distributor, while most of the hot gases pass the outside of the cup, complete the formation of the fog and then carry it away from the machine. The temperature within the fogging head operating with odourless kerosene at 95 litres/h was 265°C (Rickett and Chadwick 1972).

Fig. 11.5 TIFA fogger mounted on vehicle

Decomposition of some insecticides, including natural pyrethrins, occurs in temperatures over 230°C, but thermal degradation of certain pyrethroids was negligible in fogging machines, no doubt because they were so briefly exposed to high temperatures. Also the hot gases contain little oxygen and so are less destructive chemically. The direction of the distributor head can be set in various positions. Normally, fog is drifted across a swath of 150 m, but it can be effective for 400 m (Brown and Watson 1953). When mounted on a truck ((Fig. 11.5), some models have a self-starter and remote controls, for operating the fogger, located in the cab. The vehicle is usually equipped with a low-speed speedometer, an hour meter to record the period of fogging and fire extinguishers.

A fogging machine can be used for ULV aerosol application by restricting the flow of insecticide and removing the heating section, and utilizing the blower and distributor units: for example, Brooke et al (1974) achieved over 85 per cent reduction of adult *Aedes taeniorhynchus* applying only 1.5 g of biomesmethrin in 50 ml dieseline/ha with a modified thermal fogger. These fogging machines have also been used to treat sewers for cockroach control (Chadwick and Shaw 1974). This equipment has also been used to apply fungicides to rubber trees (Fig. 11.6).

The same machines can be used as a blower, and also for conventional spraying by fixing a hose and lance to the insecticide pump and not using the heating or air blower sections.

Fig. 11.6 Fogging fungicide in a rubber plantation in South America (photo Tifa (CI) Ltd)

Other fogging machines

A small, hand-carried, electric-powered machine has a fan which blows air over a heater so that hot air vaporizes insecticide impregnated on a special cartridge. Other machines can fog a water based formulation which is pumped into the hot gas.

Aerosols can be produced with very small droplets equivalent to a fog by mechanical devices in which a series of baffles prevent large droplets escaping from the nozzle. A cloud of droplets less than 10 μm vmd is released.

12

Seed treatment, dust and granule application

Application of dry formulations has one main advantage in that the product requires no dilution or mixing by the user. This is important in areas where the usual diluent, water, is expensive to transport. Nevertheless, the cost of transporting heavy and bulky diluent in the formulated product has to be paid for, and the relative cost of the active ingredient is increased.

Use of dusts has declined, largely because of the drift and inhalation hazards of fine particles less than 30 μm in diameter. Dusts are useful when treating small seedlings during transplanting, and in small buildings where farm produce is stored. They have also been used on certain crops such as tea, rubber and grapevines when humid conditions improve retention of dust on foliage. The main use of dusts is now for seed treatment, especially under controlled conditions and sometimes involving pelleting to produce seeds of uniform size for precision sowing (Clayton 1988).

Granular insecticides are used principally to control soil pests, aphids, stem borers and graminaceous crops and the larval stages of various flies, preferably where there is adequate rainfall or irrigation. They are sometimes added to compost used in peat blocks to raise seedlings such as brassicas

Table 12.1 Amounts of active ingredient in relation to different formulations used in rice paddy (after Kiritani 1974)

	Active ingredient in formulation (%)	Active ingredient (kg/ha)
Spray	50	0.5
Dust	2	0.6
Microgranule	3	0.9
Granule	5	1.5

(Suett 1987). An increasing number of herbicides are also formulated as granules, certain of which are used widely in rice in the Far East. Granules are very often applied by hand, especially in tropical countries, but the amount of active ingredient used is higher than with other application techniques when the granules are broadcast into irrigation water (Table 12.1) (Kiritani 1974). Accurate placement of granules at their appropriate target with precision equipment means that less active ingredient is needed than with other application methods (P. Walker 1976).

Seed treatment

The 'Rotostat' (Fig. 12.1) has been developed especially for treatment of batches of seed. It can be used to apply dusts, but is more generally used with liquid formulations. Smaller versions have been developed for use in developing countries. Fluidized-bed film coating systems are also used to deposit the pesticide in a thin durable polymer coating (Halmer, 1988; Maude and Suett 1986). Seed treatment provides a valuable method of protecting young plants with minimal quantities of toxicant (Elsworth and Harris, 1973; Middleton 1973). Seed treatment is discussed by Jeffs and Tuppen (1986) and Graham-Bryce (1988).

Equipment consists essentially of a hopper, preferably with an agitator, a metering device to feed particles at a constant rate to the discharge outlet. A blower unit to produce an airstream to convey particles towards the target is essential on a duster, and may be used also on granule applicators, unless granules are allowed to fall by gravity directly from the metering mechanism. P. Walker (1976) lists the requirements of a good applicator (Table 12.2), and Bruge (1975 1976) details characteristics of a range of granule applicators (Table 12.3), the main features of which are discussed below.

Table 12.2 Requirements of a good granule applicator (after P. Walker 1976)

1 deliver accurately amount calibrated, either continuous or intermittently
2 spread particles evenly
3 avoid damage by grinding or impaction
4 adequate mixing and feeding of material to metering device
5 easy to use, calibrate, repair, and replace worn parts
6 light hand-carried and knapsack versions need to be comfortable to carry on the back
7 robust
8 corrosion, moisture and abrasion proof
9 inexpensive
10 output directly related to distance travelled.

(a)

(b)

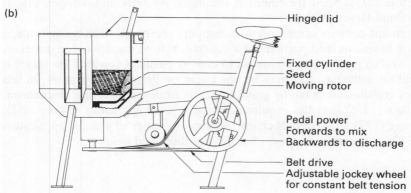

- Hinged lid
- Fixed cylinder
- Seed
- Moving rotor
- Pedal power
  Forwards to mix
  Backwards to discharge
- Belt drive
- Adjustable jockey wheel
  for constant belt tension

Fig. 12.1 (a) 'Rotostat' seed treatment machine (photo ICI Agrochemicals)
(b) Low-cost, pedal-powered seed treater

Features of dust and granule applicators

Hopper design

Ideally the hopper should have smooth sides sloping down to the outlet, thus conical-shaped hoppers are better than those which are square-box shaped, unless the floor slopes (Fig. 12.2). Conversion of spray tanks to hoppers is unsatisfactory as the floor is usually level. An agitator is useful to prevent

Table 12.3 Three examples of tractor-mounted granule applicators (some data selected from Bruge 1975)

Manufacturer	Horstine Farmery	Merriau	SMC
Model	'Microband'	Granyl	Bimigrasol
Rows/hopper	2	2	2
Hopper capacity	16 or 35 litres	25 litres	18 litres
Hopper	Galvanized steel	Polythene	Metal sheet
Indicator of level of granules	Yes, Translucent hopper	Only through translucent hopper	No
Metering device	Grooved rotors in special plastic or aluminium	Rotors with oblique lateral holes	Rubber belt
Method of regulation	By speed, using different sets of pulleys and width of rotor	By variation of speed	Variation of hopper outlet aperture by moving slide with micrometer screw
Drive	By spider wheel or by belt and pulleys	Direct from seeder by chain and sprockets	Direct from seeder or by wheel
Outlet tubing	Transparent plastic	Rubber	Transparent plastic
Internal diameter of outlet (mm)	25	20	20
Adaptability	Fit all seeders	Fit all seeders	Fit all seeders

packing of the contents and to ensure an even delivery of the contents directly to the metering device or through a constant-level device. The latter is particularly useful where friable materials such as attapulgite are damaged by an agitator. Mechanical agitators are linked to the drive shaft of the blower unit. On some machines air is ducted through the hopper from the blower unit. Certain agitators are less effective when dust particles bind together, as they merely cut a channel in the dust. Some machines have an auger in the hopper to move the contents to the metering device.

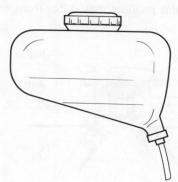

Fig. 12.2 Hopper design

The hopper should have a large opening to facilitate filling; great care is needed to avoid fine particles 'puffing' up when the hopper is filled. A sieve over the hopper opening is essential to eliminate foreign matter and large aggregates. A lid should be held in place to provide a seal to protect the contents from moisture. Ideally hoppers and components should be made from corrosion-resistant materials, so various types of plastic and light alloys are preferred to ferrous metals, although the latter are cheaper. Granules should never be left in the hopper, otherwise corrosion will occur, so the hopper should be designed to be easily emptied. One knapsack granule applicator was designed to incorporate a collapsible hopper to facilitate storage and transport.

Metering system

Various systems of metering dust and granules are used. The amount of product emitted by some machines is adjusted by altering the cross-sectional area of a chute by means of a lever or screw. For most applications the chute must be at least half open. Alternatively, particles drop through one or more holes, the size or number of which can be regulated. Both these systems are liable to block, especially if the particles are hygroscopic. Even collection of a small quantity of particles on the sides of the orifices is liable to reduce their flow and ultimately block the metering system. These systems will not give an accurate delivery unless the forward speed is constant. Metering is improved by using various types of positive-displacement rotor (Fig. 12.3) which deliver a more or less constant volume of product for each revolution. Output is varied by changing the speed of rotation or capacity of the rotors or both, as on the Horstine Farmery 'Microband' equipment. Great care must be taken in the design and construction of the metering system to avoid it acting as a very efficient grinder or compressor of granules (Amsden 1970). Variations in size, specific gravity and fluidity characteristics of particles affects the efficiency of the metering system, so each machine requires calibration for a particular product. Farmery (1976) pointed out that the characteristics of a particular product may differ from one season to the next

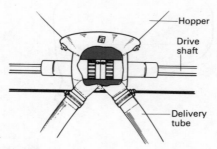

Fig. 12.3 Displacement Rotor

owing to a change in the granule base. Amsden (1970) suggested that the bulk density and flow rate of granules should be given on the label of the granule packaging as a guide to calibration.

Calibration under field conditions at the appropriate forward speed is recommended as the flow of granules can be influenced by the amount of vibration caused by passage over uneven ground. Calibration can be done by collecting granules separately from all delivery tubes in suitable receptacles while travelling over 100 m and checking their weight (Bruge 1975). The amount within each section of the target area has to be considered during calibration, as well as the total amount of product applied per hectare (Table 12.4). Thus, the metering system must provide as even a flow of particles as

Table 12.4 Amount of formulation required in small areas

Rate (kg/ha)	Area covered by 100 g (m²)	Rate per m²(g)
10	100	1.0
15	66.7	1.5
25	40	2.5

possible and avoid irregular clumping of particles. This is achieved when rotors have many small cavities to hold the particles, rather than a few larger ones. The speed of rotation can be reduced to minimize attrition of the product. Bruge (1976) emphasizes that the moving rotor must be set close to its housing to avoid particles being crushed. On some machines a scraper is positioned at the bottom of the hopper to ensure that the contents in the rotor remain intact. The metering system should not be adjusted to a lower flow rate while particles are present, as packing of the product is liable to jam the unit.

Blower unit

Small hand dusters usually have a simple piston or bellows pump. Bellows have been used in knapsack dusters as they are useful for spot treatments, but rotary blowers (Fig. 12.4) provide a more even delivery. The fan may be driven by hand through a reduction gear (about 25:1) or by a small engine. Compressed-air cylinders have also been used to discharge small quantities of dust.

Delivery system

Particles drop from the metering unit into a discharge tube connected to a blower unit, if present. When a blower unit is not used, the discharge tube

Fig. 12.4 Hand-operated blower to apply dusts or granules

should be mounted as vertical as possible to avoid impeding the fall of particles. If it must be curved, a large radius of curvature is essential. The internal diameter of the tube should be sufficiently large, ideally not less than 2 cm ID, and uniform throughout their length. Some tubes are divergent at the outlet end, or subdivided to permit treatment of two rows. At the outlet a fish-tail or deflector plate may be fitted to spread the particles. The position of the discharge tube should be fixed, especially when granules are applied in the soil, and the outlet has to be 10–30 mm from the soil at the back of a coulter. Clear plastic tubes are often used as they are less liable to condensation and blockages can be easily seen, but they are sometimes affected by static electricity. Instead of a blower and discharge tube, some machines have a spinner to throw particles over a wide swath.

Examples of equipment

Package applicators

Some dusts and granules are packaged in a container with a series of holes which are exposed on removal of a tape cover. The contents are shaken through the holes so the quantity emitted will vary, depending on the operator and amount remaining in the container. The main advantage is that the contents do not require transferring to other equipment, but the container has to be carefully disposed of after use. Similar 'pepperpot'-type applicators can be easily made by punching holes in the lid of a small tin.

Hand-carried dusters

Various types of bellows dusters are available with capacities from 20 g to 500 g. Plastic materials are used now in preference to leather or rubber which deteriorate more rapidly under hot and humid tropical conditions. They were used to apply sulphur dusts in vineyards, and Mercer (1974) used bellows dusters to treat small plots of groundnuts with fungicide. The design was similar to that described by Swaine (1954) who improvised a bellows duster by cutting a small tin in half and joining the two halves with a piece of car tyre inner tube. A metal handle was soldered to both ends and an outlet tube was fixed to the bottom half of the tin. These dusters were used to apply a puff of dust in the funnel of maize plants for stalk-borer control.

Simple plunger air-pump dusters have a bicycle-type pump which blows air into a small container. Some have double-action pumps to provide a continuous airstream. The air agitates the contents and expels a small quantity through an orifice. This type of duster was used extensively to treat humans with DDT to prevent an outbreak of typhus in the 1940s. The World Health Organization has a specification for this type of duster (WHO/EQP/4.R2). They are also useful to spot-treat small areas in gardens and around houses for controlling ants and other pests. Similar dusters with a foothold, a cut-off valve to close the dust chamber, and a flexible discharge pipe have been used to blow dust into burrows for rodent control. When sufficient dust has been emitted the valve is closed and more air blows in to drive the dust deep into the burrow (Bindra and Singh 1971). Small dusters with a rotary blower are also made for garden use.

Pest control operators sometimes use a dust applicator which is very similar in appearance to a compression sprayer. The air pump is replaced by a schrader valve and the container pre-pressurized from an air-line. Dusters

Fig. 12.5 Hand-operated granule dispenser

with an electrically powered fan are also available. A duster can be improvised by using a loosely woven linen or fabric bag, sock or stocking as a container which is shaken or struck with a stick. The amount applied is extremely variable and much of the dust is wasted.

Hand-carried granule applicators

These have a tube container (approximately 100 cm long, 1–1.5 litres capacity) with a metering outlet operated by a trigger or wrist action rotation of the container (Fig. 12.5). On one machine a small meter is positioned on each side of the outlet. The output of granules depends on the position of the cones, which can be altered by adjusting a connecting-rod. Robinson and Rutherford (1988) found that many applicators which rely on gravity flow are slow and trigger-operated systems were tiring to operate and more expensive to manufacture. They developed a 'rotary valve' using a wrist-action for granule application in transplanted tobacco. These applicators are particularly useful for spot treatment at the base of individual plants, and have been used in cabbage root fly control and in selective weed control, but are not suitable for burrowing nematode control on bananas for which larger doses are required. Granules are normally left on the soil surface but, by modifying the outlet with a spike, subsurface application is also possible.

Fig. 12.6   Horn seeder

Shoulder-slung applicator

An applicator, known as a 'horn seeder', consists of a tapered metal discharge tube containing a variable opening which is inserted into the lowest point of a rubberised or neoprene-treated cloth bag. This bag has a zipped opening and is carried by a strap over the operator's shoulder. A swath of up to 7 m can be obtained when the discharge tube is swung from side to side in a figure-of-eight pattern (Fig. 12.6).

Knapsack and chest-mounted dusters and granule applicators

A blower is usually mounted to the side and base of a hopper of 8–10 litres capacity. On hand-operated knapsack versions a crank handle is situated in front of the body and is connected to a gear-box by a driving chain which is protected by a metal case. This drive is eliminated on some machines by hanging the hopper on the operator's chest, but the front position is less comfortable, restricts the quantity of material which can be carried in the hopper, and may be hazardous if fine particles escape and contaminate the operator's face. Volume of air emitted by hand-operated machines will depend on the operator, but is at least 0.8 m^3/min at a speed of 14 m/s with the crank handle turned at 96 rpm. The discharge tube is normally on the opposite side of the hopper to the gear-case, which must be protected as much as possible from particles liable to cause wear of the gears. Compared with knapsack sprayers, dusters are relatively expensive, owing to the cost of the blower unit.

Most motorized mistblowers described in Chapter 10 can be converted for dust and granule application by removing the spray hose from the tank and inserting a wider tube to feed particles directly down from the hopper through a metering orifice into the airstream (Fig. 12.7). The outlet tube is rotated to stop the flow of material. Machines with a tank having a sloping floor are more easily adapted for application of dry materials. In Japan a 30 m long plastic tube, carried at each end, has been fitted to these machines. Dusts and microgranules are dispersed through a series of holes along its length. Tabs next to the holes improve distribution (Fig. 12.8) (Takenaga 1971). Hankawa and Kohguchi (1989) reported that good control of brown planthopper (*Nilapavata lugens*) was obtained using buprofezin or BPMC

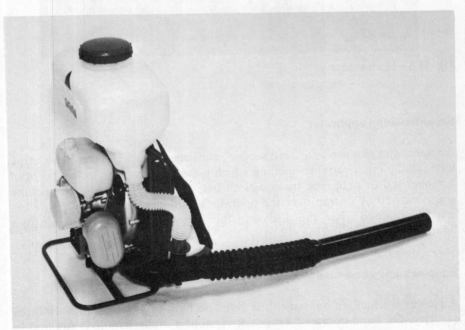

Fig. 12.7 Motorized mistblower converted for dust or granule application

(a)

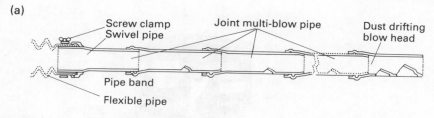

Screw clamp
Swivel pipe

Joint multi-blow pipe

Dust drifting
blow head

Pipe band

Flexible pipe

(b)

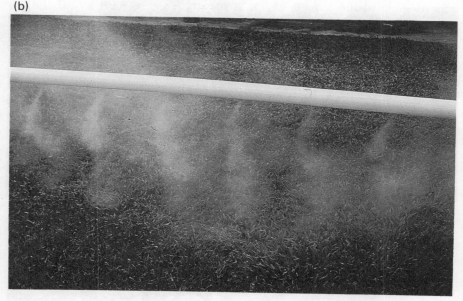

Fig. 12.8   Examples of extension tubes to apply microgranules (a) Detail of tube (from Sutherland 1980) (b) Flexible lay-flat tube inflated by the airstream applies micro-granules over a 30m swath

insecticides applied with this type of applicator fitted to a large capacity fan to increase airflow at each hole.

Some knapsack equipment is made specifically for granule application with or without an airflow. Machines designed to apply granules by gravity only can sometimes be modified to spot-treat with a measured dose. A knapsack in which a cup is moved by a lever mechanism from the hopper outlet to the discharge tube each time a dose is applied can be used for spot application on bananas.

Power dusters

Relatively few larger power-operated dusters are manufactured. Some are

Fig. 12.9 'Microband' applicator (photo Horstine Farmery Ltd)

Fig. 12.10 Airflow granule applicator (photo Horstine Farmery Ltd)

coupled to the pto of a tractor to operate a fan with an output of approximately 50 m³/min. Units with an independent 3–5 hp engine are heavy (>50 kg), but are sometimes mounted on a two-person stretcher to facilitate transport to areas inaccessible to vehicles. Such units have been used to apply fungicides, especially sulphur dust in vineyards, using machines with twin outlets. In rubber plantations dust has been projected vertically to a height of up to 24 m under calm conditions.

Tractor-mounted granule applicators

A few machines have been designed with a boom fed from a single blower unit (Palmer 1970). Some are adapted from fertilizer spreaders and have a spinner, but most have a series of separate units fixed on a horizontal frame (Fig. 12.9) which can be mounted either at the rear or front of a tractor. Normally, up to four units each with two outlets are used together, but wider booms are possible. Displacement rotors are ideally driven at 0–8 km/h from a rimless spoked land wheel designed to follow the ground contours accurately. A pto drive replaces the land wheel on high-clearance frames.

An airflow, provided by a compressor fan driven from the pto or by a hydraulic motor, increases precision when broadcasting as little as 3 kg/ha (Fig. 12.10). A deflector plate at the discharge tube outlet, if properly angled for a particular granule size and density (Goehlich 1970), will spread granules over a swath up to 1–2 m wide. Thus an applicator with four hoppers each carrying up to 30 kg granules can treat up to 40 ha without refilling, and at 8 km/h cover the ground in under 6 h. Farmery (1970) discusses the problems of achieving an even air volume and speed with airflow applicators.

Applicators can be mounted on the tractor alongside seed drills, special coulters, fertilizer applicators and other farm implements, depending on where the granules are needed (Bailey 1988; Pettifor 1988). In the 'bow-

Table 12.5 Dose required for control of cabbage root fly (after Wheatley 1976)

Application method	Amount of granules	Particle distribution		Dose/Plant	
		cm² soil/granule	cm³ soil/granule	No. of granules	mg ai
Spot surface (15 cm dia.)	0.17–1 g/plant	0.08–0.6	—	340–2 100	17–45
Band (5 × 10 cm deep)	0.7–1.2 g/m row	0.09–0.71	0.8–3.6	210–880	10
Broadcast – surface	45 kg/ha	1.1	—	160	8
Broadcast – mixed to 5 cm depth	30 kg/ha	0.71	3.5	250	2.7

wave' technique granules are metered in a 100 mm band on the soil surface in front of a coulter which then mixes them in the soil (A.G. Whitehead 1988). Dunning and Winder (1963) found that the bow-wave technique decreased phytotoxicity.

Some pesticides can be applied as granules placed as an in-seed furrow treatment at the time of sowing. The discharge tube can be positioned close to the coulter for accurate placement of the granules in relation to the seed. Wheatley (1972) studied placement and distribution on the performance of granular formulations of insecticides for carrot-fly control. He concluded that the performance of these granules could not be easily improved by minor modifications of the granule-placement equipment, provided an even distribution along the row is achieved. Granule-application criteria for certain pest-control problems is summarized in Table 12.5 (Wheatley 1976).

In contrast to 17–45 mg ai/plant required with spot treatment against cabbage root fly, the dose can be reduced to 10 mg/plant for banding treatments, and to as little as 2.7 mg ai/plant when granules are broadcast and mixed to a depth of 5 cm. Against aphids, the amount per plant depends on the size of the plant, larger plants requiring a heavier dose to maintain a similar lethal concentration in the sap (Table 12.6).

Side dressing of some insecticides is effective later in the season, especially if moisture is sufficient to redistribute the chemical.

A major problem with the development of ground equipment for granular application has been the lack of fundamental studies on particle distribution and the wide range of granule sizes available, since each manufacturer has sought the cheapest and most easily obtained supply of suitable base carriers. Sand is readily available, but is very abrasive. Heavy carriers are suitable for aerial application to reduce the risk of drift, and for soil application, but lighter and softer carriers such as the clay bentonite, are more suitable for rapid release of herbicides used in rice-weed control. A greater control of particle size within defined limits, namely controlled particle application, is perhaps needed in parallel with the development of CDA for sprayers (Table 12.7). Looking to the future, it may be possible to apply even-sized particles of 30–50 μm to foliage with the aid of electrostatic forces.

Table 12.6 Dose required for aphid control with granules on different crops (after Wheatley 1976)

Crop	Crop density (plant/ha)	Treatment (amount of granules)	Dose/Plant Number of granules	mg ai
Carrot	500 000	Band 7.5 × 1 cm (11–25 kg/ha)	32–160	1.1–2.5
Sugar beet	40 000	Soil or foliage (5.6–11 kg/ha)	38–880	7–14
Brussels sprouts	10 000	Soil or foliage (14–39 kg/ha)	3 100–12 000	33–190

Table 12.7 Size ranges for dusts and granules

Size range (μm)	
<50	fine dust
50–100	coarse dust
100–200	microgranule
200–2 000	granule
>2 000	macrogranule

13

Aerial application

Aircraft are used principally to treat large areas of crops where access is difficult for ground equipment, such as irrigated fields, as well as forests. They are also used in large-scale vector control programmes, especially for tsetse control (Allsopp 1984) and in the Onchocerciasis Control Programme in West Africa (Gratz 1985). Their use has declined in some countries due to public concern about spray drift as pesticides are released at a greater height above the crop canopy. In Europe the extensive use of tramlines to allow access for tractor mounted equipment has also reduced the need for aerial treatment. Akesson and Yates (1974) and Quantick (1985a & b) give an overall review of the use of aircraft in agriculture. Bouse (1987) discusses aerial herbicide application.

Types of aircraft

Fixed and rotary-wing aircraft (Fig. 13.1 and 13.2) are used for applying pesticides. Information on the performance for international standard atmosphere conditions at mean sea level of certain aircraft is listed in Tables 13.1 and 13.2.

Such data need to be converted to estimate correctly the performance under local operating conditions. The possibility of using microlight aircraft has also been considered for ULV application, but the conventional microlight is generally too small. The pilot is too exposed to spray contamination, and it is difficult to maintain an accurate track for adjacent swaths. Autogyros (R.F. Hill 1963; Johnstone et al 1971; Johnstone et al 1972a) and remote-controlled model aircraft (Embree et al 1976; Johnstone 1981) have also been used on a very limited scale. Larger multi-engined aircraft such as the DC–3 (Dakota) and DC–6 have been used in vector-control programmes (Lofgren

Fig. 13.1 Aircraft spraying (photo Micronair (Aerial) Ltd)

Fig. 13.2 Helicopter spraying (photo Air Lloyd)

et al 1970a; Lofgren 1970; C.W. Lee et al 1975b), and for spraying forests (A.P. Randall 1975; Quantick 1985a). Many different types of aircraft have been converted for pesticide application, but there are several which have been specifically designed for crop-spraying. The main features of these single-engined low-wing monoplanes are

Table 13.1 Data on certain fixed-wing aircraft used in agriculture (performance details are given for international standard atmosphere (ISA) conditions at mean sea level, adapted from **Agricultural Pilots' Handbook,** Anon 1973

Aircraft	'Super Cub'	'Pawnee Brave'	'AG-wagon'	'Commander'
Engine power (kW)	110	230	220	438
Capacity fuel (litres)	136	284	140	400
Capacity hopper (litres)	—*	1 000	757	1 515
Weight empty (kg)	481	930	903	1 825
Weight gross AUW (kg)	949	1 770	1 497	2 722
Weight Ag load (kg)	118	839	908	681
Wing span (m)	10.8	11.9	12.4	13.5
Wing area (m^2)	16.6	20.9	18.8	30.3
Max loading (kg/m^2)	56.7	82.3	79.8	90.2
Speed, stalling (km/h)	73	114	98	107
Speed, cruising (km/h)	145	163	182	204
Take-off ground run (m)	92	267	207	236
Take-off over 15 m (m)	290	406	332	404
Landing ground run (m)	125	213	128	244
Landing over 15 m (m)	267	503	386	412
Rate of climb (m/min)	232	241	210	274

Note: * Belly tank AUW – all-up weight

1 a high-performance engine to lift a heavy payload from earth or gravel strips to a height of 15 m in less than 400 m at sea level
2 an airframe stressed to withstand frequent landings and take-offs and to provide protection for the pilot in the event of an accident
3 an operational speed of 130–200 km/h
4 a low stalling speed of 65–100 km/h
5 a high payload to low gross weight ratio
6 light and responsive controls to reduce pilot fatigue
7 distinct separation of flight controls from application equipment
8 cockpit with good all-round visibility
9 landing gear and the canopy with sharp leading edge to minimize the hazard of hitting power lines or wires
10 a deflector cable fitted between the top of the canopy and the tail
11 a pressurized and air-conditioned cockpit to reduce the risk of contaminating the pilot with pesticide
12 a recoil-type harness and safety helmet to protect the pilot
13 a pesticide tank or hopper located in front of the cockpit and aft of the engine and over the centre of lift, so that aircraft trim is minimized by changes in weight during spraying
14 the maximum permissible weight is indicated clearly near the filler opening

Table 13.2 Data on certain rotary-wing aircraft used in agriculture (performance details are given for international standard atmosphere (ISA) conditions at mean sea level, adapted from **Agricultural Pilots' Handbook**, Anon 1973)

Aircraft	Hughes 300	Bell Ag-5	Hiller UH-12E	Mikhail Mi-45
Engine power (kW)	130	193	230	1 240
Main rotor diameter (m)	7.7	11.3	10.8	21.0
Overall height (m)	2.5	2.83	3.1	5.18
Length (m)	8.8	13.3	12.4	16.8*
Capacity fuel (litres)	114	227	174	—
Capacity hopper (litres)	304	454	635	1 600
Weight empty (kg)	433	770	770	—
Weight max AUW (kg)	755	1 293	1 220	7 800
Weight Ag load (kg)	204	544	239	1 000
Speed cruising (km/h)	97	135	140	160
Rate of climb (m/min)	350	262	463	—
Ceiling (m)	3 950	3 200	4 700	5 500
Range (km)	355	547	298	250

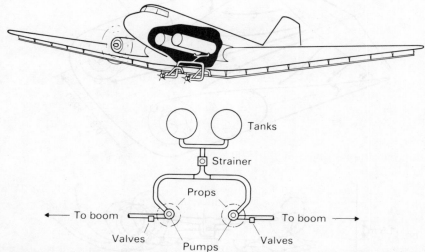

Fig. 13.3 Double air-driven pump system mounted in twin-engined aircraft (after FAO 1974)

15 a tank designed for rapid loading, easy cleaning and maintenance with provision for rapid dumping of a load in an emergency

16 provision for loading by pumping the spray into the bottom of the tank through a filler opening to the rear of the wing

17 provision of top-loading of dry particulates through large dust-tight doors

18 fuel tanks placed as far away from the pilot as possible, preferably as wing-tanks.

The basic design should facilitate inspection, cleaning and maintenance of all parts of the aircraft and application equipment. Corrosive-resistant materials and coatings should be used with readily removed panels to permit easy access to the fuselage.

Multi-engined aircraft (Fig. 13.3) are needed over populated areas and forests and swamps, where opportunities for a safe emergency landing in the event of an engine failure are limited. Multi-engined aircraft generally require well-constructed runways and can operate over long distances, even at night when it is necessary to take advantage of inversion conditions. Larger aircraft permit the use of modern electronic navigational equipment to reduce the need for ground markers (Boivin and De Camp 1976; Joyce 1985).

Helicopters provide an alternative to fixed-wing aircraft where reduced flight speed and greater manoeuvrability within fields is desirable, to increase penetration, or is necessary due to the presence of trees or other obstacles, and where landing strips are not available. Helicopters may be landed on special platforms such as the top of a vehicle. Helicopters are particularly

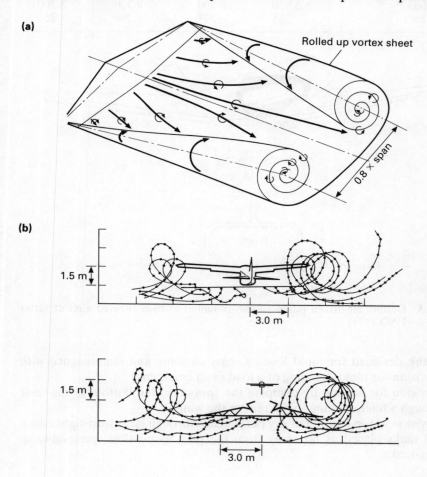

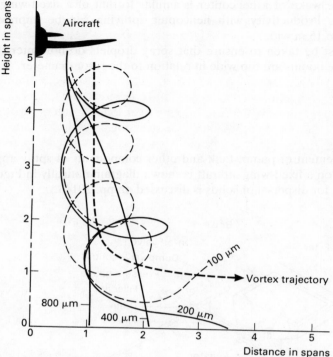

Fig. 13.4 (a) Trailing vortex system behind an aircraft (after Spillman 1977) (b) Aerodynamic trailing wake of a high-wing monoplane traced by gravitationally balanced (hydrogen-filled) balloons and of a helicopter (after FAO 1974) (c) Droplet trajectories from midspan of wing in relation to gravity, vortices, ground effects and size of droplets (after Spillman 1977)

useful where on-the-spot survey and treatment need to be combined, for example in mosquito- and black-fly control programmes. The contribution made by helicopters to agricultural aviation since the early experiments with a Sikorsky YR–4B helicopter (W.E. Ripper and Tudor 1947) has been reviewed by Voss (1976).

Improved penetration of a crop canopy with spray droplets in the strong down-wash of air created by the rotor is not achieved unless the helicopter is flown at less than 25 km/h (Parkin 1979). Unfortunately, the initial cost and maintenance costs are much greater than with fixed-wing aircraft and extra flying skills are needed by the pilot, so spray application at a low speed is not economic. Discriminative residual placement spraying of a 20 m swath, along the edge of fringing woodland and riverine forest for tsetse control at 25–40 km/h was tried in West Africa (Spielberger and Abdurrahim 1971; C.W. Lee et al 1978; Baldry et al 1978). The rotor down-wash pattern changes from a closed toroid to a horseshoe vortex pattern as the helicopter increases forward speed. At operational speeds above 40–50 km/h, distribution of spray in the wake of a helicopter is similar to that of a fixed-wing aircraft (Fig. 13.4). Productivity with helicopter spraying can be improved with booms up to 15 m wide.

Care must be taken to ensure that spray droplets do not enter the rotor vortex if the booms are too wide in relation to the rotor diameter.

Spray gear

The arrangement of pump, tank and other components of spray application equipment on a fixed-wing aircraft is shown diagrammatically in Figure 13.5. Equipment for dispersal of solids is discussed on pp. 291–3.

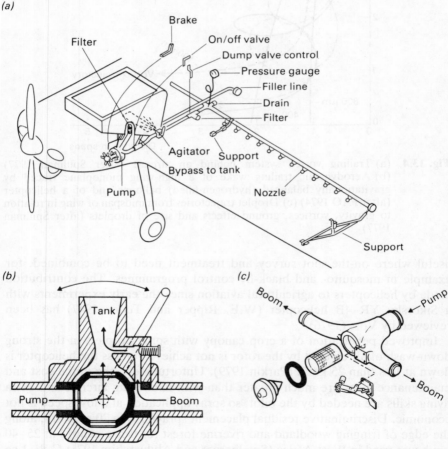

Fig. 13.5 (a) Cutaway diagram of a spraying system for a small fixed-wing aircraft (after FAO 1974) (b) Main control valve for aircraft sprayer sharing boom-vacuum position for positive shut-off; check valves are needed at each nozzle with this system (c) Liquid screen filter between pump and boom

Spray tank or hopper

The tank may be constructed with stainless steel; one aircraft has a titanium tank integral with the fuselage. Subject to government regulations concerning the structure of aircraft components, fibreglass tanks are acceptable for application of most pesticides. Such tanks have a translucent zone at the rear of the tank mounted in the fuselage to permit the pilot to check the volume of liquid remaining in the tank, otherwise a contents gauge is provided. The shape is designed so that it will drain completely, either in flight or on the ground. A dump valve is fitted so that a full load can be jettisoned within 5 s in an emergency. Small planes converted for spray work may have a belly tank fitted to the bottom of the aircraft (Fig. 13.6). In this case, the whole tank may be jettisoned if the need arises although almost all belly tanks incorporate a conventional dump and most pilots prefer not to drop the tank. Cockpit contamination and pilot exposure to pesticides is minimized with a belly tank as it is outside the fuselage. Internally mounted tanks and pumps are installed in large aircraft when required. In helicopters saddle tanks are mounted on either side of the engine with a large-diameter interconnecting pipe to maintain a level load (Fig. 13.7). On turbine helicopters a belly tank is normally used. An electrically driven agitator has been fitted to some helicopter tanks used with particulate suspensions such as *Bacillus thuringiensis* for mosquito larviciding.

The tank opening is provided with a basket filling filter, but loading is quicker and safer when the load is pumped through a bottom loading point from a ground mixing unit. A filter incorporated in this feed line usually has a sufficiently fine mesh to protect the nozzle orifices, although each nozzle

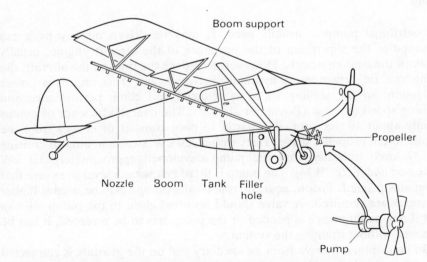

Fig. 13.6 Spray system with quick detachable belly tank on a small passenger aircraft (after FAO 1974)

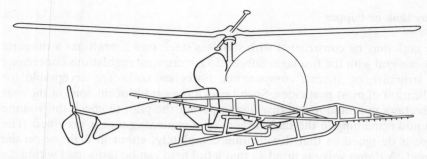

Fig. 13.7 Position of spray tanks on a helicopter with interconnecting pipe

should be provided with its own filter. The mesh size is therefore 25-100 mesh, depending on the type of nozzle used; 50 mesh is suitable for most spray work and is the first which should be used for wettable powder formulations. An in-line strainer is fitted to the outlet of the tank to protect the pump. A larger mesh size (6–8 mesh) is usually desirable to reduce pressure drop at this point. All filters should be readily accessible to allow a change of mesh size if necessary and to facilitate regular cleaning; a valve is necessary upstream of the filter to allow cleaning, even when the spray tank is full. An overflow pipe and air vent are ducted from the spray tank to the rear and bottom of the fuselage to prevent contamination of the cockpit. An air vent prevents a vacuum being created in the tank, as this would affect the flow of liquid to the pump.

Pump

A centrifugal pump is usually used. It may be driven directly by a fan mounted in the slipstream of the propeller of the aircraft engine, usually between the landing wheels. However, to reduce the drag on the aircraft, the pump can be driven by a hydraulic drive system. This improves power utilization, so the stalling speed is reduced and climb performance and cruising speed increase (Boving et al 1972). The overall efficiency of pumps is only about 10 per cent, due partly to poor transfer of energy from the engine to the propeller drive. Tests with a 220 kW Stearman aircraft cruising at 137 km/h indicated that the pump developed approximately 3.3 kW (Akesson and Yates 1974). The pump is fitted below tank level to ensure that it remains primed. Piston, gear, or roller-vane pumps may be used if higher pressures are required. A valve should be fitted close to the pump inlet so that if any maintenance is needed or the pump has to be replaced, it can be removed without draining the system.

On helicopters, a drive from an ancillary pad on the gearbox is connected through a clutch directly or via a belt drive to a pump. The addition of the belt drive is safer to protect the engine gears if the pump becomes locked.

On newer fixed-wing aircraft the pump may be similarly driven from a pto, or indirectly by means of a hydraulic system which operates at up to 200 bar and can provide 10-18 kW at the pump. Electrically operated pumps are sometimes used, particularly for LV application when only 1–2 kW or less is required.

The pump must have sufficient capacity to recirculate a proportion of its output to the tank to provide hydraulic agitation. Some agitation, even during actual spraying, is desirable.

Spray boom

On fixed-wing aircraft the boom extends for most of the wing span, but usually avoids the wing-tip area where a vortex could carry droplets upwards. Parkin and Spillman (1980) showed that the amount of spray carried off-target by wing tip vortices could be reduced by extending the wing horizontally by fitting 'sails'. These wing-tip sails, originally designed to reduce drag, have been used only in experiments. The boom is often mounted at the trailing edge of the wing, but actual positions depend on the wing structure. V.D. Young et al (1965) studied the spray-distribution patterns with different boom lengths and positions. Generally, a better distribution has been found when the spray bar is mounted below the wing. A round pipe may be used, but some booms with an internal diameter up to 50 mm to cope with high flow rates (>500 litres/min) are streamlined to reduce drag (Fig. 13.8). At lower volumes the boom can be decreased to 13–20 mm ID. A bleed line from the top of the pump may be required to remove airlocks. Larger-diameter booms (64 mm ID) have been used for very viscous materials such as invert emulsions.

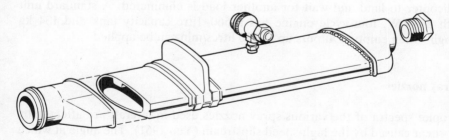

Fig. 13.8   Streamlined boom (courtesy Simplex Manufacturing Co.)

On helicopters the central section of the boom may be fixed aft of the engine, but nozzles are then in an updraught of air near the centre of the rotor. Wooley (1963) suggested nozzle positions outside the trailing vortices. Ideally it should be below and in front of the cockpit, where there is a downdraught of air. An alternative system with helicopters is to avoid fitting the spray gear directly to the aircraft by using a separate underslung unit

Fig. 13.9 Underslung unit on helicopter. (photo Simplex Manufacturing Co.)

(Fig. 13.9). The unit is attached to the cargo hook and is detachable in only a few seconds, thus permitting maximum utilization of the helicopter. Chemical contamination of the helicopter is less, thus reducing the cost of maintenance. The unit is self-contained and has an engine-driven pump controlled from the cockpit by means of an electrical connection. Ideally with two units – one spraying while the other is being refilled – the need for the helicopter to land and wait for another load is eliminated. A standard unit with a 6 kW, four-cycle engine has a 560–1litre capacity tank and 134 kg weight when empty. Outputs up to 350 litres/min can be applied.

Spray nozzles

Droplet spectra of the various spray nozzles used in aircraft are affected by air shear caused by the high-speed slipstream (Yeo 1961). The angle at which liquid is discharged from the nozzle relative to the slipstream changes the droplet spectra. Hydraulic nozzles angled forwards and downwards into the slipstream produce smaller droplets and a wider range of sizes than nozzles directed downwards or backwards (Figs 13.10 and 13.11) (Kruse et al 1949; Spillman 1982). Thus backwardly directed fan nozzles (8005) were considered suitable for herbicide application.

Fan- and cone-type hydraulic nozzles are widely used in special nozzle bodies incorporating a diaphragm check valve (Fig. 5.5, p. 104) to provide positive shut-off of the spray when a boom vacuum is provided. When

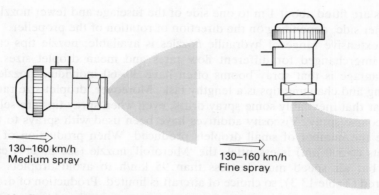

130–160 km/h
Medium spray

130–160 km/h
Fine spray

Fig. 13.10 Position of nozzles relative to aircraft slipstream

pressure along the spray boom exceeds 0.2–0.5 bar, a spring-loaded chemi-cally resistant diaphragm is lifted to allow liquid to pass through a filter to the nozzle tip. A PTFE disc should be used to protect the diaphragm when some solvents are used in the spray. Nozzles are sometimes irregularly spaced along the boom to try to counteract the effect of propeller or rotor vortex which shifts spray from one side to the other, especially when the flying altitude is less than about 3 m. At greater heights the maximum horizontal velocity of the downwash due to the wing-tip vortex is less and turbulence causes sufficient mixing of the spray droplets (Trayford and Welch 1977). Johnstone and Matthews (1965) describe experiments with a helicopter to determine the optimum nozzle arrangement. On fixed-wing aircraft extra

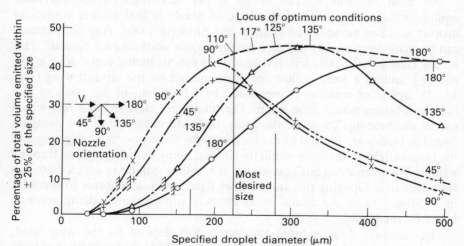

Fig. 13.11 Efficiency of droplet size control with nozzles (8005) set at different angles when aircraft is flying at approximately 46m/s) (from Spillman 1982)

nozzles are fitted about 1 m to one side of the fuselage and fewer nozzles on the other side, depending on the direction of rotation of the propeller.

An extensive range of hydraulic nozzles is available; nozzle tips can be easily interchanged for different flow rates and mean droplet sizes. The disadvantage is that spray booms often have 30–60 individual nozzles, so cleaning and changing tips is a lengthy task. Moreover, droplet-size range is so great that inevitably some spray drifts, even when spray tips are selected for a coarse spray. Viscosity additives have been used with sprays to try to reduce the number of small droplets produced. When production of large droplets (>500 μm) is essential, the 'Microfoil' nozzle (see p. 114) can be used, but air speed must be less than 95 km/h to avoid droplets being shattered (Table 13.3), so choice of aircraft is limited. Production of droplets

Table 13.3 Critical air velocity for various droplet sizes (water)

Droplet diameter (μm)	Shatter velocity (km/h)
100	322
170	241
385	161
535	137
900	105

of 250 μm is also possible with a transducer-driven, low-turbulence nozzle but this has such small orifices (125 μm) that 400-mesh filters are needed and wettable powders cannot be used (Wilce et al 1974; Yates and Akesson 1975). 'Raindrop' nozzles have been used on helicopters to apply particulate suspensions for mosquito larviciding.

The most versatile aircraft nozzle is the centrifugal-energy Micronair equipment. An advantage of this type of nozzle is that greater control of droplet size can be achieved (Parkin and Siddiqui 1990). Any adjustments can be made very rapidly as there are only a few units on each aircraft. This nozzle consists of a cylindrical, corrosion-resistant, monel metal wire gauze rotating around a fixed hollow spindle mounted on the aircraft wing (Fig. 13.12). Speed of rotation is controlled by adjustment of the pitch of five balanced blades which form a fan. The blades are clamped in a hub which carries the bearing. To adjust the angle, bolts are slackened on the clamping ring; the blades are twisted to the correct angle setting on the clamp ring and the bolts retightened evenly to nip the blades. Spray liquid is pumped through a boom via a variable restrictor unit to the hollow spindle in which there is a shut-off valve. Opening this valve allows liquid to hit a deflector to spread it in a diffuser tube. An initial break-up here provides even distribution of liquid on the gauze.

The number of units fitted to aircraft will depend on the wing span, intended swath and volume of spray being applied. A similar number of units can be fitted to helicopters, but larger propeller blades have been used. The layout of Micronair units is shown in Fig. 13.13. The earlier large AU 3000

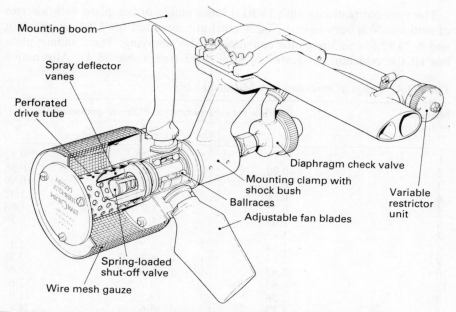

Mounting boom

Spray deflector
vanes

Perforated
drive tube

Diaphragm check valve

Mounting clamp with
shock bush

Ballraces

Adjustable fan blades

Variable
restrictor
unit

Spring-loaded
shut-off valve

Wire mesh gauze

Fig. 13.12   Micronair AU5000 with guard (courtesy Micronair (Aerial) Ltd)

unit was fitted with a hydraulically operated brake for use in an emergency
or during ferrying. The newer AU 5000 Atomiser is now preferred for normal
agricultural spraying; 6–10 of these smaller units are normally installed
instead of 4–6 AU 3000 units. Blockages are rare with these atomizers as
small orifices are not required to break up the liquid, and wettable powders
and suspensions are more easily applied than with hydraulic nozzles.

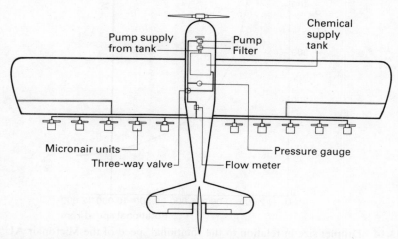

Pump supply
from tank

Pump

Filter

Chemical
supply
tank

Micronair units

Three-way valve

Flow meter

Pressure gauge

Fig. 13.13   Typical layout of Micronair AU5000 installation

The variable restrictor unit (VRU) has a single orifice plate with a series of orifices. Numbers 1–7 (0.77–2.4 mm) are intended for ULV application and 8–14 (2.65–6.35 mm) for conventional LV spraying. The standard plate has all the odd number restrictor sizes 1–13 (see Table 13.4). Alternative

Table 13.4   Range of flow rates with Micronair units

Micronair restrictor	Flow rate (litres/min) at different pressures (bar)		
	2	2.8	3
1	0.15	0.27	0.42
2	0.20	0.30	0.56
3	0.35	0.63	0.99
4	0.51	0.90	1.42
5	0.71	1.27	1.98
6	1.02	1.81	2.8
7	1.43	2.54	3.97
8	1.84	3.27	5.1
9	2.66	4.72	7.38
10	3.47	6.18	9.66
11	4.90	8.7	13.6
12	7.56	13.5	21.0
13	8.79	15.6	24.4
14	14.3	25.5	39.8

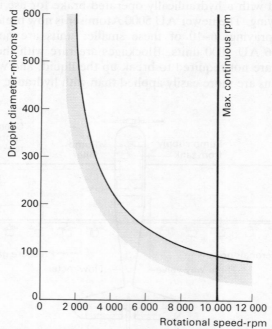

Fig. 13.14   Droplet size in relation to the rotational speed of the Micronair AU5000 Unit (courtesy Micronair (Aerial) Ltd)

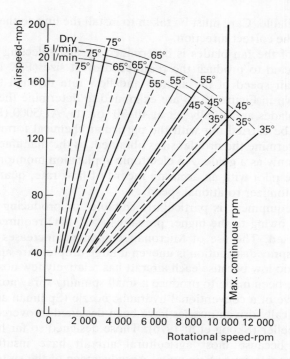

Fig. 13.15 Speed of rotation of AU5000 unit in relation to aircraft speed, blade angle and flow rate (courtesy Micronair (Aerial) Ltd)

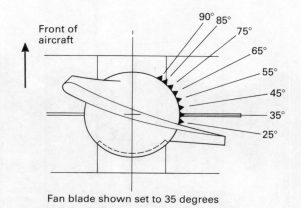

Fan blade shown set to 35 degrees

Note: This drawing is applicable to atomizers manufactured after November 1984; earlier atomizers had setting marks on the hub

Fig. 13.16 Fan blade angles for AU5000 unit

plates are available. Care must be taken to install the unit so that liquid flows through it in the correct direction.

The angle of the fan blades is determined by first selecting the speed of rotation expected to produce the required droplet size (Fig. 13.14). Then, knowing the air speed of the aircraft and flow rate through the atomizer, charts as shown in Figure 13.15 are examined to determine the angle of the blades. The blades are usually set at 40–70° on the AU5000 (Fig. 13.16) A check should be carried out with the particular chemical formulation being applied to determine the droplet sizes obtained, as the manufacturers' charts are intended only as a guide. An electronic application monitor can be fitted to provide the pilot with an accurate record of flow rate, quantity of liquid emitted and atomizer rotational speed.

Micronair equipment is particularly suitable for producing droplets less than 100 μm, owing to the higher pressures (>10 bar) required if hydraulic nozzles are used. The use of Micronair equipment increases aerodynamic drag and the spray distribution is uneven if large droplets are applied and the aircraft flies too low because each aircraft has relatively few nozzles. Various attempts have been made to produce a small spinning-spray nozzle which fits the check valve of a conventional hydraulic nozzle (Spillman and Sanderson 1983). Electrically driven nozzles have been designed to overcome the need to use a propeller (Skoog et al 1976). Those designed so far have not been widely used because most agricultural aircraft have insufficient power capacity to operate four to six units. An advantage of the system is that the pilot can control rotational speed and easily adjust droplet size when

Fig. 13.17 Vide-vite system used to apply larvicide in the onchocerciasis control programme in West africa (photo WHO by R Witlin)

required, irrespective of the forward speed of the aircraft. Such nozzles are particularly useful on the slower aircraft, including helicopters. Spraying against tsetse flies in Nigeria, a helicopter was fitted with up to six spinning discs, the speed of which is controlled by means of a computer mounted to the rear of the engine (Spielberger and Abdurrahim 1971). Multiple spinning discs have also been used to apply baculoviruses to control pine beauty moth (Entwistle 1986; Entwistle 1990) and pine sawfly in forests (Doyle 1988). To generate sufficient numbers of droplets a spray unit based on the X–15 stacked disc atomizer was developed. Six units with 30 discs driven by direct drive electric motor were used at a flying speed of 45 knots with a lane separation of 40–50m.

Prototype venturi nozzles for ULV application have been examined by Parker et al (1971) and Parkin and Newman (1977). Thermal nozzles are used to produce aerosols for vector control (Park et al 1972), but there is less control of droplet size than with centrifugal-energy nozzles. Specialized equipment for use on aircraft has been developed to release known volumes of liquid into rivers. The 'vide-vite' system (Fig. 13.17) is used to apply a larvicide, temephos, into West African rivers for the control of *Simulium damnosum*, vector of onchocerciasis (C.W. Lee et al 1975a, Baldry et al 1985).

Aerial application of dry materials

Hazardous pesticides formulated as dusts should not be applied from aircraft owing to the high drift loss potential, although fertilizers are applied. Microgranules and granules can be applied either by ram-air spreaders or spinners (Brazelton et al 1969, Spillman 1980, Bouse et al 1981). An air speed as high as 240 km/h directly from the propeller slipstream is used in the ram-air type on fixed-wing aircraft to distribute the material in the wake of the aircraft. Air enters the front of a tunnel sloped like a venturi tube, and with internal guide vanes or channels. A control gate fitted under the hopper can be opened to a preselected position determined by flight calibration, and is also the shut-off valve (Fig. 13.18). Metering can be improved by using a vaned rotor system (Bouse et al 1981). A revolving agitator may be fitted above the throat of the metering gate. This agitator may be driven via a reduction gear by a windmill placed in the propeller slipstream. Ram-air devices suffer from high drag and low spreading power. The effect of the spreader on aircraft performance has been studied by Stephenson (1976). The weight penalty is offset by gains from a wider swath, except for light aircraft. Drag is less with a tetrahedron spreader which gives a wider swath (Trayford and Taylor 1976). Ducting a limited flow of fine particles to the wing tips has been investigated by K.C. Lee (1976), who showed that wide swaths were not possible unless granules were released remote from the aircraft centre line. Experimentally, a conveyor duct was built into the

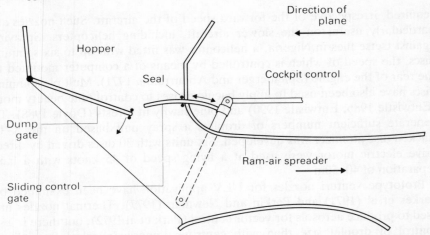

Fig. 13.18 Granule application from aircraft: simple diagram to show sliding metering gate (after FAO 1974)

trailing edge of the wing along which were a series of discharge outlets
(Harazny 1976).

On helicopters a separate blower unit driven by the engine forces air along
two ducts positioned at the base of the side tanks and out on short booms.
Each tank has a slide gate.

An alternative system to the ram-air spreader is to have two spinners, each
driven by an electric motor or hydraulically activated (Fig. 13.19). These
revolve in opposite directions, throwing granules outward from the front of
the spinner, so that swaths up to 15 m wide can be obtained. Deflector plates

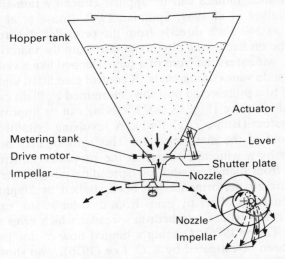

Fig. 13.19 Kawasaki Granule distributer using a spinner on a helicopter (from
Quantick 1985a)

protect the landing gear and propeller. Distribution of granular material by a spinner is reported by Courshee and Ireson (1962) and by R.F. Hill and Johnstone (1962). The metering gate is the same type as that used in ram-air spreaders. K.C. Lee and Stephenson (1969) developed a rotary-cylinder spreader. Two were mounted on an aircraft with a central spreader so that the swath was widened from 10 to 18 m, but it is not suitable for low-density or small particles.

Apart from spreading pesticides, aircraft have been used to distribute parasitoid eggs (Bouse et al 1981).

Flight planning

Aircraft flying height

Amsden (1972) has listed the factors which determine the height from which pesticides should be applied. These are

1 the velocity of the cross-wind component relative to the flight path
2 the aircraft design characteristics
3 the composition of the spray spectrum being produced
4 the specific gravity of the spray liquid (or particles)
5 the rate of evaporation from the spray droplets

All these factors may vary from one operation to the next, and even within a single flight. As discussed earlier in Chapter 4 , the relationship between spraying height and the cross-wind component can be expressed as

$$H \times U = C$$

where H = height of the wing or rotor above the crop (m) and U = cross-wind velocity in km/h.

The constant cannot be calculated, but is estimated by observing the biological effectiveness of sprays applied under a number of known conditions with a swath width determined at a particular height. Overdosing and underdosing is liable to occur if the aircraft is too low. Excessive drift is liable to occur if the aircraft is flown too high. Amsden (1972) illustrating the effect of a cross-wind on aircraft height (Table 13.5) points out that the maximum wind speed is dictated by the safety of the pilot. At speeds in excess of 25 km/h, conditions are normally so turbulent that the pilot will find it too uncomfortable to fly at the very low altitude required. If the pilot continues to spray and flies higher, excessive drift will occur. At the other extreme, the distance between the aircraft and the crop should not exceed half the wing span if full use is to be made of the downwash of turbulent air, so there is a minimum wind speed for a given HU factor. In the example in Table 13.5 these limits are 14–25 km/h for an aircraft with an 11 m wing span. According to Amsden (1972) HU values are usually between 40 and 90. The effective

Table 13.5 Variation in flying height with wind speed where **HU** = 80 (after Amsden 1972)

Height (H) (m)	Wind speed (U) (km/h)	
2.0	40	
2.67	30	
3.2	25	
4.0	20	↑
5.0	16	operating
5.7	14	limites
8	10	↓

Table 13.6 Calculation of crosswind velocity U for winds at different degrees off true cross-wind (after Amsden 1972)

Degrees	Correction factor	Effective cross-wind (km/h) where $U = 20$
0	× 1.0	20
20	× 0.94	18.8
40	× 0.77	15.4
60	× 0.5	10
80	× 0.17	3.4
90	× 0.0	—

cross-wind speed must be calculated if wind direction is at an angle to the flight path (Table 13.6).

In contrast Bache (1975) suggests that distribution of small droplets (<60 μm) is relatively insensitive to changes in windspeed, therefore consistent deposition downwind over particular crops can be obtained by adjusting flying heights with time of day (Fig. 13.20). Thus, in one example he suggests a flying height of 7 m above the crop during the morning, reducing to 5 m at dawn and dusk to achieve maximum deposition 50 m downwind. The latter technique is probably only suitable when large areas of a single crop are being treated. Johnstone and Johnstone (1977) recommend that spraying at 5 litres/ha or less with involatile droplets smaller than 120 μm vmd should cease if wind velocity exceeds 12.6 km/h.

In general, aerial sprays are more effective when there is a cross-wind and the aircraft is flying at the appropriate height.

Swath width (lane or track separation)

The swath treated by an aircraft will depend on the type of aircraft, its flying height, droplet or particle size and wind conditions prevailing at the time of application (Parkin 1979; Kuhlman 1981; Parkin and Wyatt 1982; Woods

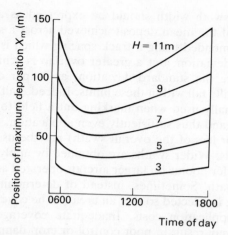

Fig. 13.20 Variation in maximum concentration with time of day and release height over a well-irrigated cotton crop in the Sudan (after Bache 1975)

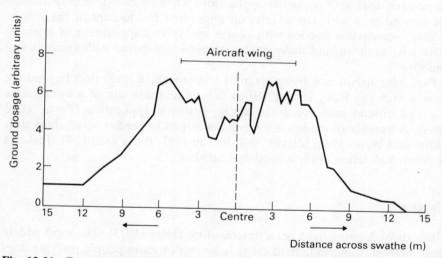

Fig. 13.21 Deposit distribution achieved with aircraft flying into wind

1986) (Fig. 13.21). The minimum swath may be determined when the aircraft flies into wind, although use is made of a cross-wind during normal commercial applications so that adjacent swaths overlap, even if there is little wind. The swath obtained with two aircraft of different dimensions can be compared by flying each one into wind with the wing at a height exactly one-half span above a line of targets. This height can also be assumed to be about the maximum height likely to be used for crop spraying, although greater heights are used, for example under carefully monitored conditions when spraying forests. Atkinson et al (1968) studied the distribution of pasture seed and recommended that swath width should be expressed as a multiple of the standard deviation of the mean deposit achieved across a single swath (Fig. 13.22). They recommended using a track spacing which is three and a half times the standard deviation, but a greater overlap is achieved if the track spacing is only twice the standard deviation, as 95 per cent of the spray deposit would normally fall within these limits. Indeed, half the overall single swath width is normally used when marking out a field to ensure adequate incremental dosing, and thus sufficiently even application. A narrower track spacing (one-third or less of the overall swath) may be used when applying unselective chemicals. Wider swaths are obtained by applying smaller droplets or particles. Wider booms on larger aircraft operated at a higher altitude will increase the swath. Sometimes, instead of determining the swath, too wide a track spacing is selected so that an area can be covered more quickly, thereby reducing application costs. Inadequate coverage or, conversely, excessive overlap, may result in poor control or crop damage, and the pilot needs guidance so that successive flights over the area being treated are correctly spaced to ensure as uniform a coverage as possible. This is particularly true with herbicide applications when an overdose may damage the treated crop and, conversely, an underdose fail to control the weeds. Greater accuracy is needed with coarse sprays and application of granular materials, as downwind movement is minimal compared with aerosols and fine sprays.

Particular attention is being given to assessments of spray drift beyond the treated area (eg Riley et al 1989) to determine the width of a buffer zone required around sensitive areas during pesticide application (Payne et al 1988). A number of models are being developed to predict aerial spray drift (Atias and Weihs 1985; Mickle 1987; Parkin 1987; Barry et al 1990; Teske et al 1990). Validation of these models is needed.

Flagging

Methods of flagging have been described by Haley (1973). The most widely used method of marking field crops is by two or more people carrying flags of a brightly coloured material. Flags are often yellow, orange or white to contrast with green vegetation and should be about 1 m square so that they are easily seen. The pilot prefers to fly along the rows of a field crop, and has little difficulty in keeping on a straight course between flagpeople standing at each end of the field, indicating the position of the swath to be treated. However, this can be done only if the wind direction is across the rows. When the land is undulating or the field is excessively long (over 3 km), additional flagpeople may be needed at intermediate positions through the field. Ideally the position for the flagpeople is measured earlier, and short pegs hammered into the ground so that flagpeople can move quickly to their next position as soon as the aircraft has flown by. The pilot may fly directly over the flagpeople but, when highly toxic materials are used, the flagpeople

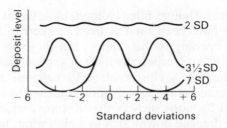

Fig. 13.22 Cumulative deposit with swaths separated by multiples of the standard deviation of a single swath

should be positioned on the upwind side of the swath, or they should move rapidly upwind as the aircraft approaches to minimize chemical contamination (Whittam 1962). If possible, they should be positioned back from the edge of the field, but a hedge or other obstruction may prevent this. The flagpeople should be well protected with suitable clothing, depending on the pesticide being applied. Other people should be kept well clear of treated fields, not only during spraying operations but also afterwards for a period, the length of which will depend on the chemical applied. When a crop is treated several times with a relatively safe pesticide, the flag may be directly under the centre of the aircraft on one application and the wing or rotor tip at the next, as this may improve the uniformity of coverage.

A bright battery-operated signal light can be used instead of a flag when pesticides are applied at night. Xenon-gas-filled tubes flashing at 60–80 times per minute have also been used in daytime. Lamps have been used in conjunction with a compass for tsetse spraying (A.W.D. Miller and Chadwick 1963).

When fields are treated many times it is often more convenient to use a series of fixed markers. In annual crops the main problem with fixed markers is the need to adjust their height in relation to crop height. Markers may be hazardous if they are set too high at the start of the season when the aircraft is flying low, whereas later in the season they may be obscured by foliage. Successive fixed markers across a field need to be painted with contrasting colours (eg orange, white, yellow, black) and shapes (eg triangular, circular, square, rectangular) so that the pilot can readily identify which is the next swath to be treated. Bright fluorescent pigment paints of different colours are ideal. Ground markers such as an upside-down white or yellow plastic laundry basket, about 0.6 m high, have also been used to warn pilots of hazards such as power and telephone lines crossing fields (Keeler 1971). A serious drawback to fixed markers is that they cannot readily be moved. This may be necessary if the wind direction has changed from the time of one application to the next. This problem may be overcome by marking the field in two directions at right angles to each other. On each visit the pilot selects the more appropriate set of markers.

Flagging is more complicated in hilly or mountainous areas and also in forest or swampy areas. Helium-filled balloons of different colours sometimes up to 1 m diameter have been used as fixed markers in forest areas where there is a sufficient opening in the canopy, but they often get entangled in the foliage, burst or deflate, especially if set at appropriate places one or two days prior to treatment. A cluster of small balloons to allow for some deflating can be mounted at the top of a pole attached to a tractor. The more aerodynamic-shaped balloons such as the 'Kytoon', which flies like a kite, are less likely to be dragged downwards in a high wind, but is impractical as it must be lowered immediately the aircraft approaches to prevent the plane's vortex from hitting it. Otherwise the Kytoon may be damaged as it gyrates wildly above the trees (Whittam 1962). Smoke flares have been used when there is radio contact between the pilot and ground crew, enabling flares to be fixed in the correct sequence and eliminating the fire hazard.

When ground marking is difficult or impossible, the pilot can use several systems from the aircraft. Coloured dyes have been sprayed and fog produced from the aircraft engine exhaust, but neither are very successful methods. Toilet rolls were dropped to mark spray lines in the early locust-control programmes. More recently a tissue-paper dispenser has been developed and is fitted to the aircraft wing, so that as the pilot flies on the swath the pilot can release pieces of paper marker weighted with a piece of cardboard. The paper forms a streamer which lies on top of the foliage to guide the pilot where the start of the next swath should be. These markers remain effective for a sufficiently long period, but eventually become torn and degraded, especially in wet weather.

Track guidance

Several navigational systems have been used for track guidance. Generally these can be divided into two categories: those using an external reference and those using only on-board equipment.

Early systems such as the Decca 'Agri-fix' (D.A. Walker 1973) rely on two mobile ground stations transmitting signals which intersect to produce a family of hyperbolic position lines. A receiver and left/right indicator in the aircraft enable the pilot to follow the hyperbolic tracks. This system has the disadvantage that the ground stations must be correctly aligned relative to the required track.

More recent systems use two or more ground-based transponders placed in or around the area to be sprayed. Equipment on the aircraft interrogates these transponders and a computer calculates its position relative to them. The computed position is compared with a grid corresponding to the required tracks and a left/right indicator guides the pilot. These systems are typically accurate to within 1 metre and have the advantage that the layout and spacing of the tracks can be defined independently of the position of the transponders. However, both transponders and airborne equipment are expensive and a high utilization is necessary for this approach to be cost effective. Commercially available systems include the Flying Flagman from Del Norte and the Maxiran from Maxiran Inc.

Established chains of ground stations have also been used for track guidance. The VLF/Omega system is primarily intended for long range navigation but has been used for track guidance in drift spraying with wide swath widths for tsetse control and similar work. The resolution and accuracy of this system is marginal even at very wide swath widths and it is difficult to achieve reliable and repeatable performance. A more satisfactory system is Loran, using chains originally intended for marine and long-range aircraft navigation. These provide much better accuracy and resolution than VLF/Omega but the chains are available only in certain areas. This system has provided effective track guidance for mosquito spraying in the United States where there are extensive Loran chains. The recently introduced Global Positioning by Satellite (GPS) system has potential for track guidance. However, commercial use will have to await the availability of all the proposed satellites and the introduction of low-cost receivers coupled to suitable navigation computers.

On-board systems use precision gyroscopes, doppler radar or a combination of both. Gyroscopes may be either mechanical (inertial) or strapdown laser types. The gyroscope senses and acceleration of the aircraft in all three axes and feeds this information to a computer. The position of the aircraft is calculated from the gyroscope data and the airspeed. Doppler systems are normally based on the Decca/Racal radar and associated computer. The movement of the aircraft is calculated from the doppler shift in the frequency of the reflections of three radar beams directed to the ground. This is combined with heading information and sometimes measured airspeed to give a computed aircraft position and distance from a pre-determined track. These systems are best suited to spray operations requiring long and widely spaced tracks such as in tsetse spraying. As with transponder based guidance systems, on board guidance equipment is very costly and has only been used in specialised operations.

Logistics

The quantity of spray applied as litres per hectare (Q) will depend on the throughput of each nozzle, the number of nozzles, swath width and flying speed thus:

$$Q = \frac{10Nq_n}{SV_s}$$

where

Q = output required (litres/ha)
N = number of nozzles
q_n = throughput of a nozzle (ml/s)
S = swath (m)
V_s = flying speed (m/s)

The output and number of nozzles is selected to give the required output for a given aircraft for which the swath width and flying speed have been determined.

The time to fly 1 ha in 1 min is 600/Sv, so the volume of spray applied per minute is:

$$Qt = \frac{QSv}{600}$$

where v = flying speed (km/h). This should be checked by putting a known quantity of liquid in the aircraft and spraying at the correct pressure for a definite time. Normally this involves a flight check, although ground checks are possible if the pump can be operated at the correct speed. The tank is then drained to determine the quantity sprayed. Alternatively, the level of the tank before and after spraying can be measured if the tank has been properly calibrated. As with tractor sprayers, small adjustments of volume can be made by changing the pressure, otherwise different nozzles will be required. The number of hectares treated with one load is:

$$\frac{Q_f}{Q}$$

where

Q_f = quantity of spray mix loaded per flight
Q = litres per hectare

The number of hectares covered for different swath widths and field lengths is indicated in Table 13.7, so the load can be adjusted to avoid the aircraft running out of spray in the middle of a swath.

The approximate time needed to spray an area can be determined by reference to Table 13.8 in which the hectares per minute is given for different combinations of swath width and flying speed.

Table 13.7 Hectares covered for given field lengths and track spacings

Field length (m)	Track spacing (m)			
	7.5	10	15	20
250	0.19	0.25	0.38	0.5
500	0.38	0.5	0.75	1.0
750	0.56	0.75	1.13	1.5
1 000	0.75	1.0	1.5	2.0
2 000	1.5	2.0	3.0	4.0
3 000	2.3	3.0	4.5	6.0
4 000	3.0	4.0	6.0	8.0
5 000	3.75	5.0	7.5	10.0

Table 13.8 Hectares/min covered with different velocities and track spacings

Velocity (km/h)	Track spacing (m)			
	7.5	10	15	20
100	1.3	1.7	2.5	3.3
120	1.5	2.0	3.0	4.0
140	1.8	2.3	3.5	4.7
160	2.0	2.7	4.0	5.3
180	2.3	3.0	4.5	6.0

Flight pattern

Pilots will normally fly a series of passes, gradually moving upwind across the area requiring treatment (Fig. 13.23). At the end of each pass pilots have to complete a procedure turn. Initially, as they approach the end of a pass, they increase power, shut off the spray, pull up sharply to 15–30 m, turn away about 45° and then brings the aircraft round to approach the next swath. The power required will depend on the load and the height of obstacles, but adequate speed and power are essential to guard against stalls or incipient spins. Sometimes a pilot may prefer to fly a 'race-track' pattern which allows a wider turn, but necessitates additional flagpeople. The race-track or round-robin pattern is useful when a number of small fields are located close to each other (Fig. 13.24) and can give a more even spray coverage as the aircraft is flying successive passes in the same direction. When obstructions are close to the edge of the field, the pilot will normally fly one or two passes along the edge and along each headland to 'finish off' after completing the main part of the field. Useful advice to pilots is given in the *Handbook for Agricultural Pilots* (Quantick 1985b).

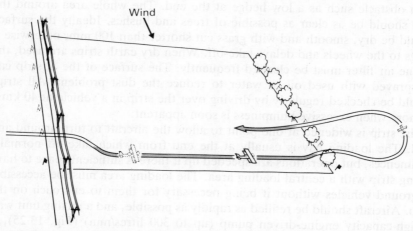

Fig. 13.23 Schematic representation of a flight pattern (after FAO 1974)

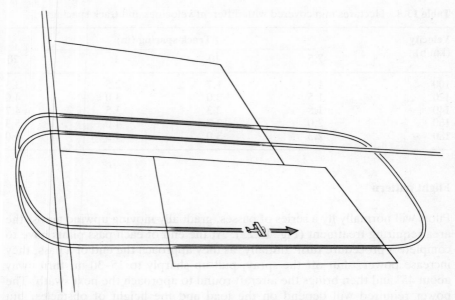

Fig. 13.24 Round-robin system of spraying a field. *Note:* Two fields are sprayed alternately

Airstrips

Agricultural flying is often from unprepared airstrips or ordinary grass fields. Some governments have specific regulations on the size and condition of airstrips which can be used. In general, a strip should be about 30 m wide with a slight camber to permit drainage and at least twice as long as the distance taken by the aircraft to take off. Longer strips are essential if there is an obstacle such as a low hedge at the end. The whole area around the strip should be as clear as possible of trees and bushes. Ideally the surface should be dry, smooth and with grass cut shorter than 100 mm, otherwise it clings to the wheels and delays take-off. When dry earth strips are used, the engine air filter must be cleaned frequently. The surface of the airstrip can be sprayed with used oil or water to reduce the dust problem. All strips should be checked regularly by driving over the strip in a vehicle at 40 km/h or more when excessive bumpiness is soon apparent.

The strip is widened at one point to allow the aircraft to turn around and load. The loading bay is usually at the end from which take-off normally commences, but operations are speeded up if there is sufficient space to have a long strip with a central loading area. The loading area must be accessible to ground vehicles without it being necessary for them to encroach on the strip. Aircraft should be refilled as rapidly as possible, and a mixing unit with a high-capacity engine-driven pump (up to 300 litres/min) (Fig. 13.25), is essential and may be mounted on the support vehicle. When open tanks are

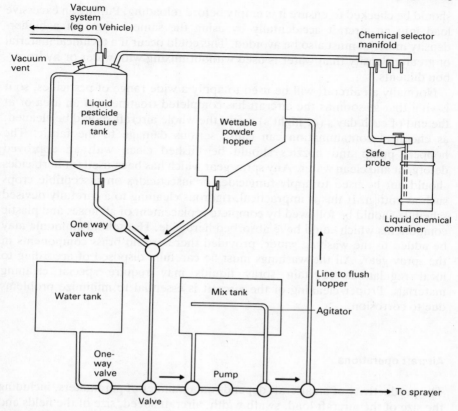

Fig. 13.25 Rapid refilling system

used, foam can be a problem if air is trapped in the spray liquid during mixing. Such foam can be reduced by adding a small quantity of a silicone anti-foam agent. Some countries have regulations concerning the mixing of pesticides (Brazelton and Akesson 1976). The hazards of handling concentrated materials can be reduced by using closed-system mixing units with which a precise quantity of pesticide is transferred from its commercial container to the mixing tank and later pumped directly into the aircraft. Dry materials are normally handled by special equipment which can be loaded through the opening on the top of the hopper.

The load will depend on the design of the aircraft, quality of the airstrip, altitude and air temperature. The pilot is responsible for determining what is a safe load for a given airstrip; the first few take-offs at a new strip will require light loads until the pilot is used to the local conditions. The pilot may need to reduce the load normally taken at a particular airstrip if the surface is softened or otherwise affected.

Great care must be taken to avoid overloading the aircraft. The hopper should be checked to ensure it is empty before reloading. Putting an excessive load into an aircraft accidentally by using the same volume of a higher-density material must also be avoided. This could occur if a technical material of greater density than water is used without mixing with water or hydrocarbon diluents.

Normally an aircraft will be used to apply a wide range of pesticides, so it is vital that as soon as the aircraft has completed treatment of an area or at the end of each day's (or night's) work, the whole aircraft should be cleaned, as chemical contamination can cause serious damage to the fabric. The hopper, pump and nozzles should be flushed clean with an approved detergent and clean water. Any spray gear which has been used for herbicides should not be used to apply fungicides or insecticides on susceptible crops such as cotton. If this is impractical, rigorous cleaning to a carefully devised schedule should be followed by complete replacement of all hoses and plastic components which could have absorbed herbicide. Household ammonia may be added to the washing water, provided there are no brass components in the spray gear. All the washings must be carefully disposed of according to local regulations. Certain spray liquids may require special cleaning materials. Proper cleaning of the aircraft is essential to minimize problems due to corrosion.

Aircraft operations

The productivity of an aircraft depends on a number of parameters, including the size of the aircraft load, swath width, aircraft speed, size of the fields and distance to the refilling point. The Baltin formula (Baltin 1959) expresses these parameters as follows:

$$t = 10^4 \left(\frac{T_r q}{Q_r} + \frac{1}{V_s S} + \frac{T_w}{SL} + \frac{2aq}{V_f Q_f} + \frac{C}{V_f F} \right)$$

where
  t = work time per hectare (s/ha)
  T_r = time for loading and taxiing (s)
  q = application rate (litres or kg/m^2)
  Q_f = quantity of chemicals loaded per flight (litres or kg)
  V_s = flying speed when spraying (m/s)
  V_f = flying speed when ferrying (m/s)
  S = swath width (m)
  T_w = time for one turn at the end of spray run (s)
  L = average length of fields (m)
  C = average distance between fields (m)
  F = average size of fields (m^2)
  a = average distance airstrip to the fields (m)

A similar equation (Baltin–Amsden formula) is available in imperial units

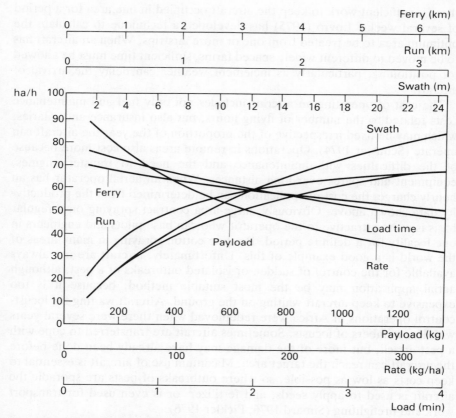

Fig. 13.26 Operational analysis for a fixed-wing aircraft at a normal application rate (after FAO 1974)

(Amsden 1959). A more detailed formula has been given by Interflug (1975). More detailed discussion on the productivity of aircraft is given by Quantick (1985a). Akesson and Yates (1974) show the effect of variations in swath width, field length, ferry distance, loading time, application rate and payload on productivity in hectares per hour (Fig. 13.26). Each factor was varied separately, while median values were used as constant values for the other factors. These values were as follows: swath 12.2 m; field length 8.05 km; ferry distance 3.2 km; loading time 2 min; application rate 112 kg/ha; payload 907 kg; flying speed 144.8 km/h; and turn time 0.5 min.

Highest productivity is obviously favoured by long fields, wide swaths, low application rates, short ferry distance and a large load. Agricultural planners should consider field shape if aerial application is anticipated and provide long runs for aircraft. Higher flying speeds favour fixed-wing aircraft in contrast to helicopters, but the ferry distance for the latter may be negligible. In most aerial application work a positioning time must be considered unless there is sufficient work to keep the aircraft occupied in one area for a period of several weeks. Lovro (1975) has developed a technique to calculate the optimum area to be treated from one or more airstrips. When an aircraft has to be moved to different widely spaced farms, sufficient time must be allowed for positioning, particularly as inclement weather can delay the arrival of aircraft.

The cost of operating an aircraft includes not only fuel and maintenance costs related to the number of flying hours, but also insurance and salaries, which may be fixed irrespective of the proportion of the year the aircraft can operate (Schuster 1974). Operations in remote areas also cost more because of the difficulties with maintenance and the need to transfer engines, equipment and spares over long distances. When an aerial operator has an hourly charge, the cost of application can be determined from the productivity data shown above. Obviously, large-area contract spraying on a regular basis is more attractive to the operator who can base pilots and engineers in one locality for a definite period. Routine cotton spraying in many areas of the world is a good example of this. Unfortunately, aircraft are not always available for the control of sudden or isolated outbreaks of a pest, although aerial application may be the most suitable method, because it is too expensive to keep aircraft waiting on the ground. Aircraft waiting for locust-control operations in Africa were redeployed when there were several years with low numbers of locusts. Sometimes aircraft are transferred to cope with a pest attack, but more of the damage may have already been done before the aircraft can reach the target area. Maximum use of aircraft is essential to keep costs as low as possible, so where outbreaks of pests are sporadic the aircraft is used to apply seeds, and fertilizer, or is even used for transport work and firefighting (Simard 1976; Pickler 1976).

Aircraft regulations

The use of aircraft for the application of pesticides is controlled by legislation. Some countries merely require aircraft to be registered with, and inspected by, a civil aviation organization which has power to issue certificates or airworthiness where appropriate and to control the period of flying between routine maintenance checks. Pilots must undergo frequent checks on their physical fitness and competence to retain a licence issued by the same organization, which also controls the number of hours a pilot is permitted to fly. Other countries have wider legislation to control which chemicals may be applied from aircraft. This legislation is essential in order to restrict the use of various herbicides, and the most toxic or most persistent pesticides. In the United Kingdom agricultural aviation operators must comply with the 'Aerial Application Permission' issued by the Civil Aviation Authority. Previously, such operators had been exempted from the relevant parts of air legislation which restricted both low flying (under 152 m) and the dropping of articles

Table 13.9 Comparison of two aerial spray treatments in a forest

Type of application	Ultra low volume application	Low volume application
Formulation	6 Parts fenitrothion 50 EC 4 Parts Butyl dioxytol	3 Parts fenitrothion 50 EC 97 Parts water
Atomizer	Two Micronair AU 3000 Standard 5″ Cage, 13.5″Flat blades Set at 25°	Six Microniar AU 3000 Standard 5″ Cage, 13.5″ Flat blades Set at 25°
Application rate	1 litre/ha	20 litres/ha
Active ingredient rate	0.3 litre/ha	0.3 litre/ha
Release height above canopy	6 m	3 m
Lane separation	50 m (two applications at 100 m on successive days)	25 m
Emission rate	15 l/min	150 l/min
Droplet sizes	vmd 97 nmd 24	vmd 104.5 nmd 22
% of volume between 10 μm and 40μm	8.0%	8.4%
Spraying speed	180 km/hour (50 m/s)	180 km/hour (50 m/s)
Wind speed	7.8 knots day 1 13 knots day 2	1.6 knots
Area sprayed	50 ha (2 km by 250 m)	100 ha
Ferry Distance	100 km	30 km
Overall work rate	309 ha/hour	88 ha/hour
Destination of active ingredient		
(a) Collected by needles or larvae	94.5%	41.7%
(b) Lost to ground within block	4.5%	38.3%
(c) Lost outside block	1%	20%
Average larval weight	28 mg	108 mg
Mean deposit on 20 needles	41.2 ng	23.6 ng
Mean deposit on single buds	18.8 ng	23.6 ng
Mean deposit on larvae per gramme of larval weight	1 285 ng/g	407 ng/g
Mortality (%)	97.5	97.5

from aircraft (Birchall 1976). The Aerial Application Permission requires comprehensive standards for all safety aspects of aerial application operations, including avoidance of spray drift, marking of fields, reconnaissance pre-flight briefing and mapping of areas requiring treatment to indicate obstructions. Aerial spraying operators may apply only pesticides selected from a 'permitted list' compiled by the Ministry of Agriculture, Fisheries and

Food under the Food and Environmental Protection legislation. Details of the regulations in the United States and other selected countries were published by the World Health Organization (1970), but considerable changes in legislation are taking place especially in the USA where the Environmental Protection Agency (EPA) has an overall responsibility for implementing the Federal Environmental Pesticide Control Act. An increase in the number of regulations and their scope can be expected, particularly in relation to the equipment which may be fitted to aircraft. Selection of nozzles and droplet size may be restricted in relation to meteorological conditions.

Where large-scale aerial spray operations are proposed, it is essential to carry out an environmental impact study. Ultra low volumes of insecticide have been applied successfully with minimal impact on non-target species by ensuring that with appropriate droplet sizes and spray concentration, most of the spray is collected on the foliage. Detailed studies have been carried out in Scotland (Holden and Bevan 1978; 1981) and in Canada (eg Sundaram et al 1988). The trials in Scotland provided an interesting comparison between LV and ULV applications, the latter doubling the recovery on pine needles and the pine beauty moth larvae with a much higher work rate (Table 13.9) (Spillman 1987).

In several countries an untreated buffer zone has been proposed to protect ecologically sensitive areas, especially ponds and streams to prevent a significant impact on fish and their food populations. Payne et al (1988) used a motorized mistblower with spinning disc to apply a synthetic pyrethroid as a fine spray to assess a worst-case scenario. Using a model they predicted that a buffer width of 20 m caused less than 0.02 per cent mortality of *Salmo gairdnei* rainbow trout in water depths greater than 0.1 m. For aerial application Riley et al (1989) have considered a 100 m buffer zone would ensure that there would be at least a 10 × decrease from the deposit observed at the edge of the target area, even when wind speeds exceed those currently recommended for agricultural sprays. As indicated earlier, much depends on the vegetation filtering out the spray droplets and, with higher wind speeds and turbulence, more of the spray will be impacted on foliage.

14

Injection, fumigation and other techniques

Injection techniques

Soil injection

Soil injection of volatile pesticides, including nematicides and herbicides, requires specialized equipment and is justified only with high-value crops, especially as the treated surface may have to be covered with a polythene sheet to retain the volatile fumigant for a sufficiently long period to be effective. Granules can be applied to the surface and incorporated into the soil with a suitable harrow of plough.

Tractor-mounted soil-injection equipment (Fig. 14.1) usually has a series of chisel tines fixed to a toolbar so that the depth to which the tines penetrate the soil can be adjusted. Pesticide is delivered through a tube mounted directly behind each tine. It may be supplied in a cylinder under pressure, otherwise a small low-pressure pump draws liquid from a spray tank through an in-line strainer, pressure regulator and distributor to a metering jet. This jet is mounted close to the tine but above the soil surface, so it is readily accessible for cleaning. It is impossible to see liquid flowing out behind the tine, so each outlet from the distributor has a flow meter which can be checked visually by the operator. Good filtration of the liquid is essential to avoid blocking the jet. Injection to the correct depth is particularly important to distribute chemical into the most appropriate zone, ie where the roots of the subsequent crop will be liable to infestation. Often soil injection is too shallow, allowing a deep-rooted crop to be attacked by soil pests below the treated zone, but injecting below 150–200 mm is extremely difficult, except on some soil types. No chemical should be left in the equipment, otherwise corrosion may occur.

When a small area needs treatment a hand-operated soil injector is used. This consists essentially of a pointed injector fitted to a piston pump, at the

Tank on front of tractor

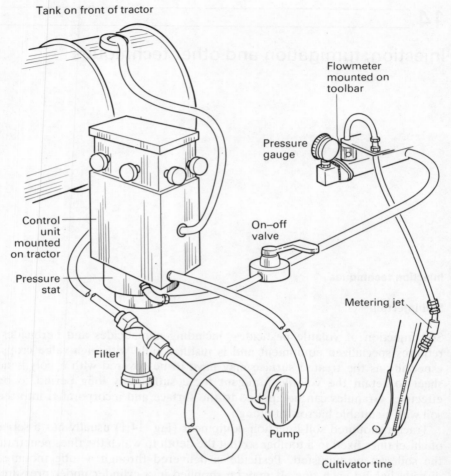

Fig. 14.1 Soil fumigant injection equipment for use on a tractor

top end of which is a handle and adjustable metering cup. The handle may be fitted into the top of a container which surrounds the pump (Fig. 14.2). The operator holding the handle thrusts the injector into the soil to the required depth. Depth of injection is determined by a plate which can be fixed at various positions to the injector head. A sharp tap on the cap or injector handle actuates the pump and forces a single measured dose, usually l-15 ml, through an orifice near the end and to the side of the injector head.

Forcing the injector into the soil is very arduous, especially on some soils, and less than 0.5 ha can be treated in one day with one injector. Injections are made at regular intervals over the whole field or along rows in which the crop will be subsequently sown or transplanted. The distance between injection points will depend on the crop spacing, soil type and dose applied.

Fig. 14.2   Hand-operated soil injector

A delay between treatment and planting may be needed to allow the chemical to disperse to avoid any phytotoxic effects on seeds or seedling.

When filling the tank, a fine-mesh filter must be used in the tank opening to reduce wear of the pump and the risk of any blockage in the system. Great care is also needed to avoid trapping air in the injector, as this will affect the dose applied. After use, a soil injector should be thoroughly cleaned, and both the inside and outside treated with a mixture of equal parts of lubricating oil and kerosene before storage.

A continuous band of an insecticide is needed below the soil surface alongside rice seedlings to protect them against stem borers and other pests.

Following investigations with paper straws and gelatine capsule containers inserted below the soil surface, a two-row liquid injector was developed for lowland rice by the International Rice Research Institute. A small platform mounted at the end of a tube is pushed along the surface of soft, wet, paddy soil between the seedlings. Insecticide is carried in a 10-litre knapsack tank and delivered to two tubes 50–100 mm underneath and at the side of the platform. By inserting the insecticide in the root zone, less is lost because of movement of the irrigation water, evaporation and oxidation. If the mud through which the injector is pushed is soft, 0.5–0.7 ha can be treated per day by one person, but as this is arduous work, the technique has not been used widely. The technique can also be used to double fertilizer efficiency compared with broadcast application.

Tree injection

Injection is most successful for treating coconuts and palm trees (Wood et al 1974; Khoo et al 1983). A simple brace and bit (Ng and Chong 1982), or preferably a power-operated drill is used to make a hole between the base of two frond butts, angled 45° downwards into the stem using a 10–15 cm long drill. Insecticide is metered into the hole as soon as possible using a small hand-operated injector (Fig. 14.3) and then the hole opening is covered with a fungicide paste to prevent loss of the insecticide. Later callus tissue will cover the hole. A systemic insecticide such as monocrotophos is used as it is readily taken by the xylem to the crown of the tree. Protection of the hands and face is needed during treatment as a concentrate is used to apply 5–7 g ai/tree. With a suitable power drill and injector about 3 ha/person-day can be treated.

Control of Dutch elm disease, caused by the fungus *Ceratocystis ulmi* with benomyl or carbendazim injected into trees has also been attempted (Gibbs and Dickinson 1975). The concentration of chemical used is usually a compromise to avoid too high a concentration which may cause phytotoxicity, and at the other extreme too low a concentration may result in an impracticable volume of liquid to be injected. The technique is slow and labour intensive, so application costs per tree are high and justified only where high-value amenity trees need to be saved.

Simpler techniques involving injection with a simple syringe to a bored hole or cut surface have been successful against some diseases. Specially adapted hatchets and secateurs have been made for repetitive treatments, for example *Trichoderma viride* has been applied to a cut shoot surface as a protection from invasion by pathogens such as silver leaf (K.G. Jones et al, 1974).

A trunk-implantation technique to apply systemic insecticides to trees was described by W.E. Ripper (1955) and used by Hanna and Nicol (1954) to control cocoa mealybug with minimal risk to pollinators, parasites and predators. Arboricides have been applied with special axes to kill trees.

(a)

(b)

Fig. 14.3 Tree injection (a) Drilling hole (b) Injection of insecticide

Dispensers into water

Certain pests live in water; for example, the snails which are the intermediate hosts of *Schistosoma* spp, the larvae of *Simulium*, vector of the disease onchocerciasis. There are also a number of important aquatic weed species.

Various types of equipment have been used to apply pesticides in sprays and granules to water, but where there is flowing water special dispensers can be used to avoid using labour involved in spraying. Complete mixing of a chemical with the water can be obtained if the dispenser is set at a narrow point or where the water is turbulent. The dose required and its distribution will depend on the volume of water and the rate of flow, as in sluggish water the chemical will be distributed only a short distance downstream. The quantity of water flowing per second past a given point can be determined from the cross-sectional area of flow and the average velocity. In many watercourses the velocity needs to be determined at various depths.

Some dispensers use a simple gravity feed, but unless the valve is adjusted or a constant-head device is used the application rate will decrease as the reservoir empties, owing to a decrease in the head of liquid.

The variation in flow can be minimized by mounting a squat tank as high as possible above the discharge point. A 200-litre drum is often used as a reservoir and can be mounted in a boat if necessary. More sophisticated systems adjust the dosage proportional to water flow (Klock 1956). Alternatively, the chemical is bled into the suction line of a pump circulating water from a stream. In general, a pump delivering 100–200 litres/s and driven by a 1 kW engine is adequate. The flow of chemical needs to be checked with a suitable flow meter and adjusted with a regulating valve.

In order to minimize effects on non-target organisms, particularly fish, great care must be exercised in the selection of chemical and dosage required for the control of aquatic pests and weeds.

Chemigation

Applying certain pesticides by injection into irrigation water using a positive displacement pump, known as chemigation, has increased in popularity in some areas as irrigation equipment, especially centre pivot systems, has been developed to improve uniformity of water distribution. Once the investment is made on the irrigation equipment, the farmer needs to utilize fully the equipment which is a viable alternative to conventional sprayers and requires less labour. Some herbicides washed into the soil by the irrigation water can be more effective, but concern about leaking of pesticides to ground water has been expressed. Care must also be taken to avoid backflow which might contaminate the water source, so safety devices are essential (Fig. 14.4). Foliar applied chemicals may be less effective due to the extreme dilution (>5000 l/ha) but herbicides may be less phytotoxic on crop foliage. Extreme

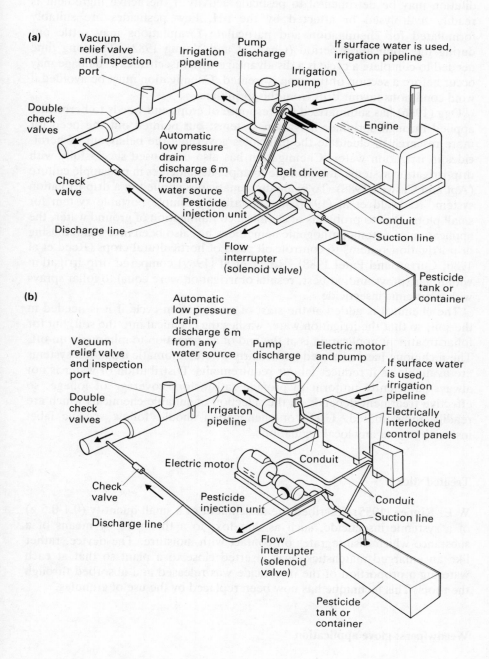

Fig. 14.4 Equipment for chemigation (a) With engine driven pump (b) With electrically driven pump

dilution may be detrimental to pesticide activity if the active ingredient is readily hydrolysed or affected by the pH. Few pesticides are suitably formulated for chemigation and particulate formulations may settle out during the application period (Chalfont and Young 1982). The long time needed to complete a cycle is a disadvantage with insecticides as damage may occur before a section of the crop is treated. Chemigation must be avoided if wind conditions favour drift from the sprinklers.

Ogg (1986) has summarized the response of crops and weeds to herbicides applied through sprinkler irrigation systems; but pointed out the need for more research to elucidate the principles governing the behaviour of herbicides in irrigation water. Chemigation has also been used successfully with drip-irrigation systems, for example to apply nematicides in pineapple culture (Apt and Caswell 1988). To evaluate nematicides applied in a drip irrigation system, Radwald et al (1986) has described a simple portable system for small plots. Due to problems of chemical contamination of ground water, the application of entomopathogenic nematodes has also been investigated using drip-irrigation systems to control soil pests of horticultural crops (Reed et al 1986; Curran and Patel 1988). Roger et al (1989) compared drip irrigation with foliar sprays and, at best, results of irrigation were equal to foliar sprays with a systemic insecticide.

The chemical is added at the start of an irrigation cycle if it is needed in the soil, so that the irrigation water washes the chemical into the soil, but for foliar treatment application is at the end of an irrigation to minimize run-off. The technique has been used particularly where automatic irrigation systems are available as it reduces labour requirements. Distribution of water is not always sufficiently uniform to provide satisfactory coverage of foliage, so effective treatment depends on the selection of suitable chemicals which are readily redistributed. Users of chemigation should follow specific label instructions and any local regulations.

Treated stick

W.E. Ripper (1955) described the attachment of a small quantity (0.1-0.5 g) of a systemic insecticide, such as dimefox, to a thin wire by means of a substance which disintegrates on contact with moisture. The device, rather like an enlarged matchstick, was inserted close to a plant so that at each watering a proportion of the insecticide was released and absorbed through the roots. This technique has now been replaced by the use of granules.

Weedwipers: glove application

Instead of hand rogueing certain weeds such as wild oats, the herbicide can be applied selectively to individual plants, using a specially designed glove. An impervious plastic glove has a foam pad in the palm covered by a porous but resistant material. A rubber pressure bulb underneath the pad is deflated when the plant is gripped by the gloved hand, and approximately 1 ml of herbicide is transferred from the liquid container to the foam pad. A non-return valve is incorporated in the unit. A dye is normally added to the herbicide to indicate which plants have been treated.

Rope wick applicator

Rope wick applicators consist of a container from which liquid is able to soak an absorbent surface without dripping excess liquid. The applicator, which can be a handheld stick (Fig. 14.5) or a rope wick attached to a horizontal boom, is used to wipe the translocated herbicide such as glyphosate on to foliage (Dale 1979). The main problems with the equipment are difficulties in avoiding dripping or conversely having too dry a wick, and accumulation of dirt on the surface. The hand carried rope wick applicator is mainly used for spot treatment.

Individual plants can also be smeared with an oily rag treated with a suitable herbicide.

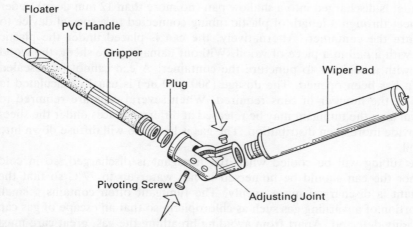

Fig. 14.5 Rope wick applicator

Fumigation

A specialized method of controlling pests is by fumigation. To be lethal, a toxic chemical is used in the gaseous state in sufficient concentration for a given time. Fumigants are used to treat plants and the soil, for example in plant quarantine work, but are particularly useful when insect and other animal pests have to be controlled inside stored grain in silos, warehouses, ships and other enclosed areas or in stacks of produce in the open. The dosage of fumigant is usually referred to as a concentration × time product ($c \times t$ product), thus a long time at a low concentration can be as effective as a short fumigation at a high concentration. Neither excessively long exposure periods nor high concentrations are practical. Some buildings are specially designed to permit fumigation of stored grain in bulk, or specially constructed fumigation chambers can be used, but if neither of these are available, a lightweight plastic sheet or gas-tight tarpaulin is used to retain the fumigant for the required exposure period. Fumigation must always be carried out by persons with proper training and equipment.

Soil fumigation

The technique is used mainly to control nematodes and weeds in seedbeds, for example tobacco seedbeds are always fumigated prior to sowing. The area to be treated is covered with a plastic sheet or tarpaulin and the edges buried and sealed with soil. The centre of the sheet is held off the soil by a suitable support such as a bag of grass. The fumigant, usually methyl bromide, is discharged into a shallow pan (no more than 12 mm deep) under the sheet through a length of plastic tubing connected to a special device to puncture the container. Alternatively, the can is placed under the plastic sheet with a nail in a piece of wood. Without damaging the sheet, the wood is hit with a hammer to puncture the container. A can cannot be resealed once it has been opened. The dosage (500 g/10 m²) is usually calculated in terms of the number of cans required. When several cans are required to treat an area the fumigant may be released at different places under the sheet to provide more even distribution. The gas is heavy so will diffuse down into the soil.

The tubing will be cooled when the fumigant is discharged, so in cold weather the can should be immersed in hot water up to 77°C so that the fumigant is discharged more evenly. The fumigant often contains a small proportion of a warning gas such as chloropicrin so that an escape of gas can be readily detected. Apart from avoiding breathing the gas, great care must be taken to avoid contact with liquid methyl bromide as it is liable to cause skin burns, although gloves should not be worn. A barrier cream can give protection to the hands. Any spillage will evaporate quickly, but hands should be washed immediately with soap and water.

The sheet can be removed after the prescribed exposure time, usually 48 h, and seed can be sown directly after removal. The soil should not be too wet, as the gas will not penetrate the soil, nor too dry, otherwise weed control will be poor. Some field crops such as strawberries have been fumigated in a similar manner with dosages of 24–48 g/m³ of air space between the sheet and the ground, for 2–3 h during early morning or late evening to avoid high temperatures while the crop is covered.

Tree fumigation

Tree fumigation is one of the oldest techniques used to control scale insects of citrus, but it also provided the first example of resistance to a pesticide. The technique is not used so extensively at the present time, except on steep hillsides where it is difficult to move spray equipment, for example in Japan.

Octagonal sheets are placed over individual trees, especially those which are dormant, with the aid of a tent puller, taking care to avoid damaging the sheet on the branches. Any dead wood should be pruned. The most practical fumigant is hydrogen cyanide (HCN) which is usually released by blowing calcium cyanide powder upwards inside the tent. Liquid HCN has also been used. Calcium cyanide is available in small briquettes (with ll.5 per cent lime) which are broken up into a powder immediately prior to use, or as a fine grey powder referred to as cyanogas. The powder releases HCN on exposure to moisture. Trees are often fumigated at night, since they may be damaged by fumigation when temperatures are higher than 26°C in moist climates or 30°C in dry areas. Also, damage may occur if the temperature is less than 7°C or if the soil is abnormally wet following heavy rainfall or irrigation. Damage has also been reported if citrus trees have been sprayed with Bordeaux mixture within the previous six months (Woglum 1923), are weakened by poor cultural conditions or if young fruit are present. Some varieties are also more susceptible than others. Great care must be taken if a number of trees are fumigated at the same time to avoid contaminating the operator, and tents should be moved ('floated') upwind so that on removal of a tent the gas is carried away from the operators. The exposure period is normally 45 min and the dosage is approximately 10 g of sodium cyanide per 1 m^3, but local recommendations should be followed.

Fumigation of stored produce

Unless a special fumigation chamber is available, the technique is basically the same for produce inside or outside buildings. Fumigation is useless if stored produce is kept in a building containing debris or has cracks and crevices in the walls and roof, harbouring pests which can reinfest produce immediately after treatment. Thorough cleaning of stores is essential prior to treatment. A residual insecticide spray may be applied to the walls, although insects flying from elsewhere can infest produce without contacting the walls. A routine post-fumigation treatment with an aerosol spray may be needed to reduce the risk of reinfestation.

Before fumigation an inspector should ensure that there is no risk of gas escaping to nearby offices, factories or living quarters, or that such areas are evacuated during treatment. Ready access to the produce is essential to permit the positioning of the cover sheets and seals. At the end of the treatment, the site must be well ventilated to clear any gas. If the floor is not gas-proof, the stack must be rebuilt on top of a gas-tight sheet. The dimensions of the stack of produce need to be checked so that the appropriate amount of fumigant needed can be calculated; for example 500 g methyl bromide per 10 tonnes of produce. An increased dose may be required for larger stacks, thus D.W. Hall (1970) indicated a dose of 5 kg/150 m^3 for larger stacks. Methyl bromide is commonly used as it can penetrate large stacks and diffuses away rapidly after the treatment is over. Large cylinders of fumigant need to be check-weighed to ensure that sufficient is available. The stack of produce in bags, boxes or cartons can be covered with sheets made of polyethylene (0.1 mm thick) which are light (100 g/m^2) and easily handled, but are liable to tear easily. Other types of sheet are made from nylon or cotton fabric and coated with neoprene, polyvinyl chloride (PVC) or butyl rubber. Pipes are laid from the gas cylinder to trays situated at the top of the stack where the fumigant is discharged. The top of the stack may need to be rearranged to accommodate the trays, and extra piping may be needed if the stack is inside a building so that the gas cylinders and scales can be kept outside. If available, fans can be placed under the sheets and operated for the first 15 min at the beginning of treatment to assist distribution of the fumigant. The fans should not be operated for a longer period as their use may tend to force fumigant out at the base of the stack, but can be used again during the subsequent ventilation period.

The sheets must be carried to the site and lifted to the top of the stack in a pre-rolled state to avoid puncturing them. Any holes must be repaired with adhesive or masking tape. The sheets are carefully opened to avoid dragging them over the surface. The edges of individual sheets are held together, overlapped by 1 m, and are rolled together to form an air-tight seal covered by bags containing dry sand, to give flexibility and to prevent the joins unrolling; G clamps are also used to secure the joints and sandbags are used to keep the edges on the ground. Plastic tubes filled with sand or water can be used as 'snakes' instead of sandbags.

Before the fumigant is released in the stack, a halide detector lamp is prepared and checked for satisfactory working. These lamps, which have a flame which turns to green and then intense blue as the concentration of methyl bromide in the air increases, are used to check any gas leakage during the fumigation. A final check is made to ensure that everyone has left the danger area, which should be cordoned off and patrolled by watchmen. Warning signs must also be placed at appropriate sites, such as entrances to the area. The operators then put on their gas masks with a correct filter-type canister and check that they are fitting correctly by placing a hand firmly over the air intake or pinching any hose connecting the canister with the face-piece. If the face-piece is fitting tightly the wearer will not be able to breathe, and the face-piece will be drawn into the face. Supervisors should also check everyone's gas-mask as well as their own to ensure that it has the correct type of canister, that the canister is not out of date or exhausted, and that a first-aid kit and torches are readily available.

While the cylinder is opened to allow fumigant into the stack, frequent checks are made with the halide detector lamp around the cylinder connections, pipes and joins in the sheet. Extra sandbags may be needed to seal the sheets at ground level. The valve on the cylinder is firmly closed as soon as the required quantity of fumigant has been released. The cylinders can then be disconnected and removed, checking that no fumigant which may be in the pipes splashes on to the operators. When the stack is inside a building, all doors are closed and locked for the entire fumigation period, usually 24 or 48 h. Then the person supervising the fumigation and the assistants replace their gas masks and inspect the premises for gas with the halide detector lamp. Gas present in the building before the sheets are removed will indicate that leakage has occurred. The sheets are then removed methodically and as quickly as possible, so that staff remain in the area with gas for as short a time as possible. Removal of sheets at the corner of the stack to allow partial aeration of the stack before returning to remove the remainder may be necessary when many sheets are involved on large stacks. Doors and windows are then left open for at least 24 h for ventilation to disperse the gas. The area is checked again with the halide detector lamp until declared safe for people without gas-masks to enter. All the warning signs can then be removed.

Bond (1984), and D.W. Hall (1970) give more details of a range of fumigants and techniques, using special fumigation chambers. The fumigant, phosphine, applied as tablets of aluminium phosphide (see Chapter 3), can be incorporated into stacks at the rate of one tablet per two bags as each layer is built. The whole stack must be covered by a gas-proof sheet, as described above, within 2 h. Tablets can also be added to a conveyor-belt moving grain into a silo, or through special probes inserted into bulk grain. Alternatively, tablets sealed in a paper envelope can be placed in individual gas-proof bags and removed by the user so that no residue of aluminium hydroxide is left in the grain. Small quantities of grain can also be fumigated inside an empty oil drum or similar container, the top of which is sealed with a polythene sheet fixed with masking tape.

15

Maintenance of equipment

Pesticides are very expensive and so must be applied efficiently. The proper maintenance of equipment is therefore essential, yet users seldom pay much attention to routine cleaning and checking of equipment. Maintenance is often considered only when a part fails and leaks occur or an engine fails to start. Instead of curative maintenance, there should be regular preventive maintenance of equipment so that all components subject to wear are replaced before they fail.

There are a number of reasons for the lack of maintenance. Most manufacturers are not sufficiently aware of the conditions under which their equipment is used in the field, so service manuals, even if available, do not always give enough detail for the user; second, operation and maintenance manuals are seldom translated into the vernacular language of the country to which the equipment is exported. Instructions given as pictograms are being used now and Groupement International des Associations Nationales de Fabricants de Produits Agrochimiques (GIFAP) have published a recommended list of internationally accepted illustrations. Manufacturers of motorised equipment often distribute a separate manual referring to the engine, instead of integrating the relevant information into a comprehensive manual describing how to use and repair the sprayer or applicator. Users of pesticides may have some training in either the biological or chemical aspects of pest control, but training in the correct and safe use calibration and storage of equipment may be totally ignored. In many areas of the world, few farmers have sufficient mechanical knowledge to maintain application equipment, so practical field training courses are essential for both individuals and supervised spray teams. Field training must be supported by the availability of suitable instruction manuals which should be well illustrated and written in simple and clear terms to facilitate translation into the vernacular. Wherever possible, government and international organizations such as FAO and WHO should prepare instruction manuals or assist manufacturers to produce relevant literature. Manufacturers or their local agents should ensure that spare parts are readily available.

The main problems that occur in the field are either those concerned with the spray delivery system from the spray tank to the nozzle or caused by the power unit. Similar problems occur, regardless of the size of the equipment and whether ground or aerial equipment is used.

Problems with the spray system

Nozzle or restrictor blockages

Water from streams and boreholes on farms inevitably contains some sediment which will block nozzles unless a filter is used at the inlet of the spray tank and in each nozzle. When a closed system is used to mix a pesticide a large filter should be positioned between the mixer and sprayer tanks. The mesh size and area of the filter must be selected to cope with the volume of liquid being used, and in relation to the nozzles used. The filter mesh at the nozzle should be smaller than the orifice diameter; for most agricultural work a 50-mesh filter is adequate. When spraying has been completed there may be several litres of spray remaining in the machine, especially in the pump and pipes (Taylor et al 1988). However, no spray mix should be left standing in the tank, hosing or nozzles, as settling out of the pesticide, especially when particulate formulations are used, is liable to cause blockages. Equally, no water should be left in the system if there is a danger of frost.

The sprayer must be washed thoroughly at the end of each day's spraying. When water-based sprays have been used, several washes using a small volume (up to 10 per cent of the spray tank capacity) are better than merely filling the spray tank once with clean water. Cleaning can be improved by using a 0.2 per cent suspension of activated charcoal, but this is expensive. Some manufacturers now market products specifically for cleaning sprayers. Household ammonia, provided there are no brass components, diluted at 10 ml per 5 litres of water, is also useful as a cleaning agent. On motorized equipment the volume of water must be sufficient to operate the agitation system. The final rinse should be with plain water.

Each nozzle should be dismantled, the individual components, filter, tip and cap cleaned and replaced. All other filters on the sprayer should also be removed, cleaned and replaced. In general, it is never possible completely to clean a sprayer, as some of the chemical becomes impregnated in hoses. If possible, separate equipment should be used for applying herbicides such as 2,4-D, which could affect other crops when different pesticides are subsequently being applied. Alternatively, equipment must be decontaminated with a charcoal suspension and the hoses replaced. The suitability of the sprayer can be checked by treating a few plants such as tomatoes, which are susceptible to 2,4-D.

Care must be taken to avoid the washings contaminating any drinking or other water supply. Some countries have issued guidelines or a code of practice concerning the cleaning of equipment, and this should be consulted. Protective clothing should not be removed until after the equipment has been cleaned and returned to the store.

If special formulations of pesticide have been used, a particular solvent may be needed to clean the equipment. Information on the suitability of solvents for cleaning should be obtained from the supplier of the pesticide or equipment to check that there is no detrimental effect on plastics and other materials used in the construction of the machinery.

If a nozzle blockage does occur while spraying, the nozzle tip and filter should be removed and replaced by clean parts. The blocked nozzle is more readily cleaned back in the workshop, and sufficient spares should be taken to the field. When spare nozzles are not available, sufficient water or solvent should be taken to the site of operations for cleaning a blockage. If washing fails to remove the obstruction, giving the nozzle a sharp tap with the inner surface downwards may be sufficient to dislodge it. Alternatively, air pressure from a car or bicycle pump can be used to blow it from the nozzle orifice. Nozzles should **never** be placed in the user's mouth to blow through the orifice as its surface is inevitably contaminated with pesticide. A hard object such as a pin, nail or stiff wire should **never** be used as the orifice can be so easily damaged. When ceramic nozzle tips are used, extra care is needed as the slightest damage to the nozzle orifice can affect the distribution of the spray liquid. If several blockages occur, the whole system should be checked to determine the source of material causing the blockage. Corrosion, especially inside the spray tank or booms, may result in small particles which can accumulate on the filter or nozzle. With some wettable powder or other particulate formulations, deterioration in storage may result in poor suspensibility so particles settle out in the boom or nozzle, even if there is agitation in the spray tank.

The flow-control valve or restrictor may become blocked on sprayers which do not have hydraulic nozzles. As mentioned above, the occurrence of blockages can be reduced by proper filtration, but if a blockage does occur it is usually quicker to replace the restrictor rather than attempt to clean it in the field.

Inefficient pumps

Piston pumps are fitted with 'O' ring seals or cup washers of leather or synthetic material. As this seal can be damaged by particles suspended in the spray liquid, it should be checked regularly to keep the pump operating well. Leather seals dry out and shrink if a sprayer is not used, so need to be kept supple with a vegetable oil. Some synthetic materials are affected by the solvents used in pesticide formulations; if swelling occurs the pump could become stiff and difficult to operate. Poor pump performance may be due to faulty valves. Ball valves and their seating can be pitted or coated with a sediment of debris or pesticide, and synthetic materials used in diaphragm or flap valves may swell and fail to seal and open properly. Apart from cleaning and replacing damaged parts, it may be necessary to change the formulation used or to improve the filtration of the water before use.

Leaks

'O' rings, washers and other types of seal are liable to wear or be damaged when hose connections, trigger valves and other components are unscrewed. Similarly, seals around the tank lid and in the pump assembly can be damaged whenever the connection is broken. The damaged part should always be replaced to avoid the occurrence of leaks. Some connections such as nozzle caps may not have a washer and rely on direct contact of smooth surfaces to seal. Any dirt on the nozzle or cap or damage to the threads may prevent a proper seal.

Proper functioning of some spray equipment, such as compression and certain motorized knapsack mistblowers, depends on an airtight seal of the container or spray tank (Fig. 15.1). For example, it is impossible to spray upwards with some mistblowers when the nozzle is above the level of the spray tank as there is insufficient pressure to force liquid to the nozzle. Small

Fig. 15.1 Checking seal on lid of spray tank of motorized knapsack mistblower

air leaks from the lid or other fittings to the tank – for example a pressure gauge on the container – can be detected by smearing a soap solution over the joint. Soap bubbles should be readily detected where air is escaping.

Problems with motorised equipment

Two-stroke engines

Users of motorized knapsack mistblowers and other equipment with a two-stroke engine frequently complain that the engine is difficult to start. Various causes for the failure to start and other problems are listed in Table 15.1, together with remedies. Many of the starting problems could be avoided if the carburettor and engine were drained of fuel after use to avoid gumming up the machine with oil when the petrol has evaporated. This can be done simply by turning off the fuel tap and allowing the engine to continue running until starved of fuel. Preferably, the fuel tank itself should also be drained to avoid the ratio of oil to petrol increasing, especially in hot climates. Starting

Table 15.1 Faults with two-stroke engines and their remedies (from Clayphon and Matthews 1973)

Fault	Remedies
Engine does not start	
Fault in fuel system	
Fuel cock not opened or blocked	Ensure fuel is present in tank. Open cock. If no flow, remove cock, clean and replace
Air vent in fuel tank filter is blocked	Clean vent
Thimble filter in carburettor is blocked	Remove filter, clean and replace
Main jet in carburettor is blocked	Remove, clean and replace
Water in carburettor float bowl	Remove and clean. Check also that fuel in tank is not contaminated with water
Float needle sticking and stopping petrol supply	Remove needle, check for burrs or rough surface. Clean off rough surface, if not possible, replace with a new one
Too much fuel in engine	Close fuel cock, remove spark plug, open throttle, pull recoil starter rope to turn engine over a few times, clean, replace
Fault in ignition system	
High-tension lead to spark plug loose or disconnected or insulation broken or burned	Fasten lead securely to plug, if badly damaged, replace

Table 15.1 *(Cont.)*

Fault	Remedies
Dirty spark plug, carbon or oil deposits on electrodes	Remove plug and clean; set gap as recommended by manufacturer. If porcelain insulation is damaged, replace with new plug
Contact breaker points dirty or pitted	Clean and adjust to correct clearance when points are open. If honing fails to remove pitting, replace with a new set
Exhaust blocked	Remove exhaust and clean or replace with a new part

Engine runs erratically or stops

Dirt or floating debris in fuel system	Clean all fuel lines, filters and carburettor bowl and check there is no air in fuel line
Main jet blocked	Remove, clean and replace. Do not use nail, pin or wire to clear obstruction
High-tension ignition lead loose or 'shorting' on metal parts of the engine	Check that lead is firmly affixed to spark plug. Where lead has been chafing on bare metal, either cover bare wire with insulation tape or replace with a new lead
Fuel running low in tank. Engine vibration or irregular movement of operator leaves outlet pipe uncovered, resulting in fuel starvation	Refill tank with correctly mixed fuel

Engine lacks power

Choke is closed	Open choke
Fuel starvation	Partially blocked pipes or filter should be removed and cleared
Air cleaner blocked with debris	Remove, clean by washing in petrol and squirt a little light oil on the cleaner element. Conform with manufacturer's recommendations
Dirty carburettor	Remove from engine, dismantle carefully, clean and examine all parts. Any worn parts such as float needle valve, etc. must be replaced with new parts
Loose or leaking joint at carburettor flange to cylinder	Check gasket. Replace if worn or damaged and tighten nuts or studs
If whistling noise is heard from cylinder when engine is running, there is a possibility of the cylinder head gasket being worn or damaged	Check carefully by feel when engine is running. If gases are escaping, remove head, fit new gasket, tighten nuts evenly. On a new machine, it may be necessary to tighten the nuts evenly without fitting a new gasket. If heavy carbon deposits are seen on piston crown or cylinder head is removed , these should be scraped away carefully. The ring of hard carbon should not be disturbed in the cylinder
Dirty exhaust	Remove exhaust, clean carbon deposits from exhaust, if possible or replace with new part

Table 15.1   *(Cont.)*

Fault	Remedies
Engine backfires	
Ignition may be badly retarded	Should be attempted only by trained or qualified personnel. Magneto should be checked and reset to manufacturer's specification
Carbon whisker bridging gap in spark plug	Remove plug, clean, adjust gap to correct clearance and replace
Overheating of engine	
Incorrect mixture of petrol and oil in fuel tank	Drain off tank. Refill with fuel in the correct ratio (see handbook or markings on tank)
Incorrect size of main jet	Remove and refit one that complies with manufacturer's specification
Ignition retarded too far	To be checked and reset by a competent person
Exhaust and silencer chocked with carbon	Remove, dismantle, clean and reassemble

problems are definitely reduced by ensuring the correct type of oil is used (see p. 231) and that the fuel is properly mixed.

The fuel line from the tank to the carburettor is often made of plastic, which is hardened by the action of the petrol and is sometimes loosened by engine vibration. This plastic tube should be regularly inspected and replaced

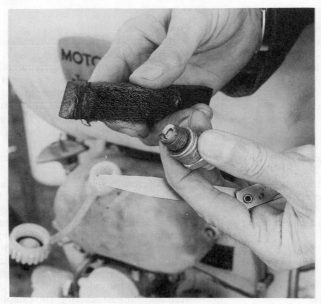

Fig. 15.2 Cleaning spark plug and checking gap

if necessary to avoid fuel leaking on to a hot engine and causing a fire. The sprayers' straps should be designed to allow the machine to be removed very quickly in case a fire starts, and manufacturers should also consider resiting the fuel tank so that it is not directly above the engine.

The spark plug should be inspected regularly if necessary (Fig. 15.2), so it should be readily accessible. The spark-plug gap may need adjusting to obtain a good spark before the plug is replaced. The plug should be replaced after 250 h as a routine. The air filter should also be examined at the end of each day's spraying or cleaned as a routine according to the manufacturer's recommendations.

Fault-finding

Some of the faults with a range of small hydraulic sprayers and possible remedies are given in Tables 15.2, 15.3 and 15.4. Similarly, faults with the small, hand-carried, spinning disc sprayers are given in Table 15.5.

Table 15.2 Faults with slide pumps (single- or double-acting continuous operation)

Fault	Remedies
No spray at nozzle	Check nozzle and clean if necessary
	Check that container is full
	Check pump, especially non-return valves
	Check for leaks on hose and connections
No suction	Check pump, especially pump seal
	Check valves and seatings
	Check strainer in container
Leaks from pump	Check gland and packing, replace if worn or damaged

Table 15.3 Faults with knapsack, lever-operated (piston or diaphragm pumps)

Fault	Remedies
No spray	If resistance is felt on downward movement of lever with cut-off valve open, check nozzle for blockage, and clean if necessary. Check and clean filter or strainer in handle of cut-off valve. If no resistance is felt, check tank contents and fill if necesary. Ensure that operating lever is tight, together with all the connections to the pump. Check that when the lever is operated, the shaft or connecting mechanism and the piston or diaphragm all move together. Pump valves and valve seat should be checked. If worn or damaged these should be replaced. Dirt and debris should be removed

Table 15.3 (*Cont.*)

Fault	Remedies
No suction	Ensure that liquid is present in the container. Check that the suction and discharge valves are not sticking. Make certain that the liquid ports that permit flow from tank to pump are not blocked. If a piston-type pump is employed, check that the piston seal is not excessively worn or damaged, as this will permit the liquid to pass between the piston and cylinder wall.
No pressure	Check liquid contents of container. Fill if necessary. After several strokes of the operating lever, look in the tank to see if air bubbles are rising to the surface. If so, this could indicate a leak in the pressure chamber. Where pressure chamber is screwed into the pump body, check that the seal is not damaged. Replace if necessary. Check both suction and discharge valves. Remove any accumulated dirt or debris from discs or balls and valve seats. If discs are worn or damaged or the rubber is perished, replace. If ball valves and seats are pitted or balls are no longer spherical, replace with new ones. If resistance is felt when pumping and no reading is seen on pressure gauge, replace gauge. If pump is of diaphragm type, check that it is seating correctly, that it is not damaged or split and that the rubber is not porous. Where a pressure-relief valve is embodied in the pressure chamber, check that it is adjusted correctly and that the spring-loaded valve is seating properly. Ensure that the openings between the pump inlet and outlet ports and the liquid container are not blocked. Check that the air vent in the filler cap is not blocked, as this could be the means of a vacuum forming in the container
Pressure drops quickly	Check pressure chamber for leaks. Air bubbles seen rising to the liquid surface are a good indication. Check valves for discharge. The discharge rate may be higher than pump capacity
Liquid leaks on to operator	Where pump is mounted in base of sprayer, a ruptured diaphragm, or one incorrectly assembled, will permit liquid under pressure to leak. For a piston type pump, a worn piston seal or deep scratches in the cylinder wall will also permit the liquid to escape and wet the operator. Check the container for cracks or leaking joints. Metal tanks can be soldered or brazed. Check that the lid of the container is fitting tightly

Table 15.4 Faults with compression sprayers

Fault	Remedies
No spray	Ensure container has liquid. If pressure gauge shows a reading and there is no spray when cut-off valve is opened, close valve and check nozzle. If nozzle is blocked, follow procedure for clearing blocked nozzles. Check strainer in cut-off valve. Clean and replace. Check hose connections and tighten. If no reading is shown on the pressure gauge, ensure that the gasket between the pump body and the liquid container is not leaking. Replace if leaks are present. Remove pump from container and check by giving a few smart strokes on the pump handle to test the valve. On each pressure stroke, the valve should 'grunt' or make a noise of escaping air. If the valve disc or ball is malfunctioning it should be replaced. Where a dip-tube is part of assembly, check that this is not blocked with debris.
Leaks from pump	After the container had been filled with spray liquid to the required level, if on the first or second downward strokes of the pump handle liquid is forced up past the shaft and out through the guide, this is a good indication that the valve requires attention. Furthermore, if strong resistance is felt on the downward stroke, again the valve is faulty and has permitted liquid to enter the pump barrel and, as liquid cannot be compressed, resistance is encountered.
Pressure drops quickly	Check that the filter cap or lid gaskets are serviceable and that the cap is properly secured. Check also where a safety valve is fitted that it is not leaking and is in a working condition. Some compression sprayers have a constant-pressure valve fitted. Check that this is adjusted correctly and that there are no leaks from the point of entry to the tank. Ensure that all connections to the tank are tight and that all gaskets and washers are serviceable. Check tank for leaking seams by pressurising and immersing completely in water. Air bubbles rising to the surface will indicate the presence of a leak. Leaking tanks cannot be repaired in the field. All repaired compression sprayers must be pressure-tested to at least twice the working pressure before being used on spraying operations.
Other faults	If nozzle dribbles with cut-off valve closed, the 'O' ring seal or the valve seat is damaged. Dismantle and check. Replace with new parts if unserviceable. With some of the plastic-type pressure gauges, the indicator pointer sometimes becomes loose on its pivot. This can

Table 15.4 (*Cont.*)

Fault	Remedies
	give a false pressure reading. By tapping the gauge against the hand it can be seen whether or not it is loose. If it is, remove the protective glass front, replace the needle on the pivot loosely and, with it pointing to zero, press it firmly on to its mounting. Replace the glass with a master gauge.

Table 15.5 Faults with spinning disc sprayers

Fault	Remedies
No spray	Restrictor may be blocked. Clean with solvent or piece of very fine wire or grass stem. Check whether air vent is blocked.
Leaks	Check that spray container is fitted correctly.
Spinning disc not rotating or rotating intermittenly or slowly	Check that enough batteries are fitted in containers. Check that all batteries are inserted the correct way. Check battery connections. Check switch (if any). Check connections to motor, clean connections with a dry cloth or sandpaper and fit new wires if necessary. Check that the '+' terminal of the batteries is connected to the '+' terminal marked on the motor. Replace batteries if necessary. Where large numbers of the sprayer are in use, it is advisable to provide a voltmeter and tachometer to check the revolutions per minute of the disc. Check whether disc is fitted correctly to motor shaft; it may be pushed on too far and touch the backing plate. If necessary, replace motor.

Maintenance in the field

One or two tools should always be taken into the field while spraying, as well as extra nozzles, washers and other spare parts. The non-mechanically minded user will find one pair of pliers, at least one screwdriver, or preferably two of differing sizes, one small adjustable wrench, a knife and a length of string invaluable. Spare washers for the trigger valve, nozzle body or even the filter caps should be available, but if not, a length of oiled or greased string can be used as a substitute (Fig. 15.3). Some washers may be cut from the inner tube of a car or cycle tyre and used temporarily until the proper spare washer can be fitted.

Quick repairs to leaking plastic containers which are not pressurized can be made by drawing the edges of a small hole with a black hot nail, and smoothing it over with a wetted finger. A 15 cm nail is suitable and can be

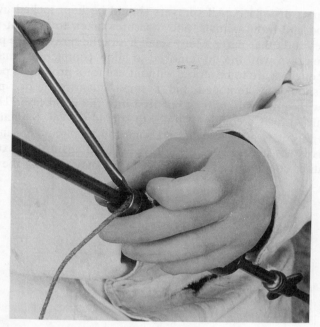

Fig. 15.3 Greased string used to make a washer

heated in a fire, even out in the field, but it must not be made too hot, otherwise the plastic may melt and the hole enlarged beyond repair.

Those using engine-driven equipment, such as a knapsack mistblower, will also need to carry a spare spark plug and a plug spanner, while those using small, battery-operated, spinning-disc sprayers also need a 'Philips' screwdriver, as well as a suitable tachometer (see Fig. 10.9, p. 232). Tools and spares can be conveniently carried in a small tool box. If the spray programme entails the use of several machines simultaneously, one or two complete machines could be taken to the field as spares so that work may continue when weather conditions are favourable, rather than delay spraying while repairs are attempted.

All stoppages and breakdowns that occur in the field should be reported to workshop personnel so that repairs and maintenance can be done without delay. Where several machines are used by a gang of operators, it is a good policy to allocate one machine to an individual who then becomes responsible for its care and maintenance.

Storage of equipment

After each day's field work, and at the end of the season, complete checks should be made of the pump and, where necessary, the engine, before storing the sprayer in a dry place. All sprayers should be kept locked away from children, food and farm animals, and measures taken to prevent rats from chewing hoses and other parts. Many small hydraulic sprayers are preferably stored upside-down with the lid removed to allow complete drainage of the container. If engines are to be stored without use for a prolonged period, the spark plug should be removed and a small quantity of oil, preferably formulated with anti-rust additives, poured into the crankcase. The engine should be turned over a couple of times to spread the oil. Similarly, at the end of each day it is advisable to add some oil to pumps on any type of sprayer. This is not necessary if the sprayer is used again the next day, but adverse weather conditions or some other factor may prolong the period of storage.

16

Safety precautions

Safety precautions depend on the hazards involved in transportation, storage and use of particular pesticides, the toxicity of which varies according to their chemical structure, purity and formulation. The risk of poisoning by the more toxic chemicals can be reduced by suitable formulation and packaging. Increasing emphasis is being given to design of application equipment to reduce the risks of operator contamination. There will also be increasing demands for the pesticide liquid to be sprayed out just before completing an application so that clean water carried in a separate tank can be used to wash the inside of the tank and fittings and then be sprayed out in the last part of the treated field. This will require greater accuracy in calculating the dosages to avoid mixing too much spray. One aspect of equipment design will be to reduce the volume of liquid that remains in the system at the end of a spray operation. In particular operator contamination can be minimized by having a closed filling system, but with existing sprayers much can be done by improved operating procedures including the use of appropriate ancillary equipment.

Pesticides, like medicines and other chemicals, must be stored and used according to instructions, so the first requirement for all users of pesticides is to '**Read the label**'.

A pesticide may be taken into the body by mouth (oral), through the skin (dermal) or through the lungs (inhalation). The uptake orally is minimal during pesticide application unless operators unwisely eat, drink or smoke before washing their hands or face. Oral poisoning has occurred when pesticides have been improperly stored in food containers, especially soft drink or beer bottles, where recently sprayed fruit has been eaten, or a person has committed suicide.

Contamination of the body is principally by absorption through the skin, which is particularly vulnerable where there is any cut or graze. The back of the hands and wrists absorb more than do the palms. Similarly, the back of the neck, feet, armpits and groin are areas which need protection, and great care must be taken to avoid contamination of the eyes. The risk of skin absorption is increased in hot weather, when sweating occurs with the minimal amount of effort and conditions are not conducive to wearing protective clothing.

A pesticide can enter the lungs by inhaling droplets or particles, principally those less than 10 μm diameter, or vapour, but the amount is usually less than 1 per cent of that absorbed through the skin. The greatest risk occurs when mixing concentrated formulations and applying dusts, fogs or smoke, especially in poorly ventilated areas. The chances that these small droplets or particles will be inhaled is reduced under field conditions as the wind usually blows them away.

The relative hazard of these routes of exposure needs to be evaluated with different operational procedures and protective clothing (Durham and Wolfe 1962).

Irrespective of how a pesticide enters the body, acute poisoning may occur after one dose or exposure, while chronic poisoning is caused by repeated small doses absorbed over a longer period of time. The latter is especially important when spray operators apply pesticides frequently, but other people, such as those scouting for pests, weeding or harvesting crops, may be at risk in treated areas. The interval before a treated crop can be entered is prescribed for the more toxic chemicals such as monocrotophos and mephosfolan.

The toxicity of a pesticide is usually quoted in milligrams of active ingredient for each kilogram of body weight (ie parts per million) of the test organism. This is measured as the dose required to kill 50 per cent of a sample of test animals in a specified time, often 24 hs, and referred to as the LD_{50} dose. This dose can be measured more accurately than the dose required to kill a higher or lower proportion of a sample of test animals. However, concern about the number of organisms needed to determine an LD_{50} has led to the development of alternative methods of assessing the risk of toxic effects. Acute toxicity is much easier to assess, but subacute toxicity is measured initially over 90-day periods and chronic toxicity subsequently over 1 or more years. Inhalation toxicity is determined as the LC_{50} (lethal concentration) measured in milligrams per litre of air.

Much of the high cost of developing a new pesticide is due to the need for extensive toxicity testing of the chemical and breakdown products and determining residue levels, before a new product can be marketed (Table 16.1). In particular extensive environmental impact studies are required (Graham-Bryce 1989).

Classification of pesticides

The World Health Organization has differentiated between acute and dermal

Table 16.1   Usual toxicological studies required before a pesticide can be marketed

1 acute oral toxicity
2 dermal toxicity
3 eye irritation
4 inhalation
5 subacute studies – 90-day and 2-year feeding tests
6 demyeliation
7 Carcinogenicity (tumour-susceptible strain)
8 teratogenicity (pregnant rats)
9 three-generation studies (mice)
10 estimation of acceptable daily intake
11 wildlife and fish studies
12 studies on metabolism in plants and mammals
13 Residue studies
14 Potentiation

toxicity data for solid and liquid formulations in its classification (Table 16.2) (Anon 1975). Thus, good-quality granular materials are less hazardous to apply than sprays of the same chemical. Unfortunately, this classification is not yet used by all countries. The classification is merely a guide, and the hazard rating for a given chemical is revised following practical experience of its use. Examples of the classification of certain pesticides are given in Table 16.3. When selecting a pesticide for a given pest problem, preference should be given to the least hazardous pesticide which is effective, and if possible the least persistent.

Users of pesticides should familiarize themselves with the appropriate legislation in their country. In the United Kingdom, the use of pesticides is included in the Food and Environmental Protection Act (FEPA) and Control of Substances Hazardous to Health (COSHH) Regulations. A Code of Practice for the Safe Use of Pesticides on Farms and Holdings sets out the responsibilities and requirements of those using pesticides to satisfy the legislation. The type of protective clothing which must be worn by law

Table 16.2   WHO classification

Class	Hazard level	Oral toxicity*		Dermal toxicity*	
		Solids†	Liquids†	Solids†	Liquids†
Ia	Extremely hazardous	<5	<20	<10	<40
Ib	Highly hazardous	5–50	20–200	10–100	40–400
II	Moderately hazardous	50–500	200–2 000	100–1 000	400–4 000
III	Slightly hazardous	>500	>2 000	>1 000	>4 000

Notes: * Based on LD_{50} for the rat (mg/kg body weight).
     † The terms 'solids' and 'liquids' refer to the physical state of the product or formulation being classified.

Table 16.3 Examined of the oral and dermal toxicity of selected technical products (Data based on WHO document VBC/88.953 and *The Pesticide Manual* 9th edition)

Common name	Trade name	Type	Toxicity (mg/kg) Oral	Dermal	Classification
aldicarb	Temik	C	0.9	2.5	Class Ia
phorate	Thimet	OP	2	70–300	pesticides
disulfoton	Disyston	OP	2.6	50	extremely
chlorfenvinphos	Birlane	OP	10	30–108	hazardous
parathion-methyl		OP	14	67	
phosphamidon	Dimecron	OP	7	125	
fonofos	Difonate	OP	8	150	
mephosfolan	Cytrolane	OP	9	10	
thionazin	Nemafos	OP	11	8–15	
azinphos methyl	Guthion	OP	16	280	Class IB
carbophenothion	Trithion	OP	32	800	pesticides
carbofuran	Furadan	C	8	120	highly
dicrotophos	Bidrin	OP	22	42	hazardous
monocrotophos	Nuvacron	OP	14	112	
dioxathion	Delnav	OP	23	350	
methidathion	Supracid	OP	25	25–400	
dichlorvos	DDVP	OP	56	75–900	
methomyl	Lannate	C	17	>1 600	
demeton S-methyl	Metasystox	OP		85	
dieldrin		OC	37	>100	
aldrin		OC	98	>200	
endosulfan	Thiodan	OC	80	74–680	
vamidothion	Kilval	OP	103	1 160	
thiometon	Ekatin	OP	120	>200	
binapacryl	Morocide	F/A	421	720–810	Class II
thiodicarb	Larvin	C	66	>2 000	moderately
propuxor	Baygon	C	95	>200	hazardous
paraquat	Gramoxone	H	150	80–480	
fenitrothion	Sumithion	OP	503	700	
pirimicarb	Aphox	C	147	>600	
gamma HCH	Lindane	OC	100	500–1 000	
dimethoate	Rogor	OP	150	700–1 150	
carbosulfan	Marshal	C	250	>2 000	
cypermethrin	Cymbush	P	250		
lambda-cyhalothrin	Karate	P	144		
deltamethrin	Decis	P	135		
permethrin	Ambush	P	500		
fenvalerate	Sumicidin	P	450		
DDT		OC	113	2 500	
diazinon		OP	300	500–1 200	
diallate	Avadex	H	395	2 000	
2,4-D		H	375	1 500	
carbaryl	Sevin	C	300	>500	
malathion		OP	1 400	>4 000	Class III
propanil	Stam F-34	H	1 400	7 080	slightly

Table 16.3 (*Cont.*)

Common name	Trade name	Type	Toxicity (mg/kg) Oral	Dermal	Classification
trichlorphon	Dipterex	OP	560	>2 800	hazardous
triadimefon	Bayleton	F	363	>1 000	
temephos	Abate	OP	8 600	>4 000	unclassified
carbendazim	Bavistin	F	>1 5000		
diafenthiuron	Polo	IGR	2 068		
atrazine	Gesaprim	H	2 000	7 500	
buprofezin	Applaud	IGR	>2 000	—	
flufenoxuron	Cascade	IGR	>3 000	—	
teflubenzuron	Nomolt	IGR	>5 000	—	
etofenprox	Trebon	I	>42 000		

OP = organophosphate insecticide   H = herbicide
OC = organochlorine insecticide   F = fungicide
C   = carbamate insecticide   A = acaricide
IGR = insect growth regulator   I = insecticide

Notes: 1 Toxicity values refer to active ingredient, but classification is dependent on toxicity of formulation. If data on the toxicity of a formulation are not available then an approximate value can be calculated from

$$\frac{LD_{50} \text{ of active ingredient x 100}}{\% \text{ active ingredient in formulation}}$$

thus, for 5% carbofuran granules,

$$\frac{8 \times 100}{5} = 160$$

2 The majority of herbicides and fungicides are safer to apply than insecticides, and only a few examples are included in the table.

depends on the chemical, its formulation and/or method of application (Table 16.4). An official leaflet explains the regulation requirements, and each year a summary of the protective clothing requirements is published in the list of approved products for farmers and growers (*The UK Pesticide Guide*). As manufacturers may not seek approval for their product for all possible uses, for example on minor crops, other organizations or individuals can seek approval for off-label use. The Guide lists these 'off label' approvals, maximum residue levels (MRLs) permitted in food crops, and occupational exposure standards (OES). The Poisons Act and subsequent rules provide for the labelling, storage and sale of scheduled poisons. Aldicarb, chloropicrin, methyl bromide, demeton-S-methyl, paraquat and several other toxic pesticides, tar oil and lead arsenate are not scheduled chemicals, but are included in the Poisons List.

The container label must have an indication of a hazard 'VERY TOXIC', 'TOXIC', 'HARMFUL' or 'CAUTION' in relation to Class Ia, Ib, II and III category pesticides, respectively, shown in Table 16.1, and the statement

Table 16.4 Summary of the protective clothing which must be worn when applying scheduled substances

Jobs for which protective clothing must be worn	Protective clothing needed
Spraying pesticides	
1 Opening container Diluting and mixing Transferring from one container to other Washing containers Washing out equipment, including aerial equipment	Overall and rubber apron* or mackintosh* Rubber boots*, rubber gloves Face shield (or respirator†)
2 Spraying ground or glasshouse crops Acting as a ground marker with aeriel spraying	As (1) above, except overalls should have a hood, and omit rubber apron and mackintosh
3 Spraying bushes and climbing plants	As (1) above, but wear a rubber coat and sou'wester and omit rubber apron
Granule application	
4 Opening container	As (1) above, but wear rubber gauntlet gloves with sleeves over their cuffs
5 Application of granules by hand or hand-operated apparatus	As (1) above, but wear sleeves of overall over cuffs or rubber gauntlet gloves and omit apron
6 Application of granules by tractor	Overall or mackintosh, but if a Part I substance, see (5) above
7 Acting as a ground marker with aerial application	As (6) above, but add hood and face shield
Other applications	
8 Sprays applied to soil Soil injection	Overall, rubber gloves rubber apron**, rubber boots and respirator
9 Bulb dipping	Overall, rubber gauntlet gloves, rubber boots and rubber apron
10 Application of nicotine to roosts, perches and other surfaces in a livestock house	Overall, rubber gloves, face shield

Notes: * Not required with part III substances
    † Respirator must be used (a) with all jobs involving Part I substances, except when diluted dimefox is applied to the soil, and (b) with any scheduled substances (Parts I, II and III) applied inside enclosed spaces, eg glasshouses, warehouses, livestock houses, as an aerosol or smoke
    ** In enclosed spaces

'Keep out of reach of children'. Such words need to be translated into the vernacular to be meaningful. Use of distinctive colours for labels is not satisfactory as some users are colour blind. Some manufacturers may use distinctive colours to advertise their products, but the significance of different colours varies in different areas of the world. Some of the information on labels is now provided as pictograms.

In the United Kingdom certain pesticides are subject to the provision of the Poisons Act 1972, the Poisons List Order 1982 and the Poisons Rules 1982. The list is published annually in the UK Pesticide Guide.

Protective clothing

Appropriate protective clothing must be worn whenever a Class I pesticide is applied or when application equipment contaminated with such pesticides is repaired (Fig. 16.1). The minimum protective clothing is a coverall defined as a single garment (or a combination of garments which offers no less protection than a single garment) with fastenings at the neck and wrists which

1 covers the whole body and all clothing other than that which is covered by other protective clothing such as faceshield, goggles, respirator, footwear or gloves
2 has its sleeves over the top of gauntlet gloves, unless elbow-length gloves are needed for dipping plants in pesticide
3 is resistant to penetration by liquid or solid particles in the circumstances in which it is being used, and minimizes thermal stress to the operator when worn.

Test methods have been devised to assist with the selection of coveralls suitable for work while applying pesticides (A.J. Gilbert and Bell 1990). The garments are rated according to the penetration of the material by solvents and by water + surfactant applied under pressure.

Obviously, great care and more protective clothing are needed when mixing concentrated formulations. This also applies to pesticides such as paraquat, which is not included as a scheduled pesticide. Irrespective of the toxicity of the product, particular attention must be given when the period of exposure is liable to be prolonged, or the concentration of chemical exceeds 10 per cent. Risks of contamination of the operator is reduced if the spray operator is upwind of the nozzle, even the power-operated equipment. Often climatic conditions are not conducive to wearing protective clothing, especially in tropical countries, but ideally, a durable cotton fabric overall is the minimum protective clothing which should be worn, even with Classes II and III pesticides, irrespective of the method of application. This should be washed regularly with soap or detergent, preferably at the end of each day's spraying after the application equipment has been cleaned and returned to the store. Unfortunately not all pesticide deposits are removed by washing, so every effort should be made to minimize contamination of clothing. Those who are unable to purchase overalls should use a spare set of clothing which is washed immediately after the day's spraying. The shirt should have long sleeves, and the long trousers should not have turn-ups where granules and dust particles can collect. A wide-brimmed, waterproof hat to protect the back of the neck as well as the face is useful, not only to reduce spray contamination but also the effects of the sun. When minimal protective clothing is used, there must be a good supply of water for washing immediately the skin which is contaminated with pesticides.

When gloves must be worn they should be made of neoprene. These should be checked previously for pin-holes by filling the glove with water, gently squeezing it and then drying it before use. Gloves should be long

(a) **(b)**

Fig. 16.1 Examples of protective clothing (photos courtesy of: Application Hazards Unit, MAFF Central Science Laboratory) (a) Garment type 1. EP Spunbonded Polypropylene boilersuit. (note: 'EP' stands for 'extra protection'.) This is a 'breathable' garment, hence no permeation resistance, hence the need for an impermeable apron (and faceshield) if handling concentrates or if expecting to get heavily contaminated by dilute liquid pesticide in the course of planned application (b) Garment type 2. PE (polyethylene) Coated Spunbonded Polypropylene boilersuit. Not breathable, but relatively low resistance to permeation. This implies the need for an impermeable apron (and faceshield) if handling concentrates. Suit is likely to withstand fairly heavy contamination by dilute liquid pesticide (c) Garment type 3. Saranex laminated 'Tyvek' boilersuit. (Note: 'Tyvek' is the proprietary name for spunbonded polyethylene made by DuPont, who also make Saranex. All garments are made up by other intermediary companies however.) Not breathable. Fairly high permeation resistance, hence no need for impermeable apron when handling most concentrates.

enough to protect the wrist, and the cuffs of coveralls should be outside the top of gloves to reduce seepage of spray down inside the gloves. Pesticide on the outside of gloves should be washed off as quickly as possible, and in any case gloves should be washed with detergent and water before removing them to avoid contaminating the hands. Handling small objects such as nozzle tips is difficult when wearing rubber gloves, but operators should not be tempted to remove the glove; this can be extremely dangerous as some pesticides are easily absorbed through skin which is wet with sweat. People

(c)   **(d)**

The operator is thus pictured wearing only the faceshield, which would be required for concentrate handling. (It would be prudent to wear an apron if handling very hazardous products, to protect the coverall from contamination and extend its protective properties) (d) Garment type 4. PVC Coated Nylon boilersuit. Not breathable. High resistance to permeation (for most liquids), hence no need for impermeable apron when handling most concentrates. The operator is thus pictured wearing only the faceshield, which would be required for concentrate handling. (It would be prudent to wear an apron if handling very hazardous products, to protect the coverall from contamination and extend its protective properties.) Please note that all subjects are wearing protective gloves and rubber boots, with coverall arm cuffs and trouser cuffs worn outside. The wearing or holding of impermeable apron and/or faceshield is intended to denote whether wearing these particular items would be necessary to supplement the coverall alone when handling concentrates

working in workshops where spraying equipment is being repaired should be particularly careful. When dismantling equipment they may touch chemical deposits which have not been removed by normal washing.

Shoes and not sandals should be worn to protect the feet unless rubber boots are specified. The legs of overalls or trousers should be outside the boot to reduce the chance of liquid or granules getting inside it.

Special protective clothing includes eye and face shields, respirators, and impermeable overalls. Two types of respirators are available: the cartridge respirator which covers the nose and mouth and the gas-mask which also covers the eyes and may be incorporated in a complete head shield. Both types have one or two 'cartridges' which absorb toxic fumes and vapour, and are suitable for use when fogging (see Chapter 11). Gas-masks usually have more efficient fittings for more prolonged use. Both types must be worn tightly so that they are sealed around the face to prevent leakage around the edge, and are generally uncomfortable to wear in hot weather. All items need to be regularly cleaned, including the inside of gloves and masks. Any special filters on respirators must be changed according to the manufacturer's instructions. One of the dangers is that some operators wear a respirator while mixing sprays, but then remove it so that the inside is liable to get contaminated. Operators are liable to inhale the poison when they replace the respirator to mix another batch of chemical. Simple disposable masks are sometimes safer to use to reduce inhalation of droplets of the less hazardous chemicals and also minimize deposition of chemical around the mouth. An eyeshield is needed when certain solvents such as isophorone are sprayed. All clothes, including protective clothing and the user's normal clothes, should be kept well away from the storage and mixing area in a separate changing room. If pesticides are used extensively, the changing room should ideally be fitted with a shower. In any case, soap and water should be available for operators to wash after work. Some tractor-mounted equipment has a separate water tank so any contamination of gloves can be washed off immediately (Fig. 16.2).

Symptoms of poisoning

Symptoms of poisoning will vary according to the pesticide involved. Where pesticides are used regularly, advice from the local Health Authorities should be sought. In the United Kingdom, the Department of Health has published a book '*Pesticide Poisoning – Notes for the Guidance of Medical Practitioners*', which gives relevant symptoms of each group. There is also a National Poisons Centre which provides a 24 h information service.

Signs of organophosphate poisoning include headache, fatigue, weakness, dizziness, anxiety, perspiration, nausea and vomiting, diarrhoea and a loss of appetite. An increase in the severity of the symptoms leads to excessive saliva and perspiration, stomach cramps, trembling with poor muscle co-

Fig. 16.2 Using extra clean water tank on sprayer to wash contaminated hands

ordination and twitching. The patient may have blurred vision, a rapid pulse and some difficulty in breathing. Severe poisoning leads to convulsions, eyes with pinpoint pupils, inability to breathe and eventually unconsciousness.

Some of these symptoms can occur with other types of poisoning or other illnesses such as heat exhaustion, food poisoning or a 'hangover'. Although a person who becomes ill after using or being near pesticides may not necessarily have been poisoned, the suspected poisoning is seldom verified by suitable tests. People using pesticides may develop dermatitis, especially on the hands. This may be due to the inside of gloves being contaminated with pesticide or may be a reaction to the solvent rather than the pesticide itself, and may also be a reaction to wearing rubber gloves and sweating.

First aid

Immediate medical attention by a doctor or at a hospital is essential, when a person using pesticides becomes ill. First-aid can be given before the patient reaches a doctor. The patient should be kept quiet and warm, away from the sprayed area and, if possible, in a sheltered place in the shade. All protective and contaminated clothing should be removed. All other clothes should be loosened and, taking care not to contaminate your own skin or clothes, the patient's contaminated skin should be washed thoroughly with soap and as much water as possible. If the person is affected by poisoning, keep the patient lying flat and at absolute rest; if conscious and able to swallow, the patient should drink as much water as possible.

When poisoning with the most toxic and rapidly acting substances has been by mouth, attempts should be made by specially trained medical personnel to induce vomiting within four hours if the patient is conscious. Administration of salt solution is not recommended now. If the patient can be attended by a doctor or nurse, the use of ipecacuanha emetic is the preferred method of inducing vomiting. Note that its use should not be recommended to first aiders. Vomiting should not be induced if the person has swallowed an acid, alkaline or petroleum product. If the chemical has got into the eye, clean water is required quickly to flush the eye several times for at least 15 min. If breathing ceases or weakens, artificial respiration must be started, making sure that the breathing passages are clear. If the patient is in convulsion, a strong piece of wood or a folded handkerchief should be placed between his teeth to prevent him biting his tongue.

The doctor must be informed of the name of the active ingredient and given as much information as possible by showing the doctor a leaflet or label about the chemical. Treatment by a doctor will depend very much on the type of poisoning. An injection of atropine is useful for organophosphate and carbamate (anticholinesterase) poisoning.

Fig. 16.3 'Packman' system of emptying containers inside a closed apparatus from which the pesticide is transferred directly to the sprayed tank and the container rinsed with clean water added to the spray

Suitable antidotes for organochlorine poisoning are not available. Large quantities of fuller's earth is used if a person is affected by paraquat. Morphine should not be given to patients affected by pesticide poisoning.

A first-aid kit should be readily available and a supply of clean water for drinking and washing any contaminated areas of the body. On large-scale spraying programmes first-aid kits should be carried in vehicles and aircraft. People regularly involved in applying pesticides should have a routine medical examination to check the cholinesterase levels in their blood plasma.

Combination of chemicals

Sometimes farmers mix two or more pesticides together to control different types of pests with a single application or to increase the effectiveness of an individual pesticide. Apart from the problem of whether the chemicals are compatible, their toxicity to humans and other organisms may be increased. For example the LD_{50} of malathion is 1 500 mg/kg and fenitrothion 400 mg/kg, but the mixture is less than 200 mg/kg. Residues of mixtures may persist longer. Because of their potential toxicity, combinations of pesticides and various additives should not be used unless the specific combination has been tested and is recommended by the appropriate authorities. There is also the danger of a cumulative effect of different pesticides used separately in a spray programme.

Pesticide packaging and labelling

Toxicity hazards can be reduced by improved methods of packaging. Most farmers in the tropics have a small acreage, and need small packets of pesticide to avoid storage of partially opened packets, for example 5-kg packets of granules are available for direct filling of hoppers, and several pesticides have been marketed in sachets containing sufficient wettable powder to mix 15 litres of spray, to fill one knapsack sprayer. Apart from reducing the risk of pesticides contaminating the farmer's store, which is often in the house, these small sachets eliminate the need to measure out the quantity required to obtain the correct spray concentration. Savings through reduction of loss by spillage and ensuring correct mixing, in addition to the improved safety aspects, more than repay the extra cost of packaging. Some products are now packaged in plastic sachets that are placed directly in the spray tank. Efforts have been made to standardize containers to facilitate the use of closed filling systems (J. Gilbert 1989) (Fig. 16.3).

Appropriate labelling is essential and should be in the vernacular language. Apart from the brand name, the label should have details of the active ingredient and inert materials used in the formulation, the intended use of

Fig. 16.4 Examples of pictograms

the product, full directions on the safe and correct procedure for mixing and application of the product (ie protective clothing required, which pests are to be controlled, dosage, time and method of application), and how to dispose of the container. Cautionary notices to protect the user, consumer (if the treated area is a food crop) and beneficial plants and animals should be clearly given on the label. The label should also indicate the minimum period between application and harvesting appropriate for the various crops on which the pesticide can be used. To assist some users important information is also given in the form of pictograms (Fig. 16.4).

Farmers should avoid storing chemicals for more than about 18 months. Containers left longer than this may corrode, or the active ingredient may be less effective.

Container and washings disposal

At the end of an application, a significant amount of pesticide liquid may be in the sprayer. Taylor et al (1988) reported that over 9 litres may remain in the pump and associated pipework of a 600 1-litre sprayer. Ideally clean water is flushed through the equipment and sprayed out in the field, as legislation in some countries covers the disposal of these washings which will contain an unknown quantity of pesticide. If washings are not sprayed in the area being treated with the pesticide, it may be necessary to install a water effluent treatment plant. Small-scale plants are now available which can remove organic substances so that the water can be discharged into a soakaway, or retained for subsequent cleaning of equipment. The small quantity of sludge produced can be buried or disposed of by a waste disposal container. These effluent plants (Fig. 16.5) are especially important for spray contractors, including aerial operators, and large-scale farmers.

Ideally all washed empty containers should be returnable to the supplier for recycling, but this has seldom, if ever, been practised.

In all the tropical countries, the empty pesticide container is valuable for other uses. Large metal containers have been flattened to provide building materials, especially for roofs, and drums have been used to collect and carry water. Fatalities have resulted because pesticide containers can never be adequately cleaned for other purposes. Strictly, all containers should be

Fig. 16.5 'Sentinel' system of decontaminating water used to wash sprayers and containers (photo E Allman and Co.)

washed out several times and the rinsings added to the spray. The containers should then be punctured before being buried in a deep pit (at least 1.5 m in depth), preferably in clay soil well away from any river or stream to avoid pollution of water supplies. Containers of special formulations which are not mixed with water, including those for ULV applications, should be cleaned with a suitable solvent such as kerosene before being buried. Boxes and other types of packaging may be burnt, provided air pollution does not become an additional significant concern. Herbicide containers should not be burnt and aerosol cans should never be punctured or burnt. Regulations concerning disposal of containers and unused chemicals have been introduced in a number of countries to minimize the risk of human poisoning and environmental pollution.

Noise

Noise ratings greater than 85 decibels (dB) in any octave band in the speech range 250–4 000 Hz can cause permanent hearing impairment. Human pain threshold is 120 dB. The noise level within a radius of 7 m from motorized sprayers often exceeds 85 dB, so the effect of noise should be considered in relation to the safe use of pesticides. Exposure to continuous noise should be restricted by interchanging spray operators or having definite rests between short periods of spraying. Ideally hearing protection should be provided, especially when sprayers are operated inside buildings.

Safety of agricultural pilots is also affected by loud engine and other noises and vibrations on an aircraft.

Table 16.5   Guidelines for safe use of a pesticide

Before applying pesticide – general instructions

---

1 Know the pest, and how much damage is really being done.
2 Use pesticides only when really needed.
3 Seek advice on the proper method of control.
4 Use only the recommended pesticide for the problem. If several pesticides are recommended, choose the least toxic to mammals and if the least persistent.
5 **READ THE LABEL**, including the small print.
6 Make sure the appropriate protective clothing is available and is used, and that all concerned with the application also understand the recommendations, and are fully trained in how to apply pesticides.
7 Commercial operators using large quantities of organophosphate pesticides should visit their doctor and have a bloood cholinesterase test, and have repeat checks during the season.
8 Check application equipment for leaks, calibrate with water and ensure it is in proper working order.
9 Check that plenty of water is available with soap and towel and that a change of clean clothing is available.

Table 16.5 (*Cont.*)

10 Check that pesticides on the farm are in the dry, locked store. Avoid inhaling pesticide mists or dusts, especially in confined spaces such as the pesticide store.
11 Warn neighbours of your spray programme, especially if they have apiaries.
12 Take only sufficient pesticide for the day's application from the store to the site of application. Do **NOT** transfer pesticides into other containers, especially beer and soft drink bottles.

While mixing pesticides and during application

1 Wear appropriate protective clothing. If it is contaminated, remove and replace with clean clothing
2 Never work alone when handling the most toxic pesticides
3 Never allow children or other unauthorised persons near the mixing
4 Recheck the instructions on the label
5 Avoid contamination of the skin, especially the eyes and mouth. Liquid formulations should be poured carefully to avoid splashing. Avoid powder formulations 'puffing up' into the face. If contaminated with the concentrate wash immediately. Use a closed system to transfer chemical to the sprayer where possible. Add washings of containers to spray tank
6 Never eat, drink or smoke when mixing or applying pesticides
7 Always have plenty of water available for washing
8 Always stand upwind when mixing
9 Make sure pesticides are mixed in the correct quantities
10 Avoid inhalation of chemical, dust or fumes
11 Start spraying near the downwind edge of the field and proceed upwind so that operators move into unsprayed areas
12 **NEVER** blow out clogged nozzles or hoses with your mouth
13 **AVOID** spraying when crops are in flower. Risk to bees is reduced if sprays are applied in evening when they are no longer foraging. Never spray if the wind is blowing towards grazing livestock or regularly used pastures
14 **NEVER** leave pesticides unattended in the field
15 Provide proper supervision of those assisting with the pesticide application, and have adequate rest periods
16 When blood tests are being conducted, do not work with pesticides if your cholinesterase level is below normal

After application

1 **RETURN** unused pesticide to the store
2 Safely dispose of all empty containers. As it may be difficult to dispose of empty containers after each day's spraying operations, they should be kept in the pesticide store until a convenient number are ready for disposal. **IT IS ABSOLUTELY IMPOSSIBLE** to clean out a container sufficiently well to make it safe for use for storage of food, water or as a cooking utensil. If any containers are burnt, **NEVER** stand in the smoke
3 **NEVER** leave pesticides in application equipment. Clean equipment and return to store
4 Remove and clean protective clothing
5 Wash well and put on clean clothing. Where there is a considerable amount of spraying, the operators should be provided with a shower room
6 Keep a record of the use of pesticides
7 Do not allow other people to enter the treated area for the required period if restrictions apply to the pesticide used

Code of conduct

Concern about the hazards of using certain pesticides led to the FAO Code of Conduct on the Distribution and Use of Pesticides. Subsequently this code has incorporated the principle of Prior Informed Consent (PIC) by which exporters of pesticides have to inform importers in developing countries about the toxicity and hazards associated with the use of products included on the PIC list (Table 16.6), and receive their authority before the products can be exported. The FAO Code is voluntary, but the requirements for PIC have been included in a European Community regulation applicable by law in the Member States. A PIC database is maintained at the International Register of Potentially Toxic Chemicals held at Geneva.

Advice on the disposal of unwanted pesticide stocks is now available in a booklet published by GIFAP, Brussels.

Table 16.6 Pesticides on the Initial Prior Informed Consent List

Category 1: Pesticides which have been banned or severely restricted in five or more countries and in active use or being phased out:

aldrin, amitrole, arsenic, BHC, (HCH), captafol, chlordane, chlorobenzilate, chloropicrin, chlordimeform, cyhexatin, DDT, demeton, -O & -S, dicofol, dieldrien, dinoseb, disulfoton, EDB, endosulfan, fluoracetamide, heptachlor, hexachlorbenzene, lindane, mercury compounds, methoxchlor, nitrofen, parathion, pentachlorophenol, phosphine generators, sodium fluoride, sodium fluoroacetate, strychnine, tepp, 2, 4, 5-T

Category 2: Candidate pesticides not yet on PIC list but now next in priority for inclusion

carbofuran dichlorvos, methamidophos, methomyl, methyl parathion, monocrotophos, phosphamidon

(Others): aldicarb, paraquat, methyl bromide

Category 3: Pesticides banned or severely restricted in five or more countries, and little or no use, or production discontinued:

chlordecone, DBCP, endrin, kelevan, leptophos, mirex, strobane, schradan, telodrin, thallium sulphate, toxaphene

17

Selection of spraying equipment

The foregoing chapters have indicated the range of equipment which can be used to apply pesticides. The choice of the equipment is normally governed by many factors, only some of which may be considered key ones (Fig. 17.1).

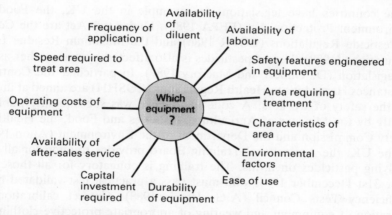

Fig. 17.1 Factors governing the selection of equipment

Environmental factors

Concern that pesticides can pollute the environment has stimulated research on methods of improving pesticide deposition, for example by using air assistance and electrostatic charges on spray droplets, but adoption of new types of equipment has been slow. The farmer seeks versatile, reliable, cost-effective equipment, while new systems of application, such as CDA and the use of electrostatic sprayers, tend to be more suited to particular applications.

Fewer pesticide products have been formulated for use in the more special-ized equipment. Nevertheless some governments, particularly in Scandinavia, have stipulated a reduction in the overall use of pesticides so efficiency of application will become a more important factor. More sophisticated equip-ment will therefore become acceptable to meet the requirements of new legislation, despite additional capital costs.

Some changes in equipment design have been used to reduce the risk of drift. Nozzles can be selected to provide a coarser spray (Rutherford et al 1989) or can be angled to reduce the distance between the nozzle and crop. An air curtain will also assist projection of spray droplets into a crop canopy and reduce drift or permit spraying with higher wind speeds (Taylor et al 1989).

If further government legislation provides stricter controls on environmen-tal contamination, farmers may have to give more serious consideration to techniques which provide greater control of droplet size, improve deposition, and allow more accurate timing of application.

Operational safety factors

Some countries have legislation, for example in the UK, the Food and Environment Protection Act (FEPA 1985). Under this Act are the Control of Pesticide Regulations (COPR 1986) and the Maximum Residue Levels Regulations (1988). Use of pesticides is also affected by many other aspects of legislation (D. Gilbert and Macrory 1989). In particular the Control of Substances Hazardous to Health Regulations (COSHH) are aimed at improv-ing the safety of workers. A code of practice was subsequently published jointly by the Ministry of Agriculture, Fisheries and Food, the Health and Safety Commission and the Department of the Environment (Anon 1990a). In the UK, the Agricultural Training Board provides training for all those applying pesticides on farms. The training is obligatory for all those born after 31st December 1964, who must pass a practical test validated by the Proficiency Tests Council (A.G. Harris 1988). Correct calibration and operation of equipment and wearing of appropriate protective clothing will undoubtedly improve the standard of application. Farmers are now much more aware of the potential problems of pesticide use (Burn and Tinsley 1990). However, in recognition that the transfer of a pesticide from its original container to the spray tank is potentially the most hazardous task, efforts have been made to improve sprayer design (see Chapter 7) to reduce the risk of operator contamination. This trend was the one most respondents agreed with in a survey carried out by A.K. Smith (1985) aimed at forecasting future development. There is no doubt that more farmers will choose equipment with safety features such as provision of closed filling systems, despite the additional costs involved (Yates et al 1981; Frost and Miller 1988).

Table 17.1   Threshold values (Spackmann and Barrie 1982)

Parameter	Threshold at which spraying is constrained
Hourly mean windspeed*	< = 4.6 m/s (hydraulic pressure sprayer)
	< = 6.2 m/s (CDA sprayer with large droplets (250 μm))
Soil moisture	Days when soil moisture deficit <5mm (conventional tractors), with no constraint on low ground pressure vehicles
Daylight	Not earlier than 0600 or later than 2100
Temperature	>1.0°C during spraying, with air temperature >7.0°C some time during the day
Precipitation	None for at least 1 hour before or at time of spraying

Notes: * measured at a height of 10 metres

Tractor Sprayers

Apart from environmental and safety factors, the selection of a tractor-mounted or trailer sprayer will be affected by the area which requires treatment and the time available for treatment. Often pest outbreaks occur when weather conditions are not ideal for pesticide application, so the number of days suitable for treating a crop can be severely limited. Spackmann and Barrie (1982) used threshold values for the UK (Table 17.1) and calculated the number of days suitable for spraying during each month, based on 10 years' weather data from 15 weather stations. In some areas of the UK, wind, rain or wet soil can reduce the number of spray days to as few as two or three, but this can be increased with a low ground pressure vehicle to allow access on wet ground. Small automatic weather stations are now available for individual farms to assist in monitoring conditions favourable for disease outbreaks and these can also be used to assist in decisions concerning spray dates.

Using the data on the number of spray days available in the period in which most pesticide treatments are required, A.K. Smith (1984) used a model (Fig. 17.2) to select ground spraying systems for arable farms. Based on the Baltin equation (see p. 304) this model allows rapid investigation of alternatives. Amsden (1962) has already pointed out the need to reduce the time taken to refill the sprayer by using a pump and by taking a water supply to the field. An increase in the tank capacity has very little effect on spraying efficiency and is more likely to cause soil compaction. Widening the boom did not significantly increase the proportion of time actually spraying, but reducing the volume applied was advantageous.

Higher vehicle speeds are not always possible (Nation 1978) due to uneven terrain and greater difficulty in controlling both the output and spray quality. Higher speeds tend to increase the proportion of small droplets in the spray and increase the risk of spray drift unless air assistance is used (Taylor el al 1989).

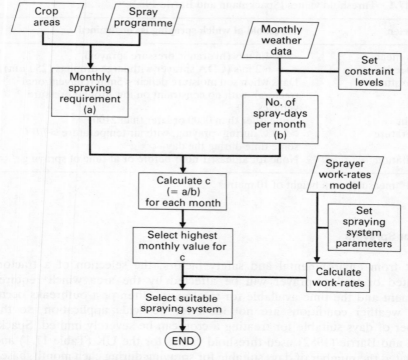

Fig. 17.2 Flow diagram of model suggesting a suitable spraying system for a farm (from Smith 1984)

Most manufacturers provide a range of machinery, the capital costs and operating costs of which can be estimated (Table 17.2) but the actual choice will be determined principally by the work rate for a critical period of the year. Costs are clearly reduced if large areas require treatment (Fig. 17.3).

Knapsack and Hand-carried Sprayers

Many areas requiring pesticide application are too small to justify tractor-mounted or aerial equipment. Sometimes access is difficult, for example around buildings or the terrain may be hilly, uneven, or too wet. Under these circumstances pesticide equipment is needed despite the operator being more exposed to the pesticide. Many different designs of lever-operated knapsack, compression and the small sprayers are available throughout the world. These have often been purchased solely on the basis of low capital costs, but increasingly there is an awareness of the need for basic design criteria to improve safety and reduce fatigue in their use. Portable equipment must be

Table 17.2 Operating costs for four power-operated boom sprayers

Sprayer	Small tractor mounted	Large tractor mounted	Trailed	Self-propelled
Initial capital cost (£)	1 060	3 680	15 000	36 000
Area sprayed annually (ha)	150	600	1 000	2 000
Tank capacity (litres)	320	1 000	2 000	1 600
Boom width (m)	6	12	18	24
Life in years	8	8	8	8
Hectares/h spraying	3.6	7.2	10.8	14.4
Overall ha/h (50% efficiency)	1.8	3.6	5.4	7.2
Use (h/annum)	83.3	166.7	185.2	277.8
Annual cost of ownership (£)	132.5	460	1 875	4 500
Repairs and maintenance† (£)	79.5	276	1 125	2 700
15% interest on half capital (£)	79.5	276	1 125	2 700
Total cost of ownership (£)	291.5	1 012	4 125	9 900
Ownership cost per hour (£)	3.50	6.07	22.27	35.64
Ownership cost per hectare (£)	1.94	1.69	2.06	4.94
Labour costs per hectare (£)	1.39	0.69	0.46	0.35
Tractor cost per hectare (£)	1.67	0.83	0.56	0.42
Total operating costs per hectare** (£)	5.0	3.21	3.08	5.71

Notes: † Based on 7.5% of capital cost, depending on sprayer
   ** Based on the overall cost of up to £3–6/ha shown above does not include various overheads, positioning of equipment, secretarial and telephone expenses and other items included in prices charged by farm contractors

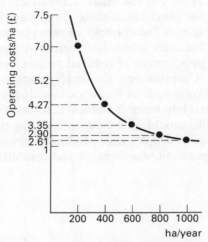

Fig. 17.3 Variation in operating costs with yearly workload, based on 12 m boom. Initial cost of tractor-mounted sprayer £4000

sufficiently durable so that frequent repairs are not needed and frustrating delays in a spray programme are avoided. It must also be comfortable and easy to carry. In some areas hand-carried battery-operated sprayers have replaced knapsack equipment because of their light weight and significant reduction in manual effort needed to use them. In particular the availability of water is often the key factor in favour of ULV/CDA equipment, which allows more rapid treatment in response to a pest infestation (Table 17.3).

Table 17.3 Comparison of spraying time for a knapsack and battery-operated CDA sprayer*

Sprayer	Knapsack			CDA sprayer		
Application rate (litres/ha)	400	200	100	10	2.5	1
Spray tank (litres)	15	15	15	0.5	0.5	0.5
Mixing and refilling time (h/ha)	2.2	1.1	0.55	0.67	0.17	0.07
Ferry time (h/ha)	14.8	7.4	3.7	0.37	0.08	0.04
Swath width (h/ha)	1	1	1	2.8	1.39	0.56
Spraying time (h/ha)	2.8	2.8	2.8	2.8	1.39	0.56
Turning time (s/row)	2	2	2	2	3	7
Total time (h/ha)	19.9	11.4	7.2	3.95	1.84	0.94
Spraying time as percentage of total	14	25	39	71	76	60

Note: * Assuming walking speed of 1 m/s, carrying 15 litres of water from supply 1 km from fields; fields 100 m long, average size 0.5 ha and separated by 150 m

The costs of operating portable equipment can be calculated using the same basic formula used for aircraft and tractor equipment, although labour input is proportionally greater, especially when water has to be carried over a long distance (Table 17.4). For the small-scale farmer, the capital cost may be the key factor, but the purchase of cheap equipment is false economy as it often fails after little use. A disadvantage of battery-operated equipment is the recurrent cost of batteries which has increased significantly in recent years, but time saved by adoption of reduced volume spraying released the farmer for other tasks. Unfortunately the battery costs increase with narrow swaths. Air-carrier sprayers such as knapsack mistblowers are needed when spray has to be projected into trees or bushes. Selection from the wide range of mistblowers available should be based on how far the spray has to be projected, and the range of droplet sizes produced, as well as how easily the motor starts, ease of operation, maintenance and comfort to the operator.

Aerial application

Farmers may prefer to choose aerial application, even when comparatively small areas require treatment, because

Table 17.4 Operational costs with knapsack and hand-carried CDA sprayer (actual costs of equipment and labour will depend on local conditions: cost of chemical, which is also affected by choice of formulation is not included)

Sprayer	Manually operated knapsack	Motorized knapsack mistblower	Hand-carried sprayer	
Initial capital cost (£)	60	350	45	45
Area sprayed annually (ha)	20	20	20	20
Tank capacity (litres)	15	10	1	1
Swath width (m)	1	3	1	3
Life in years	3	5	3	3
Hectares/h spraying*	0.36	1.08	0.36	1.08
Overall ha/h (% efficiency)*	0.18(50)	0.65(60)	0.31(85)	0.97(90)
Use (h/annum)	111	30.8	64.5	20.6
Annual cost of ownership (£)	20	70	15	15
Repairs and maintenance† (£)	6	35	4.5	4.5
15% interest on half capital (£)	4.5	26.3	3.4	3.4
Total cost of ownership (£)	30.5	131.3	22.9	22.9
Ownership cost per hour (£)	0.27	4.26	0.36	1.11
Ownership cost per hectare (£)	0.76	3.95	0.99	1.03
Labour costs per hectare‡ (£)	1.38	0.39	0.80	0.25
Operating cost including batteries** (£/ha)	—	0.68	2.2	0.74
Labour costs to collect water*** (£/ha)	1.38	0.92	0.13	0.04
Total operating costs per hectare (£)	3.52	5.94	4.12	2.06

Notes:
    * Assuming walking speed is m/s, actual efficiency will depend on how far water supply is from treated area, application rate and other factors (see Table 17.3)
    † 10% of capital cost
    ‡ Assumes labour in tropical country at £2 per 8 h day
    ** Assumes batteries cost 50p each and a set of 8 will operate for 5 h with a fast disc speed. Fuel for mistblower at 44p litre/h
    *** Water required for washing, even when special formulations are applied at ULV

Battery consumption is less on some sprayers with a single disc and smaller motors. The 'Electrodyn' sprayer uses only 4 batteries over 50+ hours, so the costs of batteries on a double row swath is 0.6 instead of 2.2

1 passage of ground equipment will damage crops such as cereals unless 'tramlines' are left for access
2 soil compaction which can be severe on certain soils, especially with repeat sprays, is eliminated
3 crops may be inaccessible to ground equipment at critical periods of pest infestation if the soil or a part of the field is too wet; poor drainage on an irrigation scheme, or arrangement of canals may restrict the use of tractor-mounted equipment

4 pests, including red spider mites, or diseases may be spread by movement of equipment or personnel through the crop
5 capital investment in equipment is not justified if pest or disease infestations are sporadic
6 access to certain crops may be difficult or impossible without specialized equipment, for example high-clearance tractors which are needed for late-season treatment of crops such as maize.

When deciding on aerial application, the higher operating costs and the availability of aircraft must be considered. The higher costs may be offset by less mechanical damage to the crop, but sufficient aircraft may not be available to meet a sudden demand when infestations of a sporadic pest reach the economic threshold over extensive areas simultaneously, or in areas remote from the aircraft operator's base.

In practice, aerial spraying is used principally over large forests and tree crops and in large irrigation schemes such as the Sudan Gezira, particularly if a pest occurs suddenly over an extensive area. Spruce budworm in North America has been sprayed with aircraft to treat large areas when outbreaks of the young larvae occur (Armstrong 1981). Similarly extensive areas infected with tsetse fly have been successfully controlled by aerial application of aerosols (Allsopp 1984). Aircraft are also very important for the control of locust swarms (Rainey 1976) to prevent crop failure. However, in general for arable crop protection penetration of a crop canopy may be inadequate to control certain pests on the lower leaves and there has been a reluctance to recommend aerial treatments in many countries due to the perceived increased risk of spray drift as droplets are released at a greater height.

General

The availability of spare parts and ease of maintenance are important criteria in selecting equipment. Certain basic spare parts such as nozzle tips, washers, 'O' ring seals and other replaceable components should be purchased wherever possible with the original equipment, to avoid any delay in supplying a spray at a crucial period during the season. The need to stock basic spares cannot be overemphasized, particularly when equipment is used in remote areas.

Routine maintenance is advisable, so preference should be given to equipment on which components subject to wear are readily accessible. Some chemicals are particularly corrosive or affect the reliability of a component, for example certain plastics such as PVC are dissolved by particular solvents, such as isophorone or cyclohexanone, used in some formulations, as 'O' ring seal may swell or parts may be abraded by some granular or wettable powder formulations. Manufacturers should be consulted regarding compatibility of their products with materials used in the construction of application equipment.

The purchase of any application equipment should be preceded by an inspection of the range of the different makes and models of the type of equipment needed when they are displayed side by side at an exhibition or agricultural show. Ideally this should be followed by a dynamic demonstration of the equipment under field conditions so that the movement of spray booms or other relevant features can be assessed.

The ultimate criterion in selecting equipment is whether the pest can be controlled economically. Despite many improvements in equipment, pesticide application remains one of the most inefficient processes practised (Himel and Moore, 1967; Graham-Bryce 1975). This need not be so, but efficiency can be achieved only if the biological target is defined in relation to the behaviour of the pest and deposition on specific targets obtained with the optimum droplet size range. This requires selecting the optimal pesticide formulation, and choosing the correct nozzle and delivery systems, sometimes air-assisted, to minimise loss of chemical and ensure that the correct dose is transferred to the target.

References

Abdalla, M. (1984) A biological study of the spread of pesticides in small droplets, PhD Thesis, University of London.

Abdelbagi, H.A. and Adams, A.J. (1987) Influence of droplet size, air-assistance and electrostatic charge upon the distribution of ultra-low volume sprays on tomatoes, *Crop Protection* **6**: 226–33.

Adams, A.J. and Lindquist, R.K. (1991) Air-assisted spraying under glass. *BCPC Monograph* **46**: 227–35.

Adams, A.J., Abdalla, M.R., Wyatt, I.J. and Palmer, A. (1987) The relative influence of the factors which determine the spray droplets density required to control the glasshouse whitefly, *Trialeurodes vaporarorium, Aspects of Appl. Biology* **14**: 257–66.

Adams, A.J., Lindquist, R.K., Adams, I.H.H. and **Hall, E.R.** (1991) Efficacy of bifenthrin against pyrethroid – resistant and – susceptible populations of glasshouse whitefly in bioassays and using three spray application methods *Crop Protection* **10**: 106–10.

Adams, A.J., Lindquist, R.K., Hall, F.R. and **Rolph, I.A.** (1989) Application, distribution and efficacy of electrostatically charged sprays on chrysanthemums *Pesticide Formulation and application Systems*: *International Aspects* 9th Volume ASTM STP 1036, Philadelphia: 179–90.

Afreh-Nuamah, K. and **Matthews, G.A.** (1987) Comparative spray distribution in a tree crop with three different spray nozzles. *Aspects of Applied Biology* **14**: 77–83.

Ahmad, S.I., Tate, R.W., Bode, L.E. and Retzer, H.J. (1980) Droplet size characteristics of the by-pass nozzle. ASAE Paper: 80–1,058.

Ahmad, S.I., Bode, L.E. and Butler, B.J. (1981) A variable-rate pesticide spraying system. *Trans. ASAE* **24**: 584–9.

Akesson, N.B. and **Yates, W.E.** (1974) The use of aircraft in agriculture. *FAO Agric. Dev. Paper* **94**, 217 pp.

Allan, J.R. McB. (1980) Development in monitoring and control systems for greater accuracy in spray application. *BCPC Monograph* **24**: 201–13.

Allen, G.G., Chopra, C.G., Friedhoff, J.F., Gara, R.I., Maggi, M.W., Neogi, A.N., Roberts, S.C. and **Wilkins, R.M.** (1973) Pesticides, pollution and polymers. *Chem. Tech.* **4**: 171–8.

Allen, J.G., Smith, A.P., Butt, D.J. and Warman, T.M. (1991) Improved performance of mistblower sprayers by electrostatic charging. *BCPC Monograph* **46**: 143–50.

Allsopp, R. (1984) Control of tsetse flies (Diptera: Glossinidae) using insecticides: a review and future prospects. *Bull. ent. Res.* **74**: 1–23.

Allsopp, R. (1990) A practical guide to aerial spraying for the control of tsetse flies (*Glossina* spp.). *Aerial Spraying Research and Development Project Final Report* **2**. Chatham, Natural Resources Institute.

Alm, S.R., Reichard, D.L. and Hall, F.R. (1987) Effects of spray drop size and distribution of drops containing bifenthrin on *Tetranychus urticae* (Acari: Tetranychidae), *J. Econ. Ent.* **50**: 517–20.

Almeida, A.A. (1967) Reactions of larvae of *Plodia interpunctella* (Hb.) (Lep. Pyralidae) to insecticidal droplets. *Bull. ent. Res.* **58**: 221–6.

Ames, B.N. (1983) Dietary carcinogens and anticarcinogens: Oxygen radicals and degenerative diseases. *Science* **221**: 1,256–64.

Amsden, R.C. (1959) The Baltin–Amsden formula. *Agric. Aviat.* **2**: 95.

Amsden, R.C. (1962) Reducing the evaporation of sprays. *Agric. Aviat.* **4**: 88–93.

Amsden, R.C. (1965) Choosing a tractor sprayer. *Congr. Prot. Cult. trop. Marseille* 485–90.

Amsden, R.C. (1970) The metering and dispensing of granules and liquid concentrates. *Br. Crop Prot. Counc. Monogr.* **2**: 124–9.

Amsden, R.C. (1972) Wind velocity in relation to aerial spraying of crops. *Agric. Aviat.* **14**: 103–7.

Amsden, R.C. (1975) A graphical method of estimating the volume median diameter of a spray spectrum. *PANS* **21**: 103–8.

Anderson, N.H., Hall, D.J. and Seaman, D. (1987) Spray retention: effects of surfactants and plant species. *Aspects of App. Biol.* **14**: 233–44.

Anon. (1971) Application and dispersal of pesticides. *Tech. Rep. Ser. Wld Hlth Org.* **465**, 66 pp. Geneva, World Health Organization.

Anon (1973) *Agricultural Pilots' Handbook*, 1st edition, The Hague, IAAC.

Anon. (1974) Microencapsulated pesticide reaches market. *Chemical and Engineering News* **52**: 15–16.

Anon. (1975) Recommended classification of pesticides by hazard. *WHO Chron.* **29**: 397–401.

Anon (1989) The use of impregnated bednets and other materials for vector-borne disease control. Geneva, WHO/VBC/89.981.

Anon (1990a) *Pesticides: Code of practice for the safe use of pesticides on farms and holdings*. Ministry of Agriculture, Fisheries and Food and Health and Safety Commission.

Anon (1990b) *Equipment for Vector Control*, 3rd edn. Geneva, World Health Organisation.

Apt, W.J. and Caswell, E.P. (1988) Application of nematicides via drip irrigation. *Annals of Appl. Nematology* **2**: 1–10.

Armstrong, J.A. (1981) ULV/CDA optimum spray droplet size for control of the eastern spruce budworm in Canada. *Outlook on Agriculture* **10**: 327–32.

Arnold, A.C. (1980) Developments in sampling and measuring techniques for monitoring small droplets. *BCPC Monograph* **24**: 233–40.

Arnold, A.C. (1983) Comparative droplets spectra for three different-angled flat fan nozzles. *Crop Protection* **2**: 193–204.

Arnold, A.C. (1985) Power sources for spinning disc sprayers and their economics. *Trop. Pest Management*, 3–10.

Arnold, A.C. (1987) The dropsize of the spray from agricultural fan spray atomisers as determined by a Malvern and the Particle Measuring System (PMS) instrument. *Atomisation and Spray Technology*, **3**: 155–67.

Arnold, A.J. (1983) UK patent 2119678.

Arnold, A.J. (1984) 'Electrostatic application with rotary atomisers' *EPPO Bull.* **13**: 451–6.

Arnold, A.J. and Pye, B.J. (1980) Spray application with charged rotary atomisers.

BCPC Monograph **24**: 109–25.

Atias, M. and **Weihs, D.** (1985) On the motion of spray drops in the wake of an agricultural aircraft. *Atomisation and Spray Technology* **1**: 21–36.

Atkinson, A.M., Filmer, P.J. and **Kinsman, K.L.** (1968) The distribution of pasture seed from light aircraft. Commonwealth of Australia, 39 pp.

Babcock, J.M., Brown, J.J. and **Tanigoshi, L.K.** (1990) Volume and coverage estimation of spray deposition using an Amino nitrogen colormetric reaction. *J. econ. Ent.* **3**: 1,633–5.

Bachalo, W.D., Houser, M.J. and **Smith, J.N.** (1987) Behaviour of sprays produced by pressure atomizers as measured using a phase/doppler instrument. *Atomisation and Spray Technology* **3**: 53–72.

Bache, D.H. (1975) Transport of aerial spray: III. Influence of microclimate on crop spraying. *Agric. Met.* **15**: 379–83.

Bache, D.H. and **Sayer, W.J.D.** (1975) Transport of aerial spray: I. A model of aerial dispersion. *Agric. Met.* **15**: 257–71.

Bache, D.H., Lawwon, T.J. and **Uk, S.** (1988) Development of a criterion for defining spray drift. *Atmos. Envir.* **22**: 131–135.

Bailey, A.G. (1986) The theory and practice of electrostatic spraying. *Atomization and Spray Technology* **2**: 95–134.

Bailey, P.W. (1988) Engineering problems associated with granule application in row crops. *BCPC Monograph.* **39**: 329–32.

Baldry, D.A.T., Kulzer, H., Bauer, S., Lee, C.W. and **Parker, J.D.** (1978) The experimental application of insecticide for a helicopter for the control of riverine populations of *Glossina tachinoides* in West Africa III Operational aspects and application techniques. *PANS* **24**: 423–34.

Baldry, D.A.T., Zerbo, D.G., Baker, R.H.A., Walsh, J.F. and **Pleszak, F.C.** (1985) Measures aimed at controlling the invasion of *Simulium damnosum* Theobald, S.I. (Diptera: Simuliidae) into the Onchocerciasis Control Programme Area 1. Experimental aerial larviciding in the Upper Sassandra Basin of South-eastern Guinea in 1985. *Trop. Pest Management* **31**: 255–63.

Bals, E.J. (1970) Rotary atomisation. *Agric. Aviat.* **12**: 85–90.

Bals, E.J. (1975a) Development of cda herbicide handsprayer. *PANS* **21**: 345–9.

Bals, E.J. (1975b) The importance of controlled droplet application (CDA) in pesticide applications. *Proc. 8th Br. Insectic. Fungic. Conf*: 153–60.

Bals, E.J. (1976) Controlled droplet application of pesticides (CDA). Paper presented at symposium 'Droplets in Air' organized by Society of Chemical Industry.

Baltin, F. (1959) The Baltin formula. *Agric. Aviat.* **1**: 104; **2**: 6.

Barlow, F. and **Hadaway, A.B.** (1947) Preliminary notes on the loss of DDT and gammexane by absorption. *Bull. ent. Res.* **38**: 335–46.

Barlow, F. and **Hadaway, A.B.** (1974) Some aspects of the use of solvents in ULV formulations. *Br. Crop Prot. Coun. Monogr.* **ll**: 84–93.

Barnett, G. (1990) The increased yield response of winter wheat to low pesticide input programmes with vegetable oil-based carrier adjuvant. *Med. Fac. Landbouww. Rijksuniv. Gent.* **55**: 1,343–47.

Barnett, M.J. and **Timbrell, V.** (1962) An apparatus for precise size measurement and analysis of particulate pharmaceutical materials. *Pharmaceutical J.* **189**: 379–81.

Barrett, P.R.F. (1978) Some studies on the use of alginates for the placement and controlled release of diquat on submerged aquatic plants. *Pesticide Science* **9**: 425–33.

Barrett, P.R.F. and **Logan, P.** (1982) The localised control of submerged aquatic weeds in lakes with diquat alginate. *Proc. EWRS 6th Symposium on Aquatic Weeds*: 193–8.

Barrett, P.R.F. and **Murphy, K.J.** (1982) The use of diquat-alginate for weed control in flowing waters. *Proc. EWRS 6th Symposium on Aquatic Weeds*: 200–8.

Barry, J.W., Ekblad, R.B., Teske, M.E. and **Skyler, P.J.** (1990) Technology transfer forest service aerial spray models. *Am. Soc. Ag. Engineers*: Paper 90–1017.

Basinski, J.J. and Wood, I.M. (1987) Kimberley Research Station In: Basinski, J.J., Wood, I.M. and Hacker, J.B. (eds) *The Northern Challenge – A History of CSIRO Crop Research in Northern Australia* Res. Rep. 3 CSIRO Division of Tropical Crops and Pastures.

Batchelor, G.K. (1967) *An Introduction to Fluid Dynamics.* Cambridge.

Bateman, R.P. (1989) Controlled droplet application of particulate suspensions of a carbamate insecticide. unpublished PhD Thesis, University of London.

Beaumont, A. (1947) The dependence on the weather of the dates of outbreaks of potato blight epidemics. *Trans. Br. Mycol. Soc.* **31**: 45–53.

Becher, P. (1973) The emulsifier. In: *Pesticide Formulations.* New York, Marcel Dekker.

Beeden, P. (1972) The pegboard – an aid to cotton pest scouting. *PANS* **18**: 43–5.

Beeden, P. (1975) Service life of batteries used in portable ULV sprayers. *Cott. Gr. Rev.* **52**: 375–85.

Beeden, P. and **Matthews, G.A.** (1975) Erosion of cone nozzles used for cotton spraying. *Cott. Gr. Rev.* **52**: 62–5.

Bell, G.A. (1989) Herbicide granules – review of processes and products. *Proc. Brighton Crop Protection Conference – Weeds*: 745–52.

Benedict, H.J., Urban, T.C., George, D.M., Seges, J.C., Anderson, D.J., McWharter, G.M. and **Zummo, G.R.** (1985) Pheromone trap thresholds for management of overwintered boll weevil. *J. econ. Ent.* **78**: 169–71.

Benedict, J.J., El-Zik, K.M., Oliver, L.R., Roberts, P.A. and **Wilson, L.T.** (1989) Economic injury levels and thresholds for pests of cotton. In: Frisbie, R.E., El-Zik, K.M. and Wilson, L.T. (eds.) *Integrated Pest Management Systems and Cotton Production.* New York, John Wiley & Sons Ltd.

Bennett, L.V. (1976) The development and termination of the 1968 plague of the desert locust *Schistocerca gregaria* (Forskil) (Orthoptera: Acrididae). *Bull. ent. Res.* **66**: 511–52.

Bergland, R.N. and **Liu, B.Y.H.** (1973) Generation of monodisperse aerosol standards. *Environ Sci. Technol* **7**: 147–53.

Beroza, M., Hood, C.S., Trefrey, D., Leonard, D.E., Knipling, E.F., Klassen, W. and **Stevens, L.J.** (1974) Large field trial with microencapsulated sex pheromone to prevent mating of the gypsy moth. *J. econ. Ent.* **67**: 661–4.

Berry, C.L. (1990) The hazards of healthy living – the agricultural component: *Proc. Brighton Crop Protection Conf. – Pest and Diseases*: 3–13.

Bindra, O.S. and **Singh, H.** (1971) *Pesticide Application Equipment*, Oxford and IBH, India.

Birchall, W.J. (1976) Developments in the regulation of agricultural aviation in the United Kingdom. *Proc. 5th Int. agric. Aviat. Congr*: 190–2.

Blair, A.M., Parker, C. and **Kasasian, L.** (1976) Herbicide protectants and antidotes – a review. *PANS* **22**: 65–74.

Bode, L.E., Langley, T.E. and **Butler, B.J.** (1979) Performance characteristics of by-pass spray nozzles. *Trans. Ann. Soc. Agr. Eng.* 1016–22.

Bode, L.E., Butler, B.J., Pearson, S.C. and **Bouse, L.F.** (1983) Characterisitcs of the Micromax rotary atomiser. *Trans. Am. Soc. Ag. Engngs*: 999–1,005.

Boivin, G. and **De Camp, S.T.** (1976) Operational aspects of inertial guidance in the 1975 province of Quebec spruce budworm control program. *Proc. 5th Int. agric. Aviat. Congr.* 350–63.

Boize, L.M. and **Dombrowski, N.** (1976) The atomization characteristics of a spinning disc ultra low volume applicator. *J. agric. Engng Res.* **21**: 87–99.

Boize, L.M., Gudin, C. and **Purdue, G.** (1976) The influence of leaf surface roughness on the spreading of oil spray drops. *Ann. appl. Biol.* **84**: 205–11.

Boize, L.M., Matthews, G.A. and **Kuntha, C.** (1975) Controlled droplet size application for residual spraying inside dwellings. *Proc. 8th Br. Insectic. Fungic. Conf*: 183–7.

Bond, C.P. (1974) Quantimet 720 for the analysis of spray droplet distribution. *Br. Crop Prot. Counc. Monogr.* **11**: 275–8.

Bond, E.J. (1984) Manual of fumigation for insect control. *FAO Plant Production and Protection* **54**, 432 pp.

Bouse, L.F. (1987) Aerial Application. In: McWhorter, C.G. and Gebhardt, M.R. (eds) *Methods of Applying Herbicides*, Monograph of Weed Science Society of America **4**: 123–36.

Bouse, L.F., Carlton, J.B. and **Merkle, M.G.** (1976) Spray recovery from nozzles designed to reduce drift. *Weed Sci.* **24**: 361–5.

Bouse, L.F., Carlton, J.B. and **Morrison, R.K.** (1981) Aerial application of insect egg parasites. *Trans. Am. Soc. Ag. Engineers* 1093–8.

Boving, P.A., Winterfeld, R.G. and **Stevens, L.E.** (1972) A hydraulic drive system for the spray system of agricultural aircraft. *Agric. Aviat*, **14**: 41–5.

Boyer, W.P., Warren, L.O. and **Lincoln, C.** (1962) Cotton insect scouting in Arkansas. *Bull. agric. expl. Stn.* Univ. of Arkansas, 656.

Brandenburg, B.C. (1974) Raindrop – drift reduction spray nozzle. *Am. Soc. Agric. Engng*, Paper 74–1595.

Brazelton, R.W. and **Akesson, N.B.** (1976) Chemical application equipment regulations in California. *Proc. 5th Int. agric. Aviat. Congr*: 206–12.

Brazelton, R.W. and **Akesson, N.B.** (1987) Principles of closed systems for handling of agricultural pesticides. *Proc. Am. Soc. Testing Materials 7th Symposium on pesticide formulations as application systems*: 16–27.

Brazelton, R.W., Roy, S., Akesson, N.B. and **Yates, W.E.** (1969) Distribution of dry materials. *Proc. 4th Int. agric. Aviat. Congr*: 166–74.

Brazzel, J.R. and **Watson, W.W.** (1966) Low volume spray patterns with three types of aerial application equipment. *Agric. Aviat.* **8**: 119–21.

Breeze, V.G. and Van Rensburg, E. (1988) Herbicide vapour as an environmental pollutant: *Aspects of Applied Biology* **17**: 339–50.

Brent, K. (1987) Fungicide resistance – practical significance. In: Brent, K. and Atkins, R.J. (eds) *Rational Pesticide Use*. Cambridge.

Brooke, J.P., Giglioli, M.E.C. and **Invest, J.** (1974) Control of *Aedes taeniorhynchus* Wied on Grand Cayman with ULV bioresmethrin. *Mosq. News* **34**: 104–11.

Brown, A.W.A. (1951) *Insect Control by Chemicals*. New York, Wiley.

Brown, A.W.A. (1971) Resistance to pesticides. In: White-Stevens, R. (ed.) *Pesticides in the Environment*. New York, Marcel Dekker.

Brown, A.W.A. and **Watson, D.L.** (1953) Studies on fine spray and aerosol machines for control of adult mosquitoes. *Mosq. News* **13**: 81–95.

Bruge, G. (1975) Les appareils localisateurs de microgranules antiparasitaires. *Phytoma* **27** (265): 9–17.

Bruge, G. (1976) Chlormephos: mutual adaptation of the formulation and the equipment used for its application. *Br. Crop Prot. Counc. Monogr.* **18**: 109–13.

Brunskill, R.T. (1956) Factors affecting the retention of spray droplets on leaves. *Proc. 3rd Br. Weed Cont. Conf.* **2**: 593–603.

Burchill, R.T. (1975) Plant pathology – fungal diseases – apple scab. *Annual Report of East Malling Research Station for 1975*.

Burges, H.D. and **Jarrett, P.** (1979) Application and distribution of *Bacillus thuringiensis* for control of tomato moth in glasshouses. *Proc. 1979 Br. Crop Prot. Conf. – Pests and Diseases*: 433–9.

Burn, A.J. and **Tinsley, M.R.** (1990) Pesticide regulation and the farmer. *Brighton Crop Protection Conf. – Pests and Diseases*: 303–12.

Burton, J. (1972) Cotton growing on Triangle Sugar Estate, Rhodesia. *Cott. Gr.*

Rev. **49**: 236–41.

Butler, B.J., Akesson, N.B. and **Yates, W.E.** (1969) Use of spray adjuvants to reduce drift. *Trans. Am. Soc. agric. Engrs.* **12**: 182–6.

Byass, J. and **Charlton, G.K.** (1968) Equipment and methods for orchard spray application research. *J. agric. Engng Res.* **13**: 280–9.

Cadou, I. (1959) Une rampe portative individuelle pour la pulvérisation a faible volume. *Cot. Fib. Trop.* **14**: 47–50.

Cammell, M.E. and **Way, M.J.** (1977) Economics of forecasting for chemical control of the black bean aphid *Aphis fabae*, on the field bean, *Vicia faba. Ann. appl. Biol.* **85**: 333–43.

Campion, D.G. (1976) Sex pheromones for the control of Lepidopterous pests using microencapsulation and dispenser techniques. *Pestic. Sci.* **7**: 636–41.

Campion, D.G. (1989) Semiochemicals for the control of insect pests. *BPCP Monograph* **43**: 119–27.

Carlton, J.B. and **Bouse, L.F.** (1980) Electrostatic spinner-nozzle for charging aerial sprays. *Trans. Am. Soc. Ag. Engrs* **23**: 1,369–73, 1,378.

Carmen, G.E. (1975) Spraying procedures for pest control on citrus in citrus. *Tech. Monogr.* 4: 28–34, Basle, Ciba-Geigy.

Carroll, M.K. and Bourg, J.A. (1979) Methods of ULV droplet sampling and analysis: effects on the size and spectrum of the droplets collected. *Mosquito News* **39**: 645–55.

Carson, A.G. (1987) Improving weed management in the draft animal-based production of early pearl millet in The Gambia. *Trop. Pest Management* **33**: 359–63.

Cauquil, J. (1987) Cotton pest control: a review of the introduction of ultra-low volume (ULV) spraying in sub-Saharan French-speaking Africa. *Crop Protection* **6**: 38–42.

Cayley, G.R., Etheridge, P., Griffiths, D.C., Phillips, F.T., Pye, B.J. and Scott, G.C. (1984) A review of the performance of electrostatically charged rotary atomisers on different crops. *Ann. appl. Biol.* **105**: 279–386.

Cayley, G.R., Griffiths, D.C., Hulme, P.J., Lewthwaite, R.J. and **Pye, B.J.** (1987) Tracer techniques for the comparison of sprayer performance. *Crop Protection* **6**: 123–9.

Chadd, E.M. (1990) Use of an electrostatic sprayer for control of anopheline mosquitoes. *Med. Vet. Ent.* **4**: 97–104.

Chadd, E.M. and **Matthews, G.A.** (1988) The evaluation of an electrostatic spraying method for mosquito control. *Trop. Pest Management* **34**: 72–5.

Chadwick, P.R. (1975), The activity of some pyrethroids, DDT and lindane in smoke from coils for biting inhibition, knockdown and kill of mosquitoes (Diptera. Culicidae). *Bull. ent. Res.* **65**: 97–107.

Chadwick, P.R. and **Shaw, R.D.** (1974) Cockroach control in sewers in Singapore using bioresmethrin and piperonyl butoxide as a thermal fog. *Pestic. Sci.* **5**: 691–701.

Chalfont R.B. and **Young J.R.** (1982) Chemigation, or application of insecticide through overhead sprinkler irrigation systems, to manage insect pests, infesting vegetable and agronamic crops *J. econ Ent.* **75**: 237–41.

Clark, C.J. and **Dombrowski, N.** (1972) On the formation of drops from the rims of fan spray sheets. *Aerosol Sci.* **3**: 173–83.

Clarke, S.E. (1990) The war against tsetse: tsetse fly eradication along the Shebelle River, Somalia. *Rural Development in Practice* **2**: 17–19.

Clayphon, J.E. (1971) Comparison trials of various motorised knapsack mistblowers at the Cocoa Research Institute of Ghana. *PANS* **17**: 209–25.

Clayphon, J.E. and **Matthews, G.A.** (1973) Care and maintenance of spraying equipment in the tropics. *PANS* **19**: 13–23.

Clayphon, J.E. and **Thornhill, E.W.** (1974a) A hand-held, drift-free herbicide applicator. *PANS* **20**: 145–9.

Clayphon, J.E. and **Thornhill, E.W.** (1974b) Spinning disc nozzle adaptation for mistblowers. *Br. Crop Prot. Counc. Monogr.* **ll**: 281.

Clayton, P.B. (1988) Seed treatment technology – the challenge ahead for the agricultural chemicals industry. *BCPC Monograph* **39**: 247–56.

Coffee, R.A. (1971a) Some experiments in electrostatic dusting. *J. agric. Engng Res.* **16**: 98–105.

Coffee, R.A. (1971b) Electrostatic crop spraying and experiments with triboelectrogasdlynamic generation system. *Proc. 3rd Conf. on Static Electrification*. Institute of Physics, Paper 17: 200–11

Coffee, R.A. (1973) Electrostatic crop spraying. *New Scientist* 25 Jan. 1973.

Coffee, R.A. (1979) Electrodynamic energy – a new approach to pesticide application. *Proc. 1979 Br. Crop Prot. Counc. Conf. Pests and Diseases* **3**: 777–89.

Coffee, R.A. (1980) Electrodynamic spraying. *BCPC Monograph* **24**: 95–107.

Coffee, R.A. (1981) Electrodynamic crop spraying. *Outlook on Agriculture* **10**: 350–56.

Coggins, S. and Baker, E.A. (1983) Micro sprayers for the laboratory application of pesticides. *Ann. Appl. Bio.* **102**: 144–54.

Combellack, J.H. and Matthews, G.A. (1981a) Droplet spectra measurements of fan and cone atomisers using a laser diffraction technique. *J. aerosol Sci.* **12**: 529–40.

Combellack, J.H. and Matthews, G.A. (1981b) The influence of atomizer, pressure and formulation on the droplet spectra produced by high volume sprayers *Weed Research* **21**: 77–86.

Comins, H.N. (1977a) The development of insecticide resistance in the presence of migration. *J. theor. Biol.* **64**: 177–97.

Comins, H.N. (1977b) The management of pesticide resistance. *J. theor. Biol.* **65**: 399–420.

Comins, H.N. (1977c) The control of adaptable pests, *Proc. of a Conf. on Pest Management* (ed. G.A. Norton and C.S. Holling), Laxenburg, Austria.

Conway, G.R. (1972) Ecological aspect of pest control in Malaysia. In: Farvar, M.T. and Milton J.P., (eds) *The Careless Technology – Ecological and International Development* New York, Natural History Press.

Cooke, B.K. and Hislop, E.C. (1987) Novel delivery systems for arable crop spraying – deposit distribution and biological activity. *Aspects of Applied Biology* **14**: 53–70.

Cooke, B.K., Herrington, P.J. and **Morgan, N.G.** (1974) Pesticide fall-out and orchard earth-worms. *Rep Long Ashton Res. Stn for 1974*; 105–6.

Cooke, B.K., Herrington, P.J., Jones, K.G. and **Morgan, N.G.** (1975a) Pest and disease control on intensive apple trees by overhead mobile spraying. *Pestic. Sci.* **6**: 571–9.

Cooke, B.K., Herrington, P.J., Jones, K.G. and **Morgan, N.G.** (1975b) Spray deposit cover and fungicide distribution obtained on intensive apple trees by overhead mobile spraying methods. *Pestic. Sci.* **6**: 581–7.

Cooke, B.K., Herrington, P.J., Jones, K.G. and **Morgan, N.G.** (1976) Spray deposit cover and fungicide distribution obtained by low and ultra low volume spraying of intensive apple trees. *Pestic. Sci.* **7**: 35–40.

Cooke, B.K., Herrington, P.J., Jones, K.G. and **Morgan, N.G.** (1977) Progress towards economical and precise fruit spraying *Proc. Br. Crop Prot. Conf. – Pests and Diseases* 323–9

Cooke, B.K., Hislop, E.C., Herrington, P.J., Western, N.M. and Humpherson-Jones, F. (1990) Air-assisted spraying of arable crops in relation to deposition, drift and pesticide performance. *Crop Protection* **9**; 303–11.

Cooper, J.F. (1991) Computer program for the analysis of droplet data. In: Allsopp, R. (ed.) *Aerial Spraying Research and Development Project Final Report* **2**: 37–38,

Chatham, Natural Resources Institute.

Cooper, S.C. and **Law, S.E.** (1985) Institute of Electrical and Electronic Engineers. *(IEEE(IAS)) Conf. Record*; 1,346–52.

Cooper, S.C. and **Law, S.E.** (1987a) Transient characteristics of charged spray deposition occurring under action of induced target coronas: space-charge polarity effect. *Inst. of Physics Conf. Series No. 85*, section 1: 21–26.

Cooper, S.C. and **Law, S.E.** (1987b) Bipolar spray charging for leaf-tip corona reduction by space-charge control *Inst. of Electrical and Electronic Engineers Transactions on Industry Applications* 1A–23: 217–23.

Courshee, R.J. (1959) Drift spraying for vegetation baiting. *Bull. ent. Res.* 50: 355–69.

Courshee, R.J. (1960) Some aspects of the application of insecticides. *Ann. Rev. Ent,* 5: 327–52.

Courshee, R.J. (1967) Application and use of foliar fungicides. In: Torgeson, D.C. (ed.) *Fungicides: An Advanced Treatise* Vol. I. New York, Academic Press.

Courshee, R.J. and **Ireson, M.J.** (1961) Experiments on the subjective assessment of spray deposits. *J. agric. Engng Res.* 6: 175–82.

Courshee, R.J. and **Ireson, M.J.** (1962) Distribution of granulated materials by helicopters – a rotating granule distributor. *Agric. Aviat.* 4: 131–2.

Courshee, R.J., Daynes, F.K. and **Byass, J.B.** (1954) A tree-spraying machine. NIAE report quoted by Ripper, W.E. in *Ann. appl. Biol.* 42: 288–324.

Coutts, H.H. (1967) Preliminary tests with the UCAR nozzle. *Agric. Aviat.* 9: 123–4.

Coutts, H.H. and **Parish, R.H.** (1967) The selection of a solvent for use with low volume aerial spraying of cotton. *Agric. Aviat.* 9: 125.

Coutts, H.H. and **Yates, W.E.** (1968) Analysis of spray droplet distribution from agricultural aircraft. *Trans. Am. Soc. agric. Engrs.* ll: 25–7.

Cowell, C. and **Lavers, A.** (1987) A laboratory examination of two prototype twin-fluid spray nozzles. *Aspects of Appl. Biol.* 14: 35–52.

Cowell, C. and **Lavers, A.** (1988) The flow rate of formulations through some typical hand-held ultra low volume spinning disc atomizers. *Trop. Pest Management* 34: 150–3.

Cowell, C., Lavers, A. and **Taylor, W.** (1988a) A preliminary evaluation of a surface deposit fluorimeter for assessing spray deposition in the field. *Proc. Int. Symp. on Pesticide Application ANPP Paris* 19–29.

Cowell, C., Lavers, A. and **Taylor, W.** (1988b) Studies to determine the fate of low volume sprays within an orchard environment. *Aspects of Applied Biology* 18: 371–83.

Crease, G.J., Ford, M.G. and **Salt, D.W.** (1987) The use of high viscosity carrier oils to enhance the insecticidal efficacy of ULV formulations of cypermethrin. *Aspects of Applied Biology* 14: 307–22.

Critchley, B.R., Campion, D.G., McVeigh, L.J., McVeigh, E.M., Cavanagh, G.G., Hosny, M.M., Nasr, E.I., Sayed, A., Khidr, A.A. and **Neguib, M.** (1985) Control of the pink bollworm *Pectinophora gossypiella* (Saunders) (hepidoptera Gelechiidae) in Egypt by mating disruption using hollow fibre, laminated flake and microencapsulated formulations of synthetic pheromone. *Bull. Ent. Res.* 45: 329–45.

Cross, J.V. (1991) Patternation of spray mass flux from axial fan air-blast sprayers in the orchard. *BCPC Monograph* 46; 15–22.

Crowdy, S.H. (1972) Translocation. In: Marsh, R.W. (ed.) *Systemic Fungicides.* London, Longman.

Crozier, B. (1976) Specifications for granules. *Br. Crop Prot. Coun. Monogr.* 18: 98–101.

Curran, J. and **Patel, V.** (1988) Use of a trickle irrigation system to distribute entomopathogenic nematodes (Nematoda: Heterorhabditidae) for the control of weevil pests (Coleoptera: Curculionidae) of strawberries. *Aust. J. Exp. Agric.* **28**: 639–43.

Curtis, C.F. (1985) Theoretical models of the use of insecticide mixtures for the management of resistance. *Bull. ent. Res.* **75**: 259–65.

Cussans, G.W. and **Taylor, W.A.** (1976) A review of research on 'controlled drop application' at the ARC. Weed Research Organisation, *13th Br. Weed Cont. Conf.* **3**: 885–94.

Dahl, G.H. and **Lowell, J.R.** (1984) Microencapsulated pesticides and their effects on non-target insects. In: Scher, H.B. (ed.) *Advances in Pesticide Formulation Technology*. ACS Symposium 254: 141–50.

Dale, J.E. (1979) A non-mechanical system of herbicide application with a rope-wick. *PANS* **25**: 431–6.

David, W.A.L. and **Gardiner, B.O.C.** (1950) Particle size and adherence of dusts. *Bull. ent. Res.* **41**: 1–61.

Davies, H. (1967) *Tsetse Flies in Northern Nigeria*. Ibadan, Ibadan UP.

Davies, H. and **Blasdale, P.** (1960) The eradication of *Glossina morsitans submorsitans* Newst. and *Glossina tachinoides* West in part of a river flood plain in Northern Nigeria by chemical means. Part III, *Bull. ent. Res.* **51**: 265–70.

Dean, H.A., Wilson, E.L., Bailey, J.C. and **Riehl, L.A.** (1961) Fluorescent dye technique for studying distribution of spray oil deposits on citrus. *J. econ. Ent.* **54**: 333–40.

De Bach, P. (1974) *Biological Control by Natural Enemies*. Cambridge, Cambridge UP.

Deutsch, A.E. (Ed.) (1976) *Manufacturers of Pesticide Application Equipment*. Oregon, Int. Plant Protect. Centre.

Deutsch, A.E. and **Poole, A.P. (ed.)** (1972) *Manual of Pesticide Application Equipment*. Oregan, Int. Plant Protect. Centre.

Doble, S.J., Matthews, G.A., Rutherford, I. and **Southcombe, E.S.E.** (1985) A system for classifying hydraulic nozzles and other atomisers into categories of spray quality. *Proc. BCPC Conference – Weeds*: 1125–33.

Dobson, C.M., Minski, M.J. and **Matthews, G.A.** (1983) Neutron activation analysis using dysprosium as a tracer to measure spray drift. *Crop Protection* **2**: 345–52.

Dombrowski, L.A. and **Schieritz, M.** (1984) Dispersion and grinding of pesticides. In: Scher, H.B. (ed.) *Advances in Pesticide Formulation Technology* ACS Symposium 254: 63–73.

Dombrowski, N. (1961) Some flow characteristics of single-orifice for spray nozzles. *J. agric. Engng Res.* **6**: 37–44.

Dombrowski, N. (1975) Improvements in and relating to liquid spray devices. Br. Pat. No. 31000/75.

Dombrowski, N. and **Lloyd, T.L.** (1974) Atomisation of liquids by spinning cups. *Chem. engng. J.* **8**: 63–81.

Dombrowski, N., Hibbert, I. and **Strechan, J.R.** (1989) Towards drift free spraying. *Proc. Brighton Crop Protection Conf. – Weeds*: 663–8.

Dorow, R. (1976) Unpublished report on Red-billed quelea *Quelea quelea* bird control in Nigeria.

Doyle, C.J. (1988) Aerial application of mixed virus formulations to control joint infestations of *Panolis flammea* and *Neodiprion sertifer* on lodgepole pine. *Ann. appl. Biol.* **113**: 119–27.

Duncombe, W.C. (1973) The acaricide spray rotation for cotton. *Rhod. agric. J.* **70**: 115–18.

Dunn, P. and **Walls, J.M.** (1978) An introduction to in-line holography and its application. *BCPC Monograph* **22**: 23–34.

Dunning, R.A. and **Winder, G.H.** (1963) Sugar beet yellows: insecticides. *Rep. Rothamsted expl Stn. for 1962*: 218–19.

Durham, W.F. and **Wolfe, H.R.** (1962) Measurement of the exposure of workers to pesticides. *Bull. Wld Hlth Org*. **26**: 75–91.

Eaton, J.K. (1959) Review of materials intended for aerial application and their properties. *Rep. 1st Int. Agric. Aviat. Conf*: 92–102.

Ebeling, W. (1971) Sorptive dusts for pest control. *Ann. Rev. Entomol*. **16**: 123–58.

Edwards, C.J. and Ripper, W.E. (1953) Droplet size, rates of application and the avoidance of spray drift. *Proc. 1953 Br. Weed Cont. Conf*: 348–67.

Elliott, J.G. and Wilson, B.J. (1983) The influence of weather on the efficiency and safety of pesticide application: the drift of herbicides. *BCPC Occasional Publication* **3**.

Elsworth, J.E. and **Harris, D.A.** (1973) The 'Rotostat' seed treater – a new application system. *Proc. 7th Brit. Insectic. Fungic. Conf*: 349–56.

Elvy, J.H. (1976) The laboratory evaluation of potential granular carriers. *Br. Crop Prot. Counc. Monogr*. **18**: 102–8.

Embree, D.C., Dobson, C.M.B. and **Kettela, E.G.** (1976) Use of radio-controlled model aircraft for ULV insecticide application in Christmas tree stands. *Commonwealth for. Rev*. **55**: 178–81.

Emden, H.F. van (1972) Plant resistance to insect pests. Developing 'risk-rating' methods. *Span* **15**: 71–4.

Emden, H.F. van (1989) *Pest Control*, 2nd ed. London, Arnold.

Entwistle, P.F. (1986) Spray droplet deposition patterns and loading of spray droplets with NPV inclusion bodies in the control of *Panolis flammea* in pine forests. In: Samson, R.A., Vlak, J.M. and Peters D (eds) *Fundamental and Applied Aspects of Invertebrate Pathology*. Foundation of the 4th International Congress of Invertebrate Pathology Wageningen, Holland.

Entwistle, P.F., Evans, H.F., Cory, J.S. and Doyle, C.(1990) Questions on the aerial application of microbial pesticides to forests. *Proc. Int. Colloquium Invert. Path. and Microbial Control*, Adelaide, Australia: 159–63.

Evans, E. (1971) Systemic fungicides in practice. *Pestic. Sci*. **2**: 192–6.

Evans, W.H. (1984) Development of an aqueous-based controlled release pheromone-pesticide system. In: Scher, H.B. (ed.) *Advances in pesticide formulation technology*. ACS Symposium **254**: 151–62.

Falcon, L.A. and **Sorensen, A.A.** (1976) Insect pathogen – ULV combination for crop pest control. *PANS* **22**: 322–6.

FAO (1974) The use of aircraft in agriculture. *FAO Agric. Dev. Paper* **94**.

Farmery, H. (1970) The mechanics of granule application. *Br. Crop Prot. Counc. Monogr*. **2**: 101–6.

Farmery, H. (1975) Controlled droplet engineering – the engineer's viewpoint. *Proc. 8th Br. Insectic. Fungic. Conf*: 171–4.

Farmery, H. (1976) Granules and their application. *Br. Crop Prot. Coun. Monogr*. **18**: 93–7.

Felber, H. (1988) Safe and efficient spraying of cereal fungicides with the Ciba-Geigy Croptilter. *ANPP International Symposium on Pesticide Application*, Paris, 249–58.

Fernando, H.E. (1956) A new design of sprayer for reducing insecticide hazards in treating rice crop. *FAO Pl. Prot. Bull*. **4**: 117–20.

Fisher, C. (1971) The new Quantimet 720. *Microscope* **19**: 1–20.

Fisher, R.W. and **Menzies, D.R.** (1979) Pick-up of phosmet wettable powder by codling moth larvae (*Laspeyresia pomonella*) (Lepidoptera: Olethreutidae) and toxicity responses of larvae to spray deposits. *Canadian Entomologist* **111**: 219–23.

Flanagan, J. (1983) Principles of pesticide formulation. In: *Formulation of Pesticides in Developing Countries*. Vienna, UNIDO.

Flint M.L. and **van den Bosch, R.** (1981) *Introduction to Integrated Pest Management*.

New York, Plenum.

Ford, M.G. and Salt, D.W. (1987) Behaviour of insecticide deposits and their transfer from plant to insect surfaces. In: Cottrell, H.J. (ed.) *Pesticides on Plant Surfaces*. London, Wiley.

Ford, M.G., Reay, R.C. and Watts, W.S. (1977) Laboratory evaluation of the activity of synthetic pyrethroids at ULV against the cotton leafworm *Spodoptera littoralis* Boisd. In: McFarlane, N.R. (ed.) *Crop Protection Agents – Their Biological Evaluation*. New York, Academic Press.

Ford, R.E. and Furmidge, C.G.L. (1966) The viscosity and size of invert emulsion droplets used in pesticidal sprays. *Proc. 3rd Int. agric. Aviat. Congr*: 172–9.

Forrester, N.W. and Cahill, M. (1987) Management of insecticide resistance in *Heliothis armigera* (Hubner) in Australia. In: Ford, M.G., Hollonan, D.W., Khambay, P.S. and Sawicki, R.M. (eds) *Biological and Chemical Approach to Combatting Resistance to Xenobiotics*. Amsterdam, Elsevier.

Fraley, R.W. (1984) The preparation of aqueous-based flowables ranging the sample size from sub-gram to several gallons. In: Scher, H.B. (ed.) *Advances in Pesticide Formulation Technology* ACS Symposium **254**: 47–62.

Franz, J.M. (1962) Definitions in biological control. *Proc. llth Int. Congr. Ent. (Vienna l960)* **2**: 670–2.

Fraser, R.P. (1958) The fluid kinetics of application of pesticidal chemicals. In: Metcalfe, R. (ed.) *Advanced Pest Control Research* vol. II. New York, Interscience.

Fraser, R.P., Dombrowski, N. and Routley, J.H. (1963) The production of uniform liquid sheets from spinning cups. *Chem. engng. Sci*, **18**: 315–21.

Frost, A.R. (1974) Rotary atomisation. *Brit. Crop Prot. Coun. Monogr.* **11**: 120–7.

Frost, A.R. (1984) Simulation of an active spray boom suspension. *J. Agric. Engng. Res.* **30**: 313–25.

Frost, A.R. (1990) A pesticide injection metering system for use on agricultural spraying machines. *J. Agric. Engng. Res.* **46**: 55–70.

Frost, A.R. and Lake, J.R. (1981) The significance of drop velocity to the determination of drop size distribution of agricultural sprays. *J. Agric. Engng. Res.* **26**: 367–70.

Frost A.R. and Law, S.G. (1981) Extended flow characterisitcs of the embedded electrode spray-charging nozzle. *J. Agric. Engng. Res.* **26**: 79–86.

Frost, A.R. and Miller, P.C.H. (1988) Closed chemical transfer system. *Aspects of Applied Biology* **18**: 345–59.

Frost, A.R. and O'Sullivan, J.A. (1988) Verification and use of a mathematical model of an active twin link boom suspension. *J. Agric. Engng. Res.* **40**: 259–74.

Fry, W.E. (1987) Advances in disease forecasting. In: Brent, K.J. and Atkins, R.J. (eds) *Rational Pesticide Use* Cambridge.

Fryer, J.D. (1977) Recent developments in the agricultural use of herbicides in relation to ecological effects. In Perring, F.H. and Mellanby, K. (eds) *Ecological Effects of Pesticides*. London, Academic Press.

Fryer J.D. and Makepeace, R.J. (1977) (eds) *Weed Control Handbook, Vol. I Principles* (6th edn.) Oxford, Blackwell.

Fuller-Lewis, P. and Sylvester, N.K. (1974) ULV in British Horticultural practice. *Br. Crop Prot. Counc. Monogr.* **11**: 262–74.

Galloway, B.T. (1891) The improved Japy knapsack sprayer. *J. Mycol.* **7**: 39.

Garnett, R. (1980) A low volume herbicide applicator for tropical small-holder farmers. *Proc. 1980 Br. Crop Prot. Counc. Conf. Weeds*: 629–36.

Gast, R.T. (1959) The relationship of weight of Lepidopterous larvae to effectiveness of topically applied insecticides. *J. econ. Ent.* **52**: 1115–17.

Gibbs, J.N. and Dickinson, J. (1975) Fungicide injection for the control of Dutch elm disease. *Forestry* **48**: 165–76.

Gilbert, A. (1989) Reducing operator exposure by the improved design and handling of liquid pesticide containers. *Proc. Brighton Crop Protection Conf. – Weeds*: 593–600.

Gilbert, A.J. and **Bell, G.J.** (1990) Test methods and criteria for selection of types of coveralls suitable for certain operations involving handling or applying pesticides. *J. of Occupational Accidents* **11**: 255–68.

Gilbert, D. and **Macrory, R. (eds.)** (1989) *Pesticide Related Law*. British Crop Protection Council Publication.

Giles, D.K. and **Blewett, T.C.** (1991) Effects of conventional and reduced volume charged spray application techniques on dislodgeable foliar residues of captan on strawberries. *J. Agric. Food Cham.* (in press)

Giles, D.K. and **Law, S.E.** (1985) Space charge deposition of pesticide sprays onto cylindrical target arrays. *Transactions of the Am. Soc. of Ag. Engineers* **28**, 658–664.

Gledhill, J.A. (1975) A review of ultra-low volume spray usage in Central Africa since 1954 and some recent developments in Rhodesia. *Proc. I Congr. ent. Soc. Sth. Afr*: 259–67.

Goehlich, H. (1970) Metering and distribution of coarse and fine granules. *Br. Crop Prot. Counc. Monogr*. **2**: 107–13.

Goehlich, H. (1979) A contribution to the demands of reduced application rates and reduced drift. *Proc. 1979 BCP Conf. – Pests and Diseases*: 767–75.

Goose, J. (1991) Mosquito control using mistblower sprayers for residual deposit of bendiocarb ULV in Mexico. *BCPC Monograph* **46**: 245–48.

Gower, J. and **Matthews, G.A.** (1971) Cotton development in the southern region of Malawi. *Cott. Gr. Rev*, **48**: 2–18.

Graham-Bryce, I.J. (1975) The future of pesticide technology: opportunities for research. *Proc 8th Br. Insectic. Fungic. Conf.* **3**: 901–14.

Graham-Bryce, I.J.(1977) Crop Protection: a consideration of the effectiveness and disadvantages of current methods and of the scope for improvement. *Phil. Trans. R. Soc. Lond. B281* 163–79.

Graham-Bryce, I.J. (1988) Pesticide application to seeds and soil: unrealised potential? *BCPC Monograph* **39**: 3–14.

Graham-Bryce, I.J.(1989) Environmental impact: putting pesticides into perspective. *Proc Brighton Crop Prot. Conf. – Weeds*: 3–20.

Gratz, N. (1985) Control of dipteran vectors. In: Haskell, P.T. (ed.) *Pesticide Application: Principles and Practice*. Oxford, Oxford University Press.

Gratz, N.G. and **Dawson, J.A.** (1963) The area distribution of an insecticide (fenthion) sprayed inside the huts of an African village. *Bull. Wld Hlth Org*. **29**: 185–96.

Greaves, M.P. and **Marshall, E.J.P.** (1987) Field margins: definitions and statistics. *BCPC Monograph*. **35**: 3–10.

Gregory, P.H. (1976) The work of the International Cocoa Black pod research project at Gambari. *Cocoa Growers' Bull*. **25**: 17–19.

Gregory, P.H. and **Maddison, A.C.** (1981) Epidemiology of *Phytophthora* on cocoa in Nigeria. *CMI Phytopathological Paper* **25**, 188 pp.

Gressel, J. (1987) Strategies for prevention of herbicide resistance in weeds. In: Brent, K.J. and Atkin, R.J. (eds) *Rational Pesticide Use*. Cambridge.

Griffiths, D.C., Cayley, G.R. Etheridge, P. Goodchild, R., Hulme, P.J., Lewthwaite, R.J., Pye, B.J., Scott, G.C. and **Stevenson, J.H.** (1984) Application of insecticides, fungicides and herbicides to cereals with charged rotary atomizers. *Proc Br. Crop Prot. Conf*: 1,021–26.

Gunn, D.L., Graham, J.F., Jaques, E.C., Perry, F.C., Seymour, W.G., Telford, T.M., Ward, J., Wright, E.N. and **Yeo, D.** (1948) Aircraft spraying against the desert locust in Kenya, 1945. *Anti-Locust Bull.* 4.

Gunning, R.V., Easton, C.S., Greenup, L.R. and **Edge, V.E.** (1984) Pyrethroid resistance in *Heliothis armiger* (Hubner) (Lepidoptera: Noctuidae) in Australia. *J. econ. Ent.* **77**: 12,837.

Hadar, E. (1991) Development criteria for an air-assisted ground crop sprayer. *BCPC Monograph* **46**: 23–6.

Hadaway, A.B. and **Barlow, F.** (1965) Studies on the deposition of oil drops. *Ann. Appl. Biol.* **55**: 267–74.

Haggar, R.J., Stent, C.J. and **Isaac, S.** (1983) A prototype hand-held patch sprayer for killing weeds activated by spectral differences in crop/weed canopies. *J. agric. Engng. Res.* **28**: 349–58.

Hale, O.D. (1975) Development of a wind tunnel model technique for orchard spray application research. *J. agric. Engng Res.* **20**: 303–17.

Haley, J. (1973) *Expert Flagging.* Univ. Grand Forks, North Dakota, North Dakota Press.

Hall, B.D. and **Inaba, D.J.** (1989) Use of water-sensitive paper to monitor the deposition of aerially applied insecticides. *J. econ. Ent.* **82**: 974–80.

Hall, D.J. and **Marr, S.G.** (1989) Microcapsules. In: Jutsum, A.R. and Gordon, R.F.S. (eds) *Insect Pheromones in Plant Protection*: 199–242. Chichester, Wiley.

Hall, D.W. (1970) Handling and storage of food grains in tropical and subtropical areas. *FAO Agric. Development Paper* **90**.

Hall, L.B. (1955) Suggested techniques, equipment and standards for the testing of hand insecticide spraying equipment. *Bull. Wld Hlth Org.* **12**: 371–400.

Halmer, P. (1988) Technical and commercial aspects of seed pelleting and film coating. *BCPC Monograph* **39**; 191–204.

Hamon, J. and **Stiles, A.R.** (1975) Global programmes for disease vector control. *Proc. 8th Brit. Insect. Fungic. Conf*: 1,045–52.

Hankawa, Y. and **Kohguchi, T.** (1989) Improvement in distribution of BPMC deposits and classification of dust particles. *J. Pesticide Sci.* **14**: 443–52.

Hanna, A.D. and **Nicol, J.** (1954) Application of a systemic insecticide by trunk implantation to control a mealybug vector of the cocoa swollen shoot virus. *Nature Lond.* **169**: 120.

Harazny, J. (1976) Agricultural equipment for the M–15 aircraft. *Proc. 5th Int. agric. Aviat. Congr*: 378–98.

Harris, A.G. (1988) The training and certification of pesticide users. *Aspects of Applied Biology* **18**: 311–15.

Harris, E.G., Cooper, J.F., Flower, L.S., Smith, S.C. and **Turner, C.R.** (1990) Toxicity of insecticide aerosol drops to tsetse flies 1. Some effects of temperature, formulation and drop size. *Trop. Pest Management* **36**: 162–5.

Hart, C.A. (1979) Use of the scanning electron microscope and cathodoluminescence in studying the application of pesticides to plants. *Pesticide Sci.* **10**: 341–57.

Hart, C.A. and **Young, B.W.** (1987) Scanning electron microscopy and cathodoluminescence in the study of interactions between spray droplets and leaf surfaces. *Aspects of Applied Biology* **14**: 127–40.

Hartley, G.S and **Graham-Bryce, I.J.** (1980) *Physical Principles of Pesticide Behaviour.* Academic Press.

Headley, J.C. (1972) The economics of pest management. In Metcalf, R.L. and Luckman, W. (eds) *Introduction to Insect Pest Management.* New York, Wiley-Interscience.

Heath, D., Knott, R.D., Knowles, D.A. and **Tadros, Th. F.** (1984) Stabilization of aqueous pesticidal suspensions by graft copolymers and their subsequent weak flocculation by addition of free polymer. In: Scher, H.B (ed.) *Advances in Pesticide Formulation Technology.* ACS Symposium Series **254**: 11–22.

Heijne, C.G. (1978) A study of the effect of disc speed and flow rate on the performance of the 'Micron Battleship'. *Proc. 1978 Br. Crop Prot. Counc. Conf. Weeds*: 673–9.

Helyer, N.L. (1985) The ecological selectivity of pesticides in integrated pest management. In: Hussey, N.W. and Scopes, N.E.A. (eds) *Biological Pest Control – The Glasshouse Experience*. Blandford.

Hetrick, W., Keim, T., Kreyling, J., Sutton, E. and Miller, T.W. (1989) An electrically controlled pesticide agitation system for rotary-winged aircraft mounted sprayers. *J. Am. Mosq. Cont. Ass.* **5**: 432–3.

Hewitt, A. (1991) Assessment of rotary atomiser attachments for motorised knapsack mistblowers. *BCPC Monograph* **46**: 271.

Hielkama, J.U. (1990) Satellite environmental monitoring for migrant pest forecasting by FAO: the Artemis system. *Phil. Trans. R. Soc. Lond. B.* **328**: 705–17.

Higgins, A.E.H. (1964) The selection of spraying equipment. *Cocoa Growers' Bull.* **3**: 21–6.

Higgins, A.H. (1967) Spread factors for technical malathion spray. *J. econ. Ent.* **60**: 280–1.

Hill, B.D. and Inaba, D.J. (1989) Use of water-sensitive paper to monitor the deposition of aerially applied insecticides. *J.econ. Ent.* **82**: 974–80.

Hill, R.F. (1963) The Beagle Wallis autogyro. *Agric. Aviat.* **5**: 48–51.

Hill, R.F. and Johnstone, D.R. (1962) Tests with a rotary granule dispenser on a Dragonfly helicopter. *Agric. Aviat.* **4**: 133–5.

Himel, C.M. (1969a) The fluorescent particle spray droplet tracer method. *J. econ. Ent.* **62**: 912–16.

Himel, C.M. (1969b) The optimum size for insecticide spray droplets. *J. econ. Ent.* **62**: 919–25.

Himel, C.M. (1974) Analytical methodology in ULV. *Br. Crop Prot. Counc. Monogr.* **11**: 112–19.

Himel, C.M. and Moore, A.D. (1967) Spruce budworm mortality as a function of aerial spray droplet size. *Science* **156**: 1,250–1.

Hinze, J.O. and Milborn, H. (1950) Atomisation of liquids by means of a rotating cup. *J. appl. Mech.* **17**: 145.

Hislop, E.C. (1987) Requirements for effective and efficient pesticide application. In: Brent, K.J. and Atkin, R.K. (eds) *Rational Pesticide Use*. Cambridge.

Hislop, E.C. (1991) Air-assisted crop spraying: an introductory review. *BCPC Monograph* **46**: 3–14.

Hobson, P.A., Miller, P.C.H., Walklate, P.J., Tuck, C.R. and Western, N. (1990) Spray drift from hydraulic nozzles: the use of a computer simulation model to examine factors influencing drift. *Agricultural Engineering Conference, Berlin*.

Holden, A.V. and Bevan, D. (eds) (1978) *Control of Pine Beauty Moth by Fenitrothion in Scotland*. Edinburgh, Forestry Commission.

Holden, A.V. and Bevan, D. (eds) (1981) *Aerial Application of Insecticide against Pine Beauty Moth*. Edinburgh, Forestry Commission.

Hollomon, D.W. (1986) Contribution of fundamental research to combating resistance. *1986 Br. Crop. Prot. Conf. – Pests and Diseases*: 801–10.

Holloway, P.J. (1970) Surface factors affecting the wetting of leaves. *Pestic. Sci.* **1**: 156–63.

Hopkinson, P.R. (1974) The prospects of enhanced impaction of fine sprays by electrostatic charging. *Br. Crop Prot. Counc. Monogr*, **11**: 166–79.

Houser, J.S. (1922) The airplane in Catapla sphinx control. *Ohio agric. expl. Stn Bull.* **7**: 126–36.

Howitt, A.J., Retzer, H.J., Himel, C.M., Hogmire, H. and Ayers, G.S. (1980) Studies on the parameters which affect spray spectra characteristics from rotary sleeve nozzles. *J. econ. Ent.* **73**: 477–83.

Hoyt, S.C. (1970) The developing program of integrated control of pest of apple in Washington. *AAAS Symposium on Theory and Practice of Biological Control*, Boston, 30–31 Dec. 1969.

Hughes, K.L. and **Frost, A.R.** (1985) A review of agricultural spray metering. *J. agric. Engng. Res.* **32**: 197–207.

Humphries, A.W. and **West, D.** (1984) The terramatic boomsprayer – automation in agriculture. *Proc. 7th Australian Weeds Conf:* 36–40.

Hunt, C.R. (1947) Toxicity of insecticide dust diluents and carriers to larvae of the Mexican bean beetle. *J. econ. Ent.* **40**: 215–19.

Hunt, G.M. and Baker, E.A. (1987) Application and fluoresence microscopy, autoradiography and energy dispersive X-ray analysis to the study of pesticide deposits. *Aspects Appl. Biol.* **14**: 113–26.

Huntington, K. and **Johnstone, D.R.** (1973) A remote recording field installation for monitoring micrometeorological variables during spray trials. *COPR Misc. Rep.* **14**.

Hussey, N.W. and **Scopes, N.E.A.** (1985) *Biological Pest Control – The Glasshouse Experience*. Blandford.

Inculet, I.I., Castle, G.S.P., Menzies, D.R. and **Frank, R.** (1981) Deposition studies with a novel form of electrostatic crop sprayer. *J. of Electrostatics* **10**: 65–72.

Ingram, W.R. and **Green, S.M.** (1972) Sequential sampling for bollworms on rain grown cotton in Botswana. *Cott. Gr. Re.* **49**: 265–75.

Interflug (1975) Method of calculating the optimum work parameters for agricultural flights. *Proc. 5th Int. agric. Aviat. Congr:* 213–22.

Isler, D.A. (1966) Atomization of low volume malathion aerial spray. *J. econ. Ent.* **59**: 688–90.

Jadhav, S.J. Sharma, R.P and Salunke, D.K. (1981) Naturally occurring toxic alkaloids in foods. *Critical Review of Toxicology* **9**: 21–104.

Jeffs, K.A. (1986) (ed.) *Seed Treatment*. BCPC Monograph, 2nd edition. Surrey, BCPC Publication.

Jeffs, K.A. and **Tuppen, R.J.** (1986) The application of pesticides to seeds. In Jeffs, K.A. (ed.) *Seed Treatment*. BCPC Monograph, 2nd edition. Surrey, BCPC Publication.

Jegatheeswaran, P. (1978) Factors concerning the penetration and distribution of drops in low growing crops. *BCPC Monograph* **22**: 91–9.

Jensen, J.A., Taylor, J.W. and **Pearce, G.W.** (1969) A standard and rapid method for determining nozzle-tip abrasion. *Bull. Wld Hlth Org.* **41**: 937–40.

Jepson, P.C., Cuthbertson, P.S., Thecker, J.R. and **Bowie, M.H.** (1987) A computerised droplet size analysis system and the measurement of non-target invertebrate exposure to pesticides. *Aspects of Applied Biology* **14**: 97–112.

Jepson, W.F. (1976) Review of granular pesticides and their use. *Br. Crop Prot. Counc. Monogr.* **18**: 1–9.

Johnstone, D.R. (1960) Assessment technique: 2, Photographic paper. TPRU, *Porton Misc. Rep:* 177.

Johnstone, D.R. (1970) High volume application of insecticide sprays in Cyprus citrus. *PANS* **16**: 146–61.

Johnstone, D.R. (1971) Droplet size for low and ultra low volume spraying. *Cott. Gr. Rev.* **48**: 218–33.

Johnstone, D.R. (1972) A differential thermistor thermometer for measuring temperature gradients in the vicinity of the ground. *E. Afr. Agric. For. J.* **37**: 300–7.

Johnstone, D.R. (1973a) Spreading and retention of agricultural sprays on foliage. In: Valkenburg, W. van (ed.) *Pesticide Formulations*. New York, Marcel Dekker.

Johnstone, D.R. (1973b) Insecticide concentration for ultra-low-volume crop spray applications. *Pestic. Sci.* **4**: 77–82.

Johnstone, D.R. (1981) Crop spraying by hovercraft, kites, RPV, ATV and other unconventional devices. *Outlook on Agriculture* **10**: 361–5.

Johnstone, D.R. (1985) Physics and meteorology. In: Haskell, P.T. (ed.) *Pesticide Application: Principles and Practice*. Oxford.

Johnstone, D.R. (1991) Variations in insecticide dose received by settled locusts – a computer model for ultra-low volume spraying. *Crop Protection* **10**: 183–94.

Johnstone, D.R. and **Cooper, J.F.** (1986) Forecasting the efficiency of the sequential aerosol technique for tsetse fly control. *Pestic. Sci.* **17**: 675–85.

Johnstone, D.R. and **Huntington, K.A.** (1970) A comparison of visual microscope methods of spray droplet size measurement using eye-pieces employing the image shearing principle and the globe and circle eye-piece graticule. *J. agric. Engng. Res.* **15**: 1–10.

Johnstone, D.R. and **Huntington, K.A.** (1977) Deposition and drift of ULV and VLV insecticide sprays applied to cotton by hand applications in N. Nigeria. *Pestic. Sci.* **8**: 101–9.

Johnstone, D.R. and **Johnstone, K.A.** (1977) Aerial spraying of cotton in Swaziland. *PANS* **23**: 13–26.

Johnstone, D.R. and **Matthews, G.A.** (1965) Evaluation of swath pattern and droplet size provided by a boom and nozzle installation fitted to a Hilter UH–12 helicopter. *Agric. Aviat.* **7**: 46–50.

Johnstone, D.R. and **Watts, W.S.** (1966) Physico-chemical assessments of cotton spraying by aeroplane and knapsack ground sprayer, Ilonga. TPRU, *Porton Rep.* **323**.

Johnstone, D.R., Lee, C.W., Hill, R.F., Huntington, K.A. and **Coles, J.S.** (1971) Ultra-low volume spray gear for installation on an ultra light weight helicopter or autogyro. *Agric. Aviat.* **13**: 57–61.

Johnstone, D.R., Huntington, K.A. and **King, W.J.** (1972a) Tests of ultra-low volume spray gear installed on the light weight Campbell Cricket autogyro. *Agric. Aviat.* **14**: 82–6.

Johnstone, D.R., Walker, P.T. and **Huntington, K.A.** (1972b) Ultra-low-volume, hand-operated motorised sprayers for insecticide application in citrus. *Int. Pest. Cont.* **14**: 8–19.

Johnstone, D.R., Huntington, K.A. and **Coutts, H.H.** (1974a) 'Penetration of spray droplets applied by helicopter into a riverine forest habitat of tsetse flies in West Africa. *Agric. Aviat.* **16**: 71–82.

Johnstone, D.R., Huntington, K.A. and **King, W.J.** (1974b) Micrometeorological and operational factors affecting ultra-low volume spray applications of insecticides on to cotton and other crops. *Agric. Met.* **13**: 39–57.

Johnstone, D.R., Huntington, K.A. and **King, W.J.** (1975) Development of hand spray equipment for applying fungicides to control *Cercospora* disease of groundnuts in Malawi. *J. agric. Engng Res.* **20**: 379–89.

Johnstone, D.R., Rendell, C.H. and **Sutherland, J.A.** (1977) The short-term fate of droplets of coarse aerosol size in ultra-low volume insecticide application onto a tropical field crop. *J. aerosol Sci.* **8**: 395–407.

Johnstone, D.R., Cooper, J.F. and **Dobson, H.M.** (1987) The availability and fall-out of an insecticidal aerosol dispersed from aircraft during operations for control of tsetse fly in Zimbabwe. *Atmos. Envir.* **21**: 2,311–21.

Johnstone, D.R., Allsopp, R., Cooper, J.F. and **Dobson, H.M.** (1988) Predicted and observed droplet deposition in tsetse flies (*Glossina morsitans*) following aerosol application from aircraft. *Pestic. Sci.* **22**: 107–21.

Johnstone, D.R., Cooper, J.F., Dobson, H.M. and **Turner, C.R.** (1989a) The collection of aerosol droplets by resting tsetse flies *Glossina morsitans* Westwood (Diptera: Glossinidae). *Bull. ent. Res.* **79**: 613–24.

Johnstone, D.R., Cooper, J.F., Gledhill, J.A. and **Jowah, P.** (1982) Preliminary trials to examine the drift of charged spray droplets. *Proc 1982 Br. Crop Prot. Conf. – Weeds*: 1,025–32.

Johnstone, D.R., Cooper, J.F., Flower, L.S., Harris, E.G., Smith, S.C. and **Turner,**

C.R. (1989b) A means of applying mature aerosol drops to insects for screening biocidal activity. *Trop. Pest Management* **35**: 65–6.

Johnstone, D.R., Cooper, J.F., Casci, F. and **Dobson, H.M.** (1990) The interpretation of spray monitoring data in tsetse control operations using insecticidal aerosols applied from aircraft. *Atmos Envir.* **24A**: 53–61.

Johnstone, K.A. and **Johnstone, D.R.** (1976) Power requirement and droplet size characteristics of a new ultra-low volume, hand carried battery operated insecticide sprayer. *COPRA Misc. Rep*: 26.

Jollands, P. (1991) Evaluation of knapsack mistblowers for the control of coffee leaf rust in Papua New Guinea. *BPCP Monograph* **46**: 177–84.

Jones, K.G., Morgan, N.G. and **Cooke, A.T.K.** (1974) Experimental application equipment. *Long Ashton Annual Report 1974*: 107.

Jones, T.R. (1966) Comparison of hand-operated machines for cotton pest control in Uganda, Part I – Description of machines under test. *E. Afr. Agric. For. J.* **31**: 409–15.

Joyce, R.J.V. (1975) Sequential aerial spraying of cotton at ULV rates in the Sudan Gezira as a contribution to synchronized chemical application over the area occupied by the pest population. *Proc. 5th Int. agric. Aviat. Congr*: 47–54.

Joyce, R.J.V. (1977) Efficiency in pesticide application with special reference to insect pests of cotton in the Sudan Gezira. Seminar organized by Rubber Research Institute of Malaysia and Agricultural Institute of Malaysia.

Joyce, R.J.V. (1985) Application from the air. In: Haskell, P.T. (ed.) *Pesticide Application: Principles and Practice.* Oxford.

Joyce, R.J.V., Marmol, L.C., Brunicardi, M.F.J. and **Kinvik, K.** (1968) Waterless spraying in East Pakistan using the Decca Navigation System. *Agric. Aviat.* **10**: 118–24.

Kao, C., Rafatjah, H. and **Kolta, S.** (1972) Replacement of spray nozzle tips based on operational considerations. *Bull. Wld Hlth Org.* **46**: 493–501.

Keeler, A.A. (1971) A note on the development of power-line markers for aerial crop spraying operations in Australia. *Agric. Aviat.* **13**: 81–6.

Keiding, J. (1974) The development of resistance to pyrethroids in field populations of Danish houseflies. *The Third International Congress of Pesticide Chemistry*, Helsinki, July 1974.

Kendrick, J.A. and **Alsop, N.** (1974) Aerial spraying with endosulphan against *Glossina morsitans morsitans* in the Okavango Delta area of Botswana. *PANS* **20**: 392–9.

Kennedy, J.S., Ainsworth, M. and **Toms, B.A.** (1948) Laboratory studies on the spraying of locusts at rest and in flight. *Anti-Locust Bull.* **2**.

Khoo, K.C., Ho, C.T., Ng, K.Y. and **Lim, T.K.** (1983) Pesticide application technology in perennial crops in Malaysia. In: Lim, G.S. and Ramasamy, S. (eds) *Pesticide Application Technology.* MAPPS, Kuala Lumpur.

Kirby, A.H.M. (1973) Progress in the control of orchard pests by integrated methods. *Hort. Abstr.* **43**: 1–16, 57–65.

Kiritani, K. (1974) The effect of insecticides on natural enemies, particular emphasis on the use of selective and low rates of insecticides. Paper submitted to *The International Rice Research Conf., IRRI*, Philippines, April 1974.

Klein, H.H. (1961) Effects of fungicides, oil and fungicide-oil-water emulsions on the development of *Cercospora* leaf spot of bananas in the field. *Phytopath.* **51**: 294–7.

Klock, J.W. (1956) An automatic molluscicide dispenser for use in flowing water. *Bull. Wld Hlth Org.* **14**: 639.

Knollenberg, R.G. (1970) The optical array: an alternative to extinction and scattering for particle size measurement. *J. appl. Meteorology* **9**: 86–103.

Knollenberg, R.G. (1971) Particle size measurements from aircraft using electro-optical techniques. *Proc. Electro-optical Systems Design Conf.* **1**: 218–33.

Knollenberg, R.G. (1976) The use of the low power lasers in particle size spectrometry. *Proc. Appl. of Low Powered Lasers* **92**: 137–52.

Koeman, J.H., Rijksen, H.D., Smies, M., Na'isa, B.K. and **Maclennan, K.J.R.** (1971) Faunal changes in a swamp habitat in Nigeria sprayed with insecticide to exterminate *Glossina*, *Ned. J. Zool.* **21**: 434–63.

Krishhan, P., Valesco, A., Williams, T.H. and **Kemble, L.J.** (1989) Spray pattern displacement measurements of TK-SS 2.5 Flood Tip Nozzles. *Trans. Am. Soc. Ag. Eng.* **32**: 1,173–6.

Kruse, C.W., Hess, E.D. and **Ludwik, G.F.** (1949) The performance of liquid spray nozzles for aircraft insecticide application. *J. Nat. Malaria Soc.* **8**: 312–34.

Kuhlman, D.K. (1981) Fly-in technology for agricultural aircraft. *World of Agricultural Aviation* **8**: 12–17.

Kummel, K., Gohlich, H. and **Westpal, O.** (1991) Development of practice-oriented control test methods for orchard spray machines by means of a vertical test stand. *BCPC Monograph* **46**: 27–33.

Lacey, A.J. (1985) Weed control. In: Haskell, P.T. (ed) *Pesticide Application: Principles and Practice*. Oxford.

Lake, J.R. (1970) Spray formation from vibrating jets. *Br. Crop Prot. Counc. Monor.* **2**: 61–8.

Lake, J.R. and **Dix, A.J.** (1985) Measurement of droplet size with a PMS optical array probe using an X–Y nozzle transporter. *Crop Protection* **4**: 464–72.

Lake, J.R. and **Marchant, J.A.** (1984) Wind tunnel experiments and a mathematical model of electrostatic deposition in barley. *J. Ag. Engng Res.* **30**: 185–95.

Lake, J.R. and **Taylor, W.A.** (1974) Effect of the form of a deposit on the activity of barban applied to *Avena fatua* L. *Weed Res.* **14**: 13–18.

Lake, J.R., Frost, A.R. and **Lockwood, A.** (1978) Drift from an ulvamast sprayer. *Proc. 1978 Br. Crop Prot. Counc. Conf. Weeds*: 681–6.

Lake, J.R., Frost, A.R. and **Lockwood, A.** (1980) The flight times of spray drops under the influence of gravitational, aerodynamic and electrostatic forces. *BCPC Monograph* **24**: 119–25.

Lake, J.R., Green, R., Tofts, M. and **Dix, A.J.** (1982) The effect of an aerofoil on the penetration of charged spray into barley. *Proc. 1982 Br. Crop Prot. Conf. – Weeds*: 1,009–16.

Landers, A.J. (1988) Closed system spraying – the Dose 2000. *Aspects of Applied Biology* **18**: 361–69.

Lane, M.D. and **Law, S.E.** (1982) Transient charge transfer in living plants undergoing electrostatic spraying. *Trans. Am. Soc. Ag. Engrs* **25**: 1,148–53, 1,159.

Langmuir, I. and **Blodgett, K.B.** (1946) A mathematical investigation of water droplet trajectories. *USAF Techn. Rep.* 5,418, pp. 65.

Last, A.J., Parkin, C.S. and **Beresford, R.H.** (1987) Low-cost digital image analysis for the evaluation of aerially applied pesticide deposits. In: *Computers and Electronics in Agriculture* **1** Amsterdam, Elsevier.

Latta, R., Anderson, L., Rogers, E., La Mer, V., Hochber, S., Lauterbach, H. and **Johnson, I.** (1947) The effect of particle size and velocity of movement of DDT aerosols in a wind tunnel on the mortality of mosquitoes. *J. Wash. Acad. Sci.* **37**: 397–407.

Lavabre, E.M. (1971) Early results of ultra low volume spraying against mirid. *Café, Cacao, Thé* **15**: 135–42.

Law, S.E. (1977) Electrostatic Spray Nozzle System, United States Patent 9004733.

Law, S.E. (1978) Embedded-electrode electrostatic-induction spray-charging nozzle: theoretical and engineering design. *Trans. Am. Soc. Ag. Engrs* **21**: 1,096–104.

Law, S.E. (1980) Droplet charging and electrostatic depositions of pesticide sprays – research and development in the USA. *BCPC Monograph* **24**: 85–94.

Law, S.E. (1986) Charge and mass flux in the radial electric field of an evaporating charged water droplet: an experimental analysis. *IEEE (IAS) Conference Record* 1,434–9.

Law, S.E. (1987) Basic phenomena active in electrostatic pesticide spraying. In: Brent, K.J. and Atkin, R.K. (eds) *Rational Pesticide Use*. Cambridge, Cambridge University Press.

Law, S.E. (1989) Electrical interaction occurring at electrostatic spraying targets. *J. Electrostatics* **23**: 145–56.

Law, S.E. and **Bailey, A.G.** (1984) Perturbations of charged-droplet trajectories caused by induced target corona-LDA analysis. *Inst. of Electrical and Electronic Engineers Trans. on Ind. Appl.* **1A–20**: 1,613–22.

Law, S.E. and **Cooper, S.C.** (1987) Induction charging characteristics of conductivity enhanced vegetable-oil sprays. *Trans. Am. Soc. Ag. Engrs* **30**: 75–9.

Law, S.E. and **Lane, M.D.** (1981) Electrostatic deposition of pesticide spray on to foliar targets of varying morphology. *Trans. Am. Soc. Ag. Engrs.* **24**:1,441–5.

Law, S.E. and **Lane, M.D.** (1982) Electrostatic deposition of pesticide sprays onto ionizing targets: charge- and mass-transfer analysis. *Inst. of Electrical and Electronic Engineers (IEEE) Transactions* **1A–19**: 673–9.

Law, S.E., **Merchant, J.A.** and **Bailey, A.G.** (1985) Charged-spray deposition characteristics within cereal crops. *Inst. of Electrical and Electronic Engineers Trans. on Ind. Appl.* **1A**: 685–93.

Lawson, T.J. and **Uk, S.** (1979) The influence of wind turbulence crop characteristics and flying height on the dispersal of aerial sprays. *Atmos. Envir.* **13**: 711–15.

Le Baron, H.M. and **Gressel, J.** (eds) (1982) *Herbicide Resistance in Plants*. New York, Wiley-Interscience.

Lee, C.W. (1974) Aerial ULV sprays for Cayman Island mosquitoes. *Br. Crop Prot. Coun. Monogr.* **11**: 190–6.

Lee, C.W. and **Miller, A.W.D.** (1966) Trials with devices for atomising insecticides by exhaust gases from a Cessna 182 aircraft. *Agric. Aviat.* **8**: 19–22.

Lee, C.W., **Parker, J.D.**, **Philippon, B.** and **Baldry, D.A.T.** (1975a) Prototype rapid release system for the aerial application of larvicide to control *Simulium damnosum* Theo. *PANS* **21**: 92–102.

Lee, C.W., **Pope, G.G.**, **Kendrick, J.A.**, **Bowles, G.** and **Wiggett, G.** (1975b) Aerosol studies using an Aztec aircraft fitted with Micronair Equipment for tsetse fly control in Botswana. *COPRA Misc. Rep.* **18**, 9 pp.

Lee, C.W., **Parker, J.D.**, **Baldry, D.A.T.** and **Molyneux, D.H.** (1978) The experimental application of insecticides from a helicopter for the control of riverine populations of *Glossina tachinoides* in West Africa. II Calibration of equipment and insecticide dispersal. *PANS* **24**: 404–22.

Lee, K.C. (1976) The design of ducted spreaders for the application of powders and granules. *Proc. 5th Int. agric. Aviat. Congr*: 328–35.

Lee, K.C. and **Stephenson, J.** (1969) The distribution of solid materials. *Proc. 4th Int. agric. Aviat. Congr*: 203–11.

Lefebvre, A.H. (1989) *Atomization and Sprays*. London, Hemisphere Publishing.

Lehtinen, J.R., **Adams, A.J.**, **Lindquist, R.K.**, **Hell, F.R.** and **Simmons, H.C.** (1989) Use of an air-assisted electrostatic sprayer to increase pesticide efficiency in greenhouses. In: Hazen, J.L. and Houde, D.A. (eds) *Pesticide Formulations and Application Systems* **9**: 165–78 (ASTM STP 1,036).

Le Patourel, G.N.J. (1986) The effect of grain moisture content on the toxicity of a sorptive silica dust to four species of grain beetle. *J. Stored Prod. Res.* **22**: 63–9.

Lewis, T. (1965) The effects of an artificial windbreak on the aerial distribution of flying insects. *Am. appl. Biol.* **55**: 503–12.

Lewis, T. (1972) Aerial baiting to control leaf cutting ants. *PANS* **18**: 71–4.

Lewis, T. (1987) Practice and progress in pest forecasting. In: Brent, K.J. and Atkin, R.K. (eds) *Rational Pesticide Use*. Cambridge.

Lloyd, G.A. Bell, G.J., **Howarth, J.A.** and **Samuels, S.W.** (1986) Rotary atomisers:

comparative spray drift studies. Unpublished report. Operator Protection Group, Central Science Laboratory, Ministry of Agriculture, Fisheries of Food, Harpenden.

Lodeman, E.G. (1896) *The Spraying of Plants*. London, Macmillan.

Lofgren, C.S. (1970) Ultra low volume applications of concentrated insecticides in medical and veterinary entomology. *Am. Rev. Entomol.* **15**: 321–42.

Lofgren, C.S., Ford, H.R., Tonn, R.J. and **Jatanasen, S.** (1970a) The effectiveness of ultra-low volume applications of malathion at a rate of 6 US fluid ounces per acre in controlling *Aedes aegypti* in a large scale test at Nakhon Sawan, Thailand. *Bull. Wld Hlth Org.* **42**: 15–25.

Lofgren, C.S., Ford, H.R., Tonn, R.J. and **Jatanasen, S.** (1970b) Equipping a multi-engined aircraft with a fuselage mounted spray system for the ultra low volume application of malathion. *Bull. Wld Hlth Org.* **42**: 157–63.

Lofgren, C.S., Anthony, D.W. and **Mount, G.A.** (1973) Size of aerosol droplets impinging on mosquitoes as determined with a scanning electron microscope. *J. econ. Ent.* **66**: 1,085–8.

Lovro, I. (1975) Optimum method of agricultural airstrips planning. *Proc. 5th Int. Agric. Aviat. Congr*: 177–89.

Maan, W.J. (1961) Fifty years of agricultural aviation. *Agric. Aviat.* **3**: 77–81.

Maas, W. (1971) *ULV Application and Formulation Techniques*. Eindhoven, NV Philips Gloeilampenfabrieken.

Mabbett, T.H. and **Phelps, R.H.** (1974) Low volume and ultra low volume spray systems in the humid tropics. *Proc. Symp. Prot. Hort. Crops in the Caribbean Univ. West Indies*, St. Augustine, Trinidad: 71–83.

Mabbett, T.H. and **Phelps, R.H.** (1976) Control of angular leaf spot of cucumber by low and ultra low volume spraying. *Trop. Agric. (Trinidad)* **53**: 105–10.

MacCuaig, R.D. (1962) The collection of spray droplets by flying locusts. *Bull. ent. Res.* **53**: 111–23.

MacCuaig, R.D. and **Watts, W.S.** (1963) Laboratory studies to determine the effectiveness of DDVP sprays for control of locusts. *J. econ. Ent.* **56**: 850–8.

McDaniel, R. and **Himel, C.M.** (1977) Standardisation of field methods for determination of insecticide spray droplet size. *Office of Naval Research* Report **2**, US Navy.

MacFarlane, R. and **Matthews, G.A.** (1978) Modifications in knapsack mistblower design to improve spray efficiency on tall tree crops. *BCPC Monograph* **22**: 151–5.

MacIver, D.R. (1963) Mosquito coils, Part I. General Description of coils, their formulation and manufacture. *Pyrethrum Post* **4**: 22–7.

MacIver, D.R. (1964a) Mosquito coils, Part II. Studies on the action of mosquito coil smoke on mosquitoes. *Pyrethrum Post* **7**: 7–14.

MacIver, D.R. (1964b) Mosquito coils, Part III. Factor influencing the release of pyrethrums from coils. *Pyrethrum Post* **7**: 15–17, 19.

Mahler, D.S. and **Magnus, D.E.** (1986) Hotwire technique for droplet measurements. *Liquid Pesticide Size Measurement Techniques*. ASTM STP 848 Philadelphia: 153–65.

Marchant, J.A. (1980) Electrostatic spraying – some basic principles. *Proc. 1980 Br. Crop Prot. Conf. – Weeds*: 987–97.

Marchant, J.A. (1985) An electrostatic spinning disc atomiser. *Trans. Am. Soc. Ag. Engrs* **28**: 386–92.

Marchant, J.A. (1987) 'Mathematical modelling in spray engineering research. In: Brent, K.J. and Atkins, R.K. (eds) *Rational Pesticide Use*. Cambridge.

Marchant J.A. and **Frost, A.R.** (1989) Simulation of the performance of state feedback controllers for an active boom suspension. *J. Agric. Engng. Res.* **43**: 77–91.

Marchant, J.A. and **Green, R.** (1982) An electrostatic charging system for hydraulic spray nozzles. *J. Agric. Engng Res*. **27**: 309–19.

Marchant, J.A., Dix, A.J. and Wilson, J.M. (1985a) The electrostatic charging of spray produced by hydraulic nozzles. Part I. Theoretical analysis. *J. Agric. Engng. Res*. **31**: 329–44.

Marchant, J.A., Dix, A.J. and Wilson, J.M. (1985b) The electrostatic charging of spray produced by hydraulic nozzles. Part II. Measurements. *J. Agric. Engng. Res*. **31**: 345–60.

Markin, G.P., Henderson, J.A. and **Collins, H.L.** (1972) Aerial application of micro encapsulated insecticide. *Agric. Aviat*. **14**: 70–5.

Marks, R.J. (1976) Field studies with the synthetic sex pheromone and inhibitor of the red bollworm *Diparopsis castamea* Hmps (Ledpidoptera: Noctiuidae) in Malawi. *Bull. ent. Res*. **66**: 243–65.

Marrs, G.J. and **Middleton, M.R.** (1973) The formulation of pesticide for convenience and safety. *Outlook on Agric*. **7**: 231–5.

Marrs, G.J. and **Scher, H.B.** (1990) Development and uses of microencapsulation. In: Wilkins, R.M. (ed.) *Controlled Delivery of Crop Protection Agent*. London, Taylor and Francis.

Martin, J.T. (1958) The comparison of high and low volume spraying techniques on fruit and ground crops. Levels and distribution of droplets. *Soc. Chem. Ind. Monogr*. **2**: 55–68.

Matthee, F.N., Thomas, A.C., Schwabe, W.F.S. and **Nel, E.W.** (1975) Low volume and ultra-low volume sprays to control apple scab (*Venturia inaequalis*). *The Deciduous Fruit Grower* **25**: 183–6.

Matthews, G.A. (1966) Investigations of the chemical control of insect pests of cotton in Central Africa. II. Tests of insecticides with larvae and adults. *Bull. ent. Res*. **57**: 77–91.

Matthews, G.A. (1969) Performance of some lever-operated knapsack sprayers. *Cott. Gr. Rev*. **46**: 134–42.

Matthews, G.A. (1971) Ultra low volume spraying of cotton – a new application technique. *Cotton Handbook of Malawi*, Amendment 2/71. Agricultural Research Council of Malawi.

Matthews, G.A. (1973) Ultra low volume spraying of cotton in Malawi. *Cott. Gr. Rev*. **50**: 242–67.

Matthews, G.A. (1975a) Determination of droplet size. *PANS* **21**: 213–25.

Matthews, G.A. (1975b) A graticule for classification of spray droplets. *PANS* **21**: 343–4.

Matthews, G.A. (1977) CDA – controlled droplet application. *PANS* **23**: 387–94.

Matthews, G.A. (1978) Spinning, the new pesticide drops to the target. *New Scientist* **78**: 753–5.

Matthews, G.A. (1981) Improved systems of pesticide application. *Phil. Trans. R. Soc. Lond. B*.: 163–73.

Matthews, G.A. (1984) *Pest Management*. London, Longman

Matthews, G.A. (1989a) *Cotton Insect Pests and their Management*. London, Longman

Matthews, G.A. (1989b) Electrostatic spraying of pesticides: a review. *Crop Protection* **8**: 3–15.

Matthews, G.A. (1990) Changes in application techniques used by the small scale cotton farmer in Africa. *Trop. Pest Management* **36**: 166–72.

Matthews, G.A. and **Clayphon, J.E.** (1973) Safety precautions for pesticide application in the tropics. *PANS* **19**: 1–12.

Matthews, G.A. and **Johnstone, D.R.** (1968) Aircraft and tractor spray deposits on irrigated cotton. *Cott. Gr. Rev*. **45**: 207–18.

Matthews, G.A. and **Mowlam, M.D.** (1974) Some aspects of the biology of cotton insects and their control with ULV spraying in Malawi. *Br. Crop Prot. Coun. Monogr*. **11**: 44–52.

Matthews, G.A. and Thornhill, E.W. (1974) A note on the use of zinc-air batteries in ULV sprayers. *Br. Crop Prot. Counc. Monogr.* **11**: 282–4.

Matthews, G.A. and Tunstall, J.P. (1966) Field trials comparing carbaryl DDT and endosulfan on cotton in Central Africa. *Cott. Gr. Rev.* **43**: 230–9.

Matthews, G.A. and Tunstall, J.P. (1968) Scouting for pests and the timing of spray applications. *Cott. Gr. Rev.* **45**: 115–27.

Matthews, G.A., Tunstall, J.P. and McKinley, D.J. (1965) Outbreaks of pink bollworm (*Pectinophora gossypiella* Saund) in Rhodesia and Malawi. *Cott. Gr. Rev.* **42**: 197–208.

Matthews, G.A., Higgins, A.,E.H. and Thornhill, E.W. (1969) Suggested techniques for assessing the durability of lever-operated knapsack sprayers. *Cott. Gr. Rev.* **46**: 143–8.

Maude, R.B. and Suett, D.L. (1986) Application of pesticide to brassica seeds using a film coating technique. *Proc. 1986 Br. Crop Prot. Conf. – Pests and Diseases*: 237–42.

Mawer, C.J. and Miller, P.C.H. (1989) Effect of roll angle and nozzle spray pattern on the uniformity of spray volume distribution below a boom. *Crop Protection* **8**: 217–22.

May, K.R. (1945) The cascade impactor: an instrument for sampling coarse aerosols. *J. Sci. Instrum.* **22**: 187–95.

May, K.R. (1950) The measurement of airborne droplets by the magnesium oxide method. *J. Sci. Instrum.* **27**, 128–30.

May, K.R. and Clifford, R. (1967) The impaction of aerosol particles, spheres, ribbons and discs. *Ann. Occp. Hyg.* **10**: 83–95.

May, M.J. (1991) Early studies on spray drift, deposit manipulation and weed control in sugar beet with two air-assisted boom sprayers. *BCPC Monograph* **46**: 89–96.

Maybank, J., Yoshida, K. and Grover, R. (1974) Droplet size spectra, drift potential and ground deposition pattern of herbicide sprays. *Can. J. Plant Sci.* **54**: 541–6.

Maybank, J., Yoshida, K. and Grover, R. (1978) Spray drift for agricultural pesticide application. *J. Air Pollution Control Assoc.* **28**: 1,009–14.

Mayes, A.J. and Blanchard, T.W. (1978) The performance of a prototype Microdrop (CDA) sprayer for herbicide application. *BCPC Monograph* **22**: 171–8.

Mboob, S.S. (1975) Preliminary assessment of the effectiveness of two droplet sizes of insecticide for the control of glasshouse whitefly *Trialeurodes vaporariorum* (Westwood). *Pl. Path.* **24**: 158–62.

Mercer, P.C. (1974) Disease control of groundnuts in Malawi. *Wld Crops* **26**: 162–4.

Mercer, P.C. (1976) Ultra low volume spraying of fungicides for the control of *Cercospora* leaf spot of groundnuts in Malawi. *PANS* **22**: 57–60.

Merritt, C.R. (1976) The interaction of surfactant type and concentration with controlled drop application of MCPA and difenzoquat. *Proc. 1976 Br. Crop Prot. Conf. – Weeds*: 413–18.

Merritt, C.R. (1980) The influence of application variables on the biological performance of foliage applied herbicides. *BCPC Monograph* **24**: 35–43.

Merritt, C.R. and Taylor, W.A. (1977) Glasshouse trials with controlled drop application of some foliage applied herbicides. *Weed Research* **17**: 241–5.

Mickle, R.E. (1987) A review of models for ULV spraying scenarios. In. Green G.W. (ed.) *Symposium on the Aerial Application of Pesticides in Forestry*: 179–188. Ottawa, Canada.

Middleton, M.R. (1973) Assessment of performance of the 'Rotostat' seed treater. *Proc. 7th Br. Insectic. Fungic. Conf.*: 357–64.

Miller, A.W.D. and Chadwick, P.R. (1963) Swath marking in aerial spraying. *Agric. Aviat.* **5**: 114–20.

Miller, P.C.H. (1988) Engineering aspects of spray drift control. *Aspects of Applied Biology* **17**: 377–84.

Miller, P.C.H. (1989) The field performance of electrostatically charged hydraulic nozzle sprayers. *Proc. 4th European Weed RS Mediterranean Symposium*: 324–33.

Miller, P.C.H. and **Hobson, P.A.** (1991) Methods of creating air-assisting flows for use in conjunction with crop sprayers. *BCPC Monograph* **46**: 35–43.

Miller, P.C.H., Mawer, C.J. and **Merritt, C.R.** (1989) Wind tunnel studies of the spray drift from two types of agricultural spray nozzle. *Aspects of Applied Biology* **24**: 237–8.

Miller, P.C.H., Juck, C.R., Gilbert, A.J. and **Bell, G.J.** (1991) The performance characteristics of a twin-fluid nozzle sprayer. *BCPC Monograph* **46**: 97–106.

Ministry of Agriculture, Fisheries and Food (1976) The utilization and performance of field crop sprayers. *Farm Mechanization Studies* **29**.

Morel, M. (1985) Field trials with the Girojet. *BPCP Monograph* **28**: 107–12.

Morgan, N.G. (1964) Gallons per acre of sprayed area – an alternative standard term for the spraying of plantation crops. *Wld Crops* **16**: 64–5.

Morgan, N.G. (1972) Spray application in plantation crops. *PANS* **18**: 316–26.

Morgan, N.G. (1981) Minimizing pesticide waste in orchard spraying. *Outlook on Agriculture* **10**: 342–4.

Morrill, A.W., Hahl, R.G. and **Lotze, K.A.** (1955) Studies on insecticidal fog generation for military use. *Mosq. News* **15**: 85–90.

Morton, N. (1977) The wind leaf orientation and ULV spray coverage on cotton plants. *FAO Pl. Prot. Bull.* **25**: 29–37.

Morton, N. (1982) The 'Electrodyn' sprayer: first studies of spray coverage in cotton. *Crop Prot.* **1**: 27–54.

Moser, E. and **Schmidt, K.** (1983) Einige grundlagen der electrostatik im chemischen. *Pflanzenschultz Landtechnik* **33**: 96–100.

Mount, G.A. (1970) Optimum droplet size for adult mosquito control with space sprays or aerosols of insecticides. *Mosq. News* **30**: 70–5.

Mount, G.A. and **Pierce, N.W.** (1972) Droplet size of ultra low volume ground aerosols as determined by three collection methods. *Mosq. News* **32**: 586–9.

Mount, G.A., Pierce, N.W. and **Baldwin, K.F.** (1975) Comparison of two aerosol generator nozzle systems: estimates of droplet size and caged mosquito assays. *Mosq. News* **35**: 501–4.

Mowlam, M.D. (1974) Aerial spraying research on cotton in Malawi. *Agric. Aviat.* **16**: 36–40.

Mowlam, M.D., Nyirenda, G.K.C. and **Tunstall, J.P.** (1975) Ultra-low volume application of water-based formulations of insecticides to cotton. *Cott. Gr. Rev.* **52**: 360–70.

Muller, J.R. and **Stovell, F.R.** (1966) Development of the biflon system. *Proc. 3rd Int. agric. Aviat. Congr*: 55–9.

Munthali, D.C. (1976) Studies on choice of applicator that enables integration of chemical control with *Encarsia formosa* (Gahan) for the control of *Trialeurodes vaporariarum* (Westwood) in glasshouses. Unpublished M.Sc. manuscript.

Munthali, D.C. (1984) Biological efficiency of small dicofol droplets against *Tetranychus urticae* (Koch) eggs, larvae and protonymphs. *Crop Prot.* **3**: 327–34.

Munthali, D.C. and **Scopes, N.E.A.** (1982) A technique for studying the biological efficiency of small droplets of pesticide solutions and a consideration of the implications. *Pestic. Sci.* **13**: 60–2.

Munthali, D.C. and **Wyatt, I.J.** (1986) Factors affecting the biological efficiency of small pesticide droplets against *Tetranychus urticae* eggs. *Pestic. Sci.* **17**: 155–64.

Nation, H.J. (1978) Logistics of spraying with reduced volume of spray and higher vehicle speeds. *Proc. 1978 Br. Crop Prot. Conf. – Weeds*: 641–8.

Nation, H.J. (1982) The dynamic behaviour of field sprayer booms. *J. agric. Eng. Res.* **27**: 61–70.

Nation, H.J. (1985) Construction and evaluation of a universal links spray boom suspension. *NIAE Divisional Note DN 1299, Wrest Park, Silsoe.*

Nettleton, D.M. (1991) Field experiences with an 'Airtec' twin fluid spraying system. *BCPC Monograph* **46**: 107–12.

Ng, K.Y. and **Chong, Y.W.** (1982) Studies on some aspects of the control of *Darna trima* in oil palm. *Malay Peninsula Agric. Ass. Yr. Bk.* 1982: 39–45.

Nguyen, T. and **Jarvis, L.R.** (1982) The use of a computerised digitiser for the analysis of spray droplet sizes on oat leaf surfaces. *Pestic. Sci.* **13**: 463–46.

Nordby, A. and **Skuterud, R.** (1975) The effects of boom height, working pressure and wind speed on spray drift. *Weed Res.* **14**: 385–95.

Nyirenda, G.K.C. (1991) Effect of swath width, time of application and height on the efficacy of very low volume (VLV) water-based insecticides on cotton in Malawi. *Crop Protection* **10**: 111–16.

Ogborn, J. (1972) The control of *Striga hermonthica* in peasant farming. *Proc. 11th Br. Weed Cont. Conf:* 1,068–77.

Ogborn, J. (1977) Herbicides and hoe farmers. *Wld Crops* **29**: 9–11.

Ogg, A.G. (1986) Applying herbicides in irrigation water – a review. *Crop Protection* **5**: 53–65.

Oliver-Bellasis, H.R. and **Southerton, N.W.** (1986) The cereals and gamebirds research project – An independent viewpoint. *Proc. 1986 BCPC Conf. – Pests and Diseases* **3**: 1,225–8.

Omar, D. and **Matthews, G.A.** (1987) Biological efficiency of spray droplets of permethrin ULV against diamond back moth. *Aspects of Applied Biology* **14**: 173–9.

Onstad, D.W. (1987) Calculation of economic-injury levels and economic thresholds for pest management. *J. econ. Ent.* **80**: 299–303.

O'Sullivan, J.A. (1988) Verification of passive and active versions of a mathematical model of a pendulum spray boom suspension. *J. Agric. Engng. Res.* **40**: 89–101.

Palmer, J.E. (1970) Dry formulations for selective weed control in cereals. *Br. Crop Prot. Counc. Monogr.* **2**: 114–23.

Park, P.O., Gledhill, J.A., Alsop, N. and **Lee, C.W.** (1972) A large-scale scheme for the eradication of *Glossina morsitans morsitans* Westw. in the Western Province of Zambia by aerial ultra low volume application of endosulfan. *Bull. ent. Res.* **61**: 373–84.

Parker, J.D., Collings, B.G.P. and **Kahumbura, J.M.** (1971) Preliminary tests of a suction spray nozzle for use with aircraft spraying systems. *Agric. Aviat.* **13**: 24–8.

Parkin, C.S. (1979) Rotor induced air movements and their effects on droplet dispersal. *J. Royal Aero Soc.* (May issue) 183–7.

Parkin, C.S. (1987) Factors affecting the movement of spray droplets above a forest canopy. In. Green G.W. (ed.) *Symposium on the Aerial Application of Pesticides in Forestry*, pp. 69–79. Ottawa, Canada.

Parkin, C.S. and **Newman, B.W.** (1977) The bifoil atomiser – a variable geometry venturi atomiser. *Agric. Aviat.* **18**: 15–23.

Parkin, C.S. and **Siddiqui, H.A.** (1990) Measurement of drop spectra from rotary cage aerial atomizers. *Crop Protection* **9**: 33–8.

Parkin, C.S. and **Spillman, J.J.** (1980) The use of wing-tip sails on a spraying aircraft to reduce the amount of material carried off-target by a crosswind. *J. agric. Engng. Res.* **25**: 65–74.

Parkin, C.S. and **Wyatt, J.C.** (1982) The determination of flight-lane separations for the aerial application of herbicides. *Crop Protection* **1**: 309–21.

Parkin, C.S., Wyatt, J.C. and **Warner, R.** (1980) The measurement of drop spectra in agricultural sprays using a Particle Measuring Systems Optical Array Spectrometer. *BCPC Monograph* **24**: 241–9.

Parkin, C.S., Outram, I., Last, A.J. and **Thomas, A.P.W.** (1985) An evaluation of aerially applied ULV and LV sprays using a double spray system and two tracers. *BCPC Monograph* **28**: 211–20.

Parnell, F.R., King, H.E. and **Ruston, D.F.** (1949) Jassid resistance and hairiness of the cotton plant. *Bull. ent. Res.* **39**: 539–75.

Parr, W.J., Gould, H.J., Jessop, N.H. and **Ludlam, F.A.B.** (1976) 'Progress towards a biological control programme for glasshouse whitefly (*Trialeurodes vaporariorum*) on tomatoes. *Ann. appl. Biol.* **83**: 349–63.

Parrott, W.N., Jenkins, J.N. and **Smith, D.B.** (1973) Frego bract cotton and normal bract cotton: how morphology affects control of boll weevils by insecticides. *J. econ. Ent.* **66**: 222–5.

Pasquill, F. (1974) *Atmospheric Diffusion*, 2nd edn. Horwood, UK.

Patterson, D.E. (1963) The effect of saturn yellow concentration on the visual inspection and photographic recording of spray droplets. *J. agric. Engng Res.* **8**: 342–4.

Payne, N.J. and **Schaefer, G.W.** (1986) An experimental quantification of coarse aerosol deposition on wheat. *Atomisation and Spray Technology* **2**: 45–71.

Payne, N.J., Helson, V.B., Sunderam, K.M.S. and **Fleming, R.A.** (1988) Estimating buffer zone widths for pesticide application. *Pestic. Sci.* **24**: 147–61.

Pearson, A.J.A. and **Masheder, S.** (1969) Bi-flon. *Agric. Aviat.* **11**: 126–9.

Peat, J.E. and **McKinstry, A.H.** (1938) Cotton pests. *Prog. Rep. expl. Stns*, Southern Rhodesia 1936–37, Emp. Cott. Gr. Corp.

Pedigo, L.P., Hutchins, S.H. and **Hegley, L.G.** (1986) Economic injury levels in theory and practice. *Ann. Rev. Entomol.* **31**: 341–68.

Peregrine, D.J. (1973) Toxic baits for the control of pest animals. *PANS* **19**: 523–33.

Pereira, J.L. (1967) Uses of fluorescent tracer for assessment of spray efficiency. *Kenya Coffee* **32**: 461–4.

Pereira, J.L. (1970) Aerial spraying of coffee for disease control. *Agric. Aviat.* **12**: 17–20.

Pereira, J.L. (1972) Modifications of a hydraulic sprayer (Hardi) for improved coffee spraying. *E. Afr. Agric. For. J.* **37**: 318–24.

Pereira, J.L. (1985) Chemical control of phytophthora pod rot of cocoa in Brazil. *Cocoa Growers' Bulletin* **36**: 23–38.

Pereira, J.L. and **Mapother, H.R.** (1972) Overhead application of fungicide for the control of coffee berry disease. *Expl Agric.* **8**: 117–22.

Perich, M.J., Tidwell, M.A., Williams, D.C., Sardelis, M.R., Pena, C.J., Mandeville, D. and **Boobar, L.R.** (1990) Comparison of ground and aerial ultra-low-volume applications of malathion against *Aedes aegypti* in Santo Domingo, Dominican Republic. *J. Am. Mos. Cont. Assoc.* **6**: 1–6.

Pettifor, M.J. (1988) Practical problems in achieving recommended placement of granules. *BCPB Monograph* **39**: 333–5.

Phillips, F.T. (1968) Microencapsulation, a method for increasing specificity and controlling persistence of insecticides. *PANS* **14**: 407–10.

Phillips, F.T. and **Gillham, E.M.** (1971) Persistence to rain washing of DDT wettable powders. *Pestic. Sci.* **2**: 97–100.

Phillips, F.T. and **Gillham, E.M.** (1973) A comparison of sticker performance against rain washing of microcapsules on leaf surfaces. *Pestic. Sci.* **4**: 51–7.

Phillips, F.T. and **Lewis, T.** (1973) Current trends in the development of baits against leaf cutting ants. *PANS* **19**: 483–7.

Pickler, R.A. (1976) Expanding the applications of firefighting aircraft to improve cost effectivnesss. *Proc. 5th Int. agric. Aviat. Congr*: 79–85.

Planas, S. and **Pons, L.** (1991) Practical considerations concerning pesticide application in intensive apple and pear orchards. *BCPC Monograph* **46**: 45–52.

Polles, S.G. and **Vinson, S.B.** (1969) Effect of droplet size on persistence of ULV malathion and comparison of toxicity of ULV and EC malathion to tobacco budworm larvae. *J. econ. Ent.* **62**: 89–94.

Polon, J.A. (1973) Formulation of pesticidal dusts, wettable powders and granules. In: Walkenburg, E. van (ed.) *Pesticide Formulations*. New York, Marcel Dekker.

Poston, F.L., Pedigo, L.P. and **Welch, S.M.** (1983) Economic injury levels: reality and practicality. *Bull. Entomol. Soc. Am.* **29**: 49–53.

Potts, S.F. (1946) Particle size of insecticides and its relation to application, distribution and deposit. *J. econ. Ent.* **39**: 716–20.

Potts, S.F. (1958) *Concentrated Spray Equipment, Mixtures and Application Methods*, New Jersey, Dorland Books.

Potts, S.F. and **Garman, P.** (1950) Concentrated sprays for application by mistblowers for control of forest, shade and fruit tree pests. *Conn. agric. expl. Stn Circ*: 177.

Prandtl, L. (1952) *The Essentials of Fluid Dynamics*. London, Blackie.

Purdy, L.H. (1967) Application and use of soil and seed treatment fungicides. In: Torgeson, D.C. (ed.) *Fungicides: An Advanced Treatise*, vol. 1. New York, Academic Press.

Quantick, H.A. (1985a) *Aviation in Crop Protection, Pollution and Insect Control*. London, Collins.

Quantick, H.A. (1985b) *Handbook for Agricultural Pilots*. London, Collins.

Radwald, J.D., Shibuya, F., McRae, N. and **Platzer, E.G.** (1986) A simple, inexpensive portable apparatus for injecting experimental chemicals in drip-irrigation systems. *J. Nematology* **18**: 423–5.

Raheja, A.K. (1976) ULV spraying for cowpea in Northern Nigeria. *PANS* **22**: 327–32.

Rainey, R.C. (1958) The use of insecticide against the desert locust. *J. sci. Fd Agric.* **9**: 677–92.

Rainey, R.C. (1974) Flying insects as targets for ultra low volume spraying. *Br. Crop Prot. Counc. Monogr.* **11**: 20–8.

Rainey, R.C. (1976) New prospects for the use of aircraft in the control of flying insects and in the development of semi-arid regions. *5th Int. Agric. Aviat. Congr*: 229–33.

Raisigl, V., Felber, H., Siegfried, W. and **Krebs, C.** (1991) Comparison of different mistblowers and volume rates for orchard spraying. *BCPC Monograph* **46**: 185–96.

Randall, A.P. (1975) Application Technology In: Prebble, M.L. (ed.) *Aerial Control of Forest Insects in Canada*. Dept of Environment, Ottawa.

Randall, J.M. (1971) The relationship between air volume and pressure on spray distribution in fruit trees. *J. agric. Engng. Res.* **16**: 1–31.

Ras, M.C.D. (1986) Effective application of chemical agents for the control of pests and diseases by the correct use of spraying machinery in orchards and vineyards. *Deciduous Fruit Grower* (Nov.) 467–77.

Rayleigh, Lord (1882) On the equilibrium of liquid conducting masses charged with electricity. *Philosophical Magazine* **14**: 184–86.

Reay, R.C. and **Ford, M.G.** (1977) Toxicity of pyrethroids to larvae of the Egyptian cotton leafworm *Spodoptera littoralis* (Boisd). II Factors determining the effectiveness of permethrin at ULV. *Pestic Sci.* **8**: 243–53.

Reed, D.K., Reed, G.L. and **Creighton, C.S.** (1986) Introduction of entomogenous nematodes into trickle irrigation systems to control striped cucumber beetle (Coleoptera: Chrysomelidae). *J. econ. Ent.* **79**: 1330–3.

Reichard, D.L., Retzer, H.J., Liljedahl, L.A. and **Hall, F.R.** (1977) Spray droplet size distributions delivered by airblast orchard sprayers. *Trans. Am. Soc. Ag. Eng.* **20**: 232–42.

Reichard, D.L., Alm, S.R. and **Hall, F.R.** (1987) Equipment for studying effects of spray drop size, distribution and dosage on pest control. *J. econ. Ent.* **80**: 540–43.

Renne, D.S. and **Wolf, M.A.** (1979) Experimental studies of 2, 4-D herbicide drop characteristics. *Ag. Meteorology* **20**: 7–24.

Rennison, B.D. (1962) A method of sampling *Antestiopsis* in arabica coffee in chemical control schemes. *E. Afr. Agric. For J*. **27**: 197–200.

Rice, B. (1967) Spray distribution from ground crop sprayers. *J. agric. Engng Res*. **12**: 173.

Rice, B. (1970) A review of procedures and techniques for testing ground crop sprayers. *Br. Crop Prot. Counc. Monogr*. **2**: 1–11.

Rice, B. and **Connolly, J.** (1969) Quality control limits for the distribution patterns of ground crop sprayer nozzles. *J. agric. Engng. Res*. **14**: 313–18.

Richardson, E.G. (1960) Introduction and historical survey. In: Richardson, E.G. (ed.) *Aerodynamic Capture of Particles* Oxford, Pergamon Press.

Richardson, L.F. (1920) The supply of energy from and to atmospheric eddies. *Proc. R. Soc*. **A97**: 354–73.

Richardson, R.G., Combellack, J.H. and **Andrew, L.** (1986) Evaluation of a spray nozzle patternator. *Crop Protection* **5**: 8–11

Rickett, F.E. and **Chadwick, P.R.** (1972) Measurements of temperature and degradation of pyrethroids in two thermal fogging machines, the Swingfog and Tifa. *Pestic. Sci*. **3**: 263–9.

Ridgway, R.L. (1967) Research on systemic insecticides and methods of applying them for control of cotton insects. *Cotton Gr. Rev*. **44**: 39–50.

Riley, C.M., Wiesner, C.J. and **Ernst, W.R.** (1989) Off-target deposition and drift of aerially applied agircultural sprays. *Pesticide Sci*. **26**: 159–66.

Ripper, R.A. (1967) The use of extended booms to increase efficiency of crop-spraying aircraft. *Agric. Aviat*. **9**: 77–9.

Ripper, W.E. (1944) Biological control as a supplement to chemical control of insect pests. *Nature* **153**: 448–52.

Ripper, W.E. (1955) Application methods for crop protection chemicals. *Ann. appl. Biol*. **42**: 288–324.

Ripper, W.E. (1956) Effect of pesticide on balance of arthropod populations. *Ann. Rev. Ent*, **1**: 403–38.

Ripper, W.E. and **Tudor, P.** (1947) The development of a helicopter spraying machine. *Bull. ent. Res*. **39**: 1–13.

Robinson, R.C. (1978) The field performance of some herbicides applied by rotary atomiser in spray volumes 5–50 l/ha. *BPCP Monograph* **22**: 185–91.

Robinson, R.C. and **Rutherford, S.J.** (1988) A hand-held precision spot-applicator for granular insecticide. *BCPC Monograph* **39**: 341–7.

Rogers, E.V. (1974) The selection and development of equipment and methods for ULV herbicide spraying in forestry. *Br. Crop Prot. Counc. Monogr*. **11**: 226–31.

Rogers, E.V. (1975) Ultra low volume herbicide spraying. HMSO Forestry Commission Leaflet **62** 20 pp.

Rogers, R.B. and **Ford, R.J.** (1985) The windproof sprayer: its progress and prospects. *Ag. Eng*. **66**: 11–13.

Rose, G.J. (1963) *Crop Protection*. London, Leonard Hill.

Roger, T.A., Edelson, J.V., Bogle, C.R. and **McCrate, S.** (1989) Insecticide applicator and insect control using a drip irrigation delivery system. *Pestic Sci*. **25**: 231–40.

Rozendaal, J.A. (1989) Self protection and vector control with insecticide-treated mosquito nets. WHO Unpublished document WHO/VBC89.96.

Ruscoe, C.N.E. (1987) Pesticide resistance: Strategies and co-operation in the agrochemical industry. In: Brent, K.H. and Atkin, R.K. (eds) *Rational Pesticide Use*. Cambridge.

Rutherford, I. (1976) An ADAS survey on the utilization and performance of field crops sprayers. *Proc. 1976 Br. Crop Prot. Conf. – Weeds*. **2**: 357–61.

Rutherford, I., Bell, G.J., Freer, J.B.S., Herrington, P.J. and **Miller, P.C.H.** (1989)

An evaluation of chemical application systems. *Brighton Crop Protection Conf. – Weeds*. **3**: 601–13.

Sander, T.P.Y. (1991) Development and evaluation of rotary cage atomiser conversion for orchard sprayers. *BPCP Monograph* **46**: 203–10.

Sawicki, R.M. and **Denholm, I.** (1987) Management of resistance to pesticides in cotton pests. *Trop. Pest Man.* **33**: 262–72.

Sawyer, K.F. (1950) Aerial curtain spraying for locust control: a theoretical treatment of some of the factors involved. *Bull. ent. Res.* **41**: 439–57.

Sayer, H.J. (1959) An ultra low volume spraying technique for the control of the desert locust *Schistocerca gregaria*. *Bull. ent. Res.* **50**: 371–86.

Sayer, H.J. (1969) Ultra low volume spraying systems – comparison and assessment. *Agric. Aviat.* **11**: 78–85.

Sayer, H.J. and **Rainey, R.C.** (1958) *An Exhaust Nozzle Sprayer for Ultra Low Volume Application of a Persistent Insecticide*. London, Anti-Locust Research Centre

Scher, H.B. (1984) Advances in pesticide formulation technology: an overview. In: *Advances in Pesticide Formulation Technology* ACS Symposium **254**: 1–7.

Scholz, E., Spielberger, U. and **Ali, J.** (1976) The night resting sites of the tsetse fly *Glossina palpalis palpalis* (Robineau Desooidy) (Diptera, Glossinidae) in northern Nigeria. *Bull. ent. Res.* **66**. 443–52.

Schroeder, W.T. and **Provvidenti, R.** (1969) Resistance to benomyl in powdery mildew of cucurbits. *Pl. Dis. Reptr.* **53**: 271–5.

Schuster, W. (1974) 'Selection standards for agricultural aircraft. *Agric. Aviat.* **16**: 98–104.

Scopes, N.E.A. and **Biggerstaff, S.M.** (1973) Progress towards integrated pest control on year round chrysanthemums. *Proc. 7th Brit. Insectic. Fungic. Conf*: 227–34.

Seaman, D. (1990) Trends in the formulation of pesticides – an overview. *Pestic. Sci.* **29**: 437–49.

Shang, H. and **Li, W.** (1990) Study on property of Model 50E electrostatic sprayer and its application. Poster presented at Shenyang Conf. organized by UNIDO.

Sharkey, A.J., Salt, D.W. and **Ford, M.G.** (1987) Use of simulation to define an optimum deposit for control of a sedentary pest. *Aspects of Appl. Biology* **14**: 267–80.

Sharp, R.B. (1973) A rapid method of spray deposit measurement and its use in new apple orchards. *Proc. 7th Brit. Insectic. Fungic. Conf*: 637–42.

Sharp, R.B. (1974) A method for tracing the initial placement of soil applied herbicides. *J. agric. Engng. Res.* **19**: 93–5.

Sharp, R.B. (1984) Comparison of drift from charged and uncharged hydraulic nozzles. *Proc. 1984 British Crop Prot. Conf. – Pests and Diseases*: 1,027–32.

Shorey, H.H. (1973) Behavioural responses to insect pheromones. *Ann. Rev. Entomol.* **18**: 349–80.

Shreenivwasan, T.N., Pettit, T.R. and **Rudgard, S.A.** (1990) An alternative method of applying copper to control cocoa pod diseases. *Brighton Crop Prot. Conf. – Pests and Diseases*: 583–8.

Simard, A.J. (1976) Air tanker utilization and wildland fire management. *Proc. 5th Int. agric. Aviat. Congr*: 71–8.

Skaf, R., Popov, G.B. and **Roffey, J.** (1990) The desert locust – an international challenge. *Phil. Trans. R. Soc. Land. B.* **328**: 525–38.

Skoog, F.E., Hanson, T.L., Higgins, A.H. and **Onsager, J.A.** (1976) Ultra low volume spraying: systems evaluation and meteorological data analysis. *Trans. Am. Soc. Agric. Engrs*, **19**: 2–6.

Slatter, R., Stewart, D.C., Martin, R. and **White, A.W.C.** (1981) An evaluation of Pestigas, B.B. – a new system for applying synthetic pyrethroids as space sprays using pressurised carbon dioxide. *Int. Pest Control*: 162–4.

Smith, A.K. (1984) A model to aid decision making in choosing a suitable crop spraying system. *1984 Br. Crop Prot Conf. – Pests and Diseases*: 621–6.

Smith, A.K. (1985) Forecasting future developments in application technology. *Crop Protection* **4**: 121–9.

Smith, C.M. and **Goodhue, L.D.** (1942) Review of particle size and toxicity. *Ind. Engng Chem.* **34**: 490–3.

Smith, D.B. and **Burt, E.C.** (1970) Effects of the size of ULV droplets on deposits within cotton foliage both inside and immediately downwind from a treated swath. *J. econ. Ent.* **63**: 1400–5.

Smith, R. (1988) The 'Electrodyn' sprayer: matching the technology to contrasting areas of smallholder agriculture. *Chemistry and Industry* **6**: 196–199.

Smith, R. (1989) The 'Electrodyn' sprayer as a tool for rational pesticide management in smallholder cotton. In: Green, M.B. and Lyon, D. (eds) *Pest Management in Cotton*. Chichester, Ellis Horwood.

Smith, R.F. (1970) Pesticides: their use and limitations in pest management. In: Rabb, R.L. and Guthrie, F.E. (eds) *Concepts of Pest Management* Raleigh, NC, North Carolina State Univ.

Smith, R.F. and **van den Bosch, R.** (1967) Integrated control. In: Kilgore, W.W. and Doutt, R.L. (eds) *Pest Control – Biological, Physical and Selected Chemical Methods* New York and London, Academic Press

Sopp, P.I. and **Palmer, A.** (1990) Deposition patterns and biological effectiveness of spray deposits on pot plants applied by the Ulvafan and three prototype electrostatic sprayers. *Crop Protection* **9**: 295–302.

Sopp, P.I., Gillespie, A.T. and **Palmer, A.** (1989) Application of *Verticillium lecani* for the control of *Aphis gossypii* by a low-volume electrostatic rotary atomiser and a high-volume hydraulic sprayer. *Entomophaga* **34**: 417–28.

Southwood, T.R.E. (1966) *Ecological Methods*. London, Methuen.

Southwood, T.R.E. (1977) Entomology and mankind. *Am. Scientist* **65**: 30–9.

Spackmann, E. and **Barrie, I.A.** (1982) Spray occasions determined from meteorological data during the 1980–81 season at 15 stations in the UK and comparison with 1971–80. *Meterological Office Agricultural Memorandum No. 933*.

Speelman, L. (1971) A fluorescent tracer technique for determination of the liquid distribution of field crop sprayers. *J. agric. Engng Res.* **16**: 301–6.

Spencer, D.M. (1972) Results in practice. II. Glasshouse crops. In: Marsh, R.W. (ed.) *Systemic Fungicides* London, London.

Spielberger, U. and **Abdurrahim, U.** (1971) Pilot trial of discriminative aerial application of persistent dieldrin deposits to eradicate *Glossina morsitans submorsitans* in the Anchau and Ikara forest reserves, Nigeria. In: *International Scientific Committee for Trypanosomiasis Research 13th meeting*, Lagos, Sept. 1971. OAU/STRC Publication **105**: 271–91.

Spillman, J.J. (1976) Optimum droplet sizes for spraying against flying targets. *Agric. Aviat.* **17**: 28–32.

Spillman, J.J. (1977) Air velocities induced by aircrafts. Cranfield Insititute of Technology Course on Aerial Application of Pesticides quoted in Elliott and Wilson (1983).

Spillman, J.J. (1980) The SB-1 aircraft spreader. Paper presented at the 6th IAAC Congress, Turin, Italy, 22–26 Sept. quoted in Quantick (1985a).

Spillman, J.J. (1982) Atomizers for the aerial application of herbicides – ideal and available. *Crop Protection* **1**: 473–82.

Spillman, J.J. (1984) Evaporation from freely falling droplets. *Aeronautical Journal of the Royal Aeronautical Society*: 181–4.

Spillman, J.J. (1987) Improvements required in spray droplet formation to improve application technology. *Proc. Symp Aerial Appl. of Pesticides in Forestry, Ottawa, Oct. 1987.*

Spillman, J.A. and **Sanderson, R.** (1983) Design and development of a disc-windmill atomiser for aerial applications. *European and Mediterranean Plant Protection Organisation Bulletin* **13**: 265–70.

Spuybrock, P.H.G. (1959) Airworthiness of agricultural airplanes. *Agric. Aviat.* **17**: 28–32.

Stafford, E.M., Byass, J.B. and **Akesson, N.B.** (1970) A fluorescent pigment to measure spray coverage. *J. econ. Ent.* **63**: 769–76.

Staniland, L.N. (1959) Fluorescent tracer techniques for the study of spray and dust deposits. *J. agr. Eng. Res.* **4**: 110–25.

Staniland, L.N. (1960) Field testing of spraying equipment by means of fluorescent tracer techniques. *J. agr. Eng. Res.* **5**: 42–81.

Stent, C.J., Taylor, W.A. and **Shaw, G.B.** (1981) A method for the production of uniformly sized drops using electrostatic dispersion. *Trop. Pest Management* **27**: 262–4.

Stephenson, J. (1976) Assessment of the performance penalty of spreaders on agricultural aircraft. *Proc. 5th Int. agric. Aviat. Conf*: 321–7.

Stern, V.M. (1966) Significance of the economic threshold in integrated pest control. *Proc. FAO Symp. on Integrated Pest Control* **2**: 41–56. Rome, Italy, 11–15 Oct. 1965.

Stern, V.M., Smith, R.F., van des Bosch, R. and **Hagen, K.S.** (1959) The integrated control concept. *Hilgardia* **29**: 81–101.

Suett, D.L. (1987) Accuracy and uniformity of insecticide treatment of commercially produced plant propagation modules. *Crop Protection* **6**: 179–84.

Suett, D.L. (1987) Influence of treatment of soil with carbofuran on the subsequent performance of insecticides against cabbage root fly and carrot fly. *Crop Protection* **6**: 371–8.

Sundaram, K.M.S., Milliken, R.L. and **Sundaram, A.** (1988) Assessment of canopy and ground deposit of fenitrothion following aerial and ground application in a Northern Ontario forest. *Pesticide Science* **25**: 59–69.

Sutherland, J.A. (1979) *Non-Motorised Hydraulic Energy Sprayers*. London, COPR.

Sutherland, J.A. (1980) *Mistblowers*. COPR, London.

Sutherland, J.A., King, W.J., Dobson, H.M., Ingram, W.R. Attique, M.R. and **Sanjani, W.** (1990) Effect of application volume and method on spray operator contamination by insecticide during cotton spraying. *Crop Protection* **9**: 343–50.

Sutton, O.G. (1953) *Micrometeorology*, New York, McGraw-Hill.

Sutton, T.B. and **Unrath, C.R.** (1984) Evaluation of the tree-row-volume concept with density adjustments in relation to spray deposits in apple trees *Plant Disease.* **68**: 480–4.

Swaine, G. (1954) A simple and inexpensive insecticide duster. *E. Afr. agric. J.* **20**: 38–9.

Swift D.L. and **Proctor, D.F.** (1982) Human respiratory deposition of particles during aronasal breathing. *Atmos. Env.* **16**: 2,279–82.

Swithenbank, J., Beer, J.M., Taylor, D.S., Abbot, D. and **McCreath, G.C.** (1975) *A laser diagnostic technique for the measurement of droplet and pesticide size distribution. Report HIC 245* University of Sheffield, Dept of Chemical Engineering and Fuel Technology.

Swithenbank, J., Beer, J.M., Taylor, D.S., Abbot, D. and **McCeath, G.C.** (1977) Laser diagnostic technique for the measurement of droplet and particle size distribution. *Prog. Astron. and Aeron.* **53**.

Symmons, P.M., Boase, C.J., Clayton, J.S. and **Garta, M.** (1989) Controlling desert locust nymphs with bendiocarb applied by a vehicle-mounted spinning disc sprayer. *Crop Protection* **8**, 324–31.

Tadros, Th.F. (1989) Colloidal aspects of pesticidal and pharmaceutical formulations – an overview. *Pestic. Sci.* **26**: 51–77.

Taft, H.M., Hopkins, A.R., Jernigan, C.E. and **Webb, J.C.** (1969) A new 8-row ground sprayer with auxiliary air for ULV application of pesticide to cotton. *J. econ. Ent.* **62**: 570–4.

Takenaga, T. (1971) Pesticide applicator used by the granular boom type blow head. *Japan agric. Res. Q.* **6**: 92–6.

Takenaga, T. (1972) Knapsack type LV concentrate (ULV) sprayer. *Japan Pestic. Inf.* **13**: 5–10.

Takenaga, T. (1976) Evaluation of ultra low volume and low volume sprayers for the agricultural pest control as ground application', *Toyo Memka Kaisha Agric. Rep. Ser.* **100**.

Taylor, W.A. and **Andersen, P.G.** (1991) Enhancing conventional hydraulic nozzle use with the Twin Spray System. *BPCP Monograph* **46**: 125–36.

Taylor, W.A. and **Merritt, C.R.** (1975) Some physical aspects of the performance of experimental equipment for controlled drop application with herbicides. *Proc. 8th Br. Insectic. Fungic. Conf*: 161–70.

Taylor, W.A., Merritt, C.R. and **Drinkwater, J.A.** (1976) An experimental tractor-mounted machine for applying herbicides to field plots at very low volumes and varying drop sizes. *Weed Res.* **16**: 203–8.

Taylor, W.A., Pretty, S. and **Oliver, R.W.** (1988) Some observations quantifying and locating spray remnants within an agricultural field crop sprayer. *Aspects of Applied Biology* **18**: 385–93.

Taylor, W.A., Andersen, P.G. and **Cooper, S.** (1989) The use of air assistance in a field crop sprayer to reduce drift and modify drop trajectories. *Brighton Crop Prot. Conf. – Weeds* **3**: 631–

Teske, M.E., Barry, J.W. and **Ekblad, R.B.** (1990) Canopy penetration and deposition in a Douglas-fir seed orchard. *Am. Soc. Ag. Engineers* Paper 90–1,019.

Thompson, N. (1983a) Diffusion and uptake of chemical vapour volatilising from a sprayed target area. *Pestic. Sci.* **14**: 33–9.

Thompson, N. (1983b) Estimating the hazard to sensitive plants from herbicide drift in vapour form. *Aspects of Applied Biology* **3**: 181–90.

Thong, K.C. and **Weinberg, F.J.** (1971) Electrical control of the combustion of solid and liquid particulate suspensions. *Proc. Roy. Soc. Lond.* A. **324**: 201–15.

Thornhill, E.W. (1974a) The adaptation of a stainless steel container for use as a compression sprayer. *PANS* **20**: 241–5.

Thornhill, E.W. (1974b) 'Adaptation of knapsack mistblowers for ULV concentrate spraying. *Br. Crop Prot. Counc. Monogr.* **11**: 279–80.

Thornhill, E.W. (1979) A rotary droplet sampler. *PANS* **25**: 68–70.

Thornhill, E.W. (1982) A summary of methods of testing pesticide application equipment. *Trop. Pest Management* **28**: 335–46.

Thornhill, E.W. (1984) Maintenance and repair of spraying equipment. *Trop. Pest Management* **30**: 266–81.

Thornhill, E.W. (1985) A guide to knapsack sprayer selection. *Trop. Pest Management* **31**: 11–17.

Thornton, M.E. and **Kibble-White, R.** (1974) Apparatus used for spray nozzle evaluation at the Weed Research Organisation. *PANS* **20**: 465–75.

Trayford, R.S. and **Taylor, P.A.** (1976) Development of the tetrahedron spreader. *Proc. 5th Int. Agric. Aviat. Congr*: 294–300.

Trayford, R.S. and **Welch, L.W.** (1977) Aerial spraying: a simulation of factors influencing the distribution and recovery of liquid droplets. *J. agric. Engng Res.* **22**: 183–96.

Tsuji, K. (1990) Preparation of microencapsulated insecticides and their release mechanisms. In: Wilkins, R.M. (ed.) *Controlled Delivery of Crop Protection Agents*. London, Taylor and Francis.

Tu, Y.Q. (**1990**) Implications of biological and pesticidal behaviour in chemical control of pests. In: *Proceedings of International Seminar – Recent developments in* the field of Pesticides and their application to Pest Control in China and other Developing Countries of the Region. *UNIDO*. In press.

Tu, Y.Q., Lin, Z.M. and Zhang, J.Y. (1986) The effect of leaf shape on the deposition of spray droplets in rice. *Crop Protection* **5**: 3–7.

Tunstall, J.P. (1962) The biology of the cotton bollworms. *Proc. First fed. Sci. Congr*. Salisbury, Rhodesia.

Tunstall, J.P. and Matthews, G.A. (1961) Cotton insect control recommendations for 1961–62 in the Federation of Rhodesia and Nyasaland. *Rhod. agric. J*. **58**: 289–99.

Tunstall, J.P. and Matthews, G.A. (1965) Contamination hazards in using knapsack sprayers. *Cott. Gr. Rev*. **42**: 193–6.

Tunstall, J.P., Matthews, G.A. and Rhodes, A.A.K. (1961) A modified knapsack sprayer for the application of insecticide to cotton. *Cott. Gr. Rev*. **38**: 22–6.

Tunstall, J.P., Matthews, G.A. and Rhodes, A.A.K. (1965) Development of cotton spraying equipment in Central Africa. *Cott. Gr. Rev*. **42**: 131–45.

Turner, C.R. and Huntington, K.A. (1970) The use of water sensitive dye for the detection and assessment of small spray droplets. *J. Agric. Engng Res*. **75**: 385–7.

Turner, D.J. and Loader, M.P.C. (1974) Studies with solubilized herbicide formulations. *Proc. 12th Br. Weed Cont. Conf*: 177–84.

Uk, S. (1977) Tracing insecticide spray droplets by sizes on natural surfaces – the state of the art and its value. Paper presented at SCI Symposium 'Droplets in Air' Part 2, 'Capture by natural surfaces', *Pestic. Sci*. **8**: 501–9.

Uk, S. and Parkin, C.S. (1983) New instruments for rapid spray deposit assessment. Unpublished paper presented at AAB meeting held at Imperial College (Instruments made at Cranfield Inst. of Technology).

Van Valkenburg, W. (ed.) (1973) The stability of emulsions. In: *Pesticide Formulations*, New York, Marcel Dekker.

Vliet, M.W. van and Picot, J.J.C. (1987) Drop spectrum characterization for the Micronair Au 4000 aerial spray atomizer. *Atomisation and Spray Technology* **3**: 123–34.

Voss, C.M. (1976) The helicopter's contribution to agricultural aviation at present and in the future. *Proc. 5th Int. Agric. Aviat. Congr*: 223–8.

Walker, A., Farrant, D.M., Bryant, J.H. and Brown, P.A. (1976) The efficiency of herbicide incorporation into soil with different implements. *Weed Res*. **16**: 391–7.

Walker, D.A. (1973) Agri-Fix, a track guidance system for aerial application. *Agric. Aviat*. **15**: 99–104.

Walker, P. (1971) The use of granular pesticides from the point of view of residues. *Residue Rev*. **40**: 65–131.

Walker, P. (1976) Pesticide granules: development overseas, and opportunities for the future. *Br. Crop Prot. Counc. Monogr*. **18**: 115–22.

Wall, C. and Greenway, A.R. (1981) An effective line for use in pheromone traps for monitoring pea moth *Cydia nigricana* (F). *Plant Pathology* **30**: 75–6.

Walton, W.H. and Prewett, W.C. (1949) Atomization by spinning discs. *Proc. Phys. Soc*. **B62**: 341–50.

Ware, G.W., Cahill, W.P. and Estesen, B.J. (1975) Pesticide drift: aerial applications comparing conventional flooding vs raindrop nozzles. *J. econ. Ent*. **68**: 329–30.

Watkins, T.C. and Norton, L.B. (1955) *Handbook of Insecticide Dust Diluents and Carriers*, 2nd edn revised by D.E. Weidhaas and J.L. Brann Jr. Caldwell, NJ, Dorland Books

Watts, W.S., Thornhill, E.W., Davies, A.L. and Matthews, G.A. (1976) The primary evaluation of the Evers and Wall Mk II exhaust nozzle sprayer. *COPR Misc. Rep*. **28**.

Way, M.J. (1972) Objectives, methods and scope of integrated control. In Geier,

P.W., Clark, L.R., Anderson, D.J. and Nix, H.A. (eds) *Insects, Studies in Population Management*. Ecol. Soc. Aust. (memoirs I), Canberra.

Way, M.J. and Cammell, M.E. (1974) The problem of pest and disease forecasting – possibilities and limitations as exemplified by work on the bean aphid, *Aphis fabae. Proc. 7th Br. Insectic. Fungic. Conf.* **3**: 933–54.

Way, M.J., Bardner, R., van Baer, R. and Aitkenhead, P. (1958) A comparison of high and low volume sprays for control of the bean aphid *Aphis fabae* Scop. on field beans. *Ann. Appl. Biol.* **46**: 399–410.

Way, M.J., Cammell, M.E., Alford, D.V., Gould, H.J., Graham, C.W., Lane, A., Light, W.I. St. G., Rayner, J.M., Heathcote, G.D., Fletcher, K.E. and Seal, K. (1977) Use of forecasting in chemical control of black bean aphid *Aphis fabae* Scop., on spring sown field beans *Vicia faba* L. *Plant Path.* **26**: 1–7.

Way, M.J., Cammell, M.E., Taylor, L.R. and Waiwood, I.P. (1981) The use of egg counts and suction trap samples to forecast the infestation of spring-sown field beans *Vicia faba* by the black bean aphid *Aphis fabae, Ann. Appl. Biol* **98**: 21–34.

Weick, F.E. (1960) Design philosophy of agricultural airplanes. *Agric. Aviat.* **2**: 84–91.

Werken, J. van de (1991) The development of an unmanned air assisted tunnel sprayer for orchards. *BCPC Monograph* **46**: 211–17.

Western, N.M. and Hislop, E.C. (1991) Drift of charged and uncharged spray droplets from an experimental air-assisted sprayer. *BCPC Monograph* **46**: 69–73.

Western, N.M. and Woodley, S.E. (1987) Influence of drop size and application volume on the effectiveness of two herbicides. *Aspects of Applied Biology* **14**: 181–92.

Western, N.M., Hislop, E.C., Herrington, P.J. and Jones, E.I. (1989) Comparative drift measurements for BPCP Reference hydraulic nozzles and for an Airtec twin-fluid nozzle under controlled conditions. *Proc. Brighton Crop Prot. Conf. – Weeds* 641–8.

Wheatley, G.A. (1972) Effects of placement distribution on the performance of granular formulations of insecticides for carrot fly control. *Pestic. Sci.* **3**: 811–22.

Wheatley, G.A. (1976) Granular pesticides: some developments and opportunities for the future (developed countries). *Br. Crop Prot. Counc. Monogr.* **18**: 131–9.

Whitehead, A.G. (1988) Principles of granular nematicide placement for temperate field crops. *BCPC Monograph* **39**: 309–18.

Whitehead, D. (1976) The formulation and manufacture of granular pesticides. *Br. Crop Prot. Coun. Monogr.* **18**: 81–92.

Whittam, D. (1962) Aircraft guidance methods for pest control in the United States. *Agric. Aviat.* **4**: 8–15.

Whittam, D. (1965) New spinning nozzle for low volume aerial application. *Agric. Aviat.* **7**: 51–2.

Wilce, S.E., Akesson, N.B., Yates, W.E., Christensen, P., Lowden, R.E., Hudson, D.C. and Weigt, G.I. (1974) Drop size control and aircraft spray equipment. *Agric. Aviat.* **16**: 7–16.

Wilkins, R.M. (1990) Biodegradable polymers. In: Wilkins, R.M. (ed.) *Controlled Delivery of Crop Protection Agents* Taylor and Francis.

Wilkins, R.M., Batterby, S. Heinrichs, E.A. Aquino, G.B. and Valencia, S.L. (1984) Management of the rice tungro virus vector *Nephotettix virescens* (Homoptera: Cicadellidae) with controlled-release formulations of carbofuran. *J.econ. Ent.* **77**, 495–99.

Wilson, A.G.L., Basinski, J.J. and Thomson, N.J. (1972) Pests, crop damage and control practices with irrigated cotton in a tropical environment. *Cott. Gr. Rev.* **49**: 308–40.

Wilson, J.M. (1982) A linear source of electrostatically charged spray. *J. Agric. Eng. Res.* **27**: 181–92.

Wilson, L.T., Sterling, W.L., Rumual, D.R. and De Vay, J.E. (1989) Quantification sampling principles in cotton IPM. In: Frisbie, R.E., Elizik, K.M. and Wilson, L.T. (eds) *Integrated Pest Management Systems and Cotton Production*. New York, John Wiley & Sons Ltd.

Wodagenah, A. and **Matthews, G.A.** (1981a) The addition of oil to pesticide sprays. 2. Down-wind movement of droplets. *Trop. Pest Management* **27**: 501–4.

Wodagenah, A. and **Matthews, G.A.** (1981b) The addition of oil to pesticide sprays – effect on droplet size. *Trop. Pest Management* **27**: 121–4.

Wofford, J.T., Luttrall, R.G. and **Smith, D.B.** (1987) Relative effect of dosage, droplet size, deposit density and droplet concentration on mortality of *Heliothis virescens* (Lepidoptera: Noctuidae) larvae treated with vegetable oil and water sprays containing permethrin. *J. econ. Ent.* **80**, 460–4.

Woglum, R.S. (1923) Fumigation of citrus trees for control of insect pests. *USDA Farmers Bull.* **1321**.

Wolfenbarger, D.A., Lukefahr, M.J. and **Graham, H.M.** (1973) LD_{50} values of methyl parathion and endrin to tobacco budworms and bollworms collected in the Americas and hypothesis in the spread of resistance in these lepidopterans to these insecticides. *J. econ. Ent.* **66**: 211–16.

Wood, B.J. (1971) Development of integrated control programs for pests of tropical perennial crops in Malaysia. In: Huffaker, C.B. (ed.) *Biological Control*. New York, Plenum Press.

Wood, B.J., Liau, S.S. and **Knecht, J.C.X.** (1974) Trunk injection of systemic insecticides against the bagworm *Metisa plana* (Lepidoptera Psychidae) on oil palm *Oleagineux*. **29**: 499–505.

Woods, N. (1986) Agricultural aircraft spray performance; calibration for commercial operations. *Crop Protection* **5**: 417–21.

Wooff, W.R. (1964) The eradication of *Glossina morsitans* Westw. in Ankile, Western Uganda, by dieldrin application. Report; 10th Meeting ISCTR, Kampala, Publ. Comm. Lech. Co-op. Afr. **97**: 157–66.

Wooley, D.H. (1963) A note on helicopter spray distribution', *Agric. Aviat.* **5**: 43–7.

World Health Organization (1970) *Control of Pesticides: A Survey of Existing Legislation*, Geneva.

World Health Organization (1973) *Specifications for Pesticides used in Public Health*, 4th edn. Geneva.

Wright, J.F. and **Ibrahim, N.I.** (1984) Steps of water dispersible granule development. In; Scher, H.B. (ed.) *Advances in Pesticide Formulation Technology* ACS Symposium **254**: 185–92.

Wrigley, G. (1973) Mineral oils as carriers for ultra-low volume spraying. *PANS* **19**: 54–61.

Wyatt, I.J., Abdalla, M.R., Atkey, P.T. and **Palmer, A.** (1984) Activity of discrete permethrin droplets against whitefly scales. *Proc. BCPC Conf. – Pests and Diseases*: 1,045–8.

Wyatt, I.J., Abdalla, M.R., Palmer, A. and **Munthali, D.C.** (1985) Localized activity of ULV pesticide droplets against sedentary pests. *BPCP Monograph* **248**: 259–64.

Yates, W.E. and **Akesson, N.B.** (1963) Fluorescent tracers for quantitative microresidue analysis. *Trans. Am. Soc. Agric. Engrs* **6**: 104–7, 114.

Yates, W.E. and **Akesson, N.B.** (1973) Reducing pesticide chemical drift. In: Valkenburg, W. van (ed.) *Pesticide Formulations*. New York, Marcel Dekker.

Yates, W.E. and **Akesson, N.B.** (1975) Systems for reducing airborne spray losses and contamination downwind from aerial pesticide applications. *Proc. 5th Int. Agric. Aviat. Congr*: 146–56.

Yates, W.E., Akesson, N.B. and **Brazelton, R.W.** (1981) Systems for safe use of pesticides. *Outlook on Agriculture* **10**: 321–6.

Yates, W.E., Cowden, R.E. and **Akesson, N.B.** (1982) Procedure for determining in situ measurements of pesticide size distribution produced by agricultural aircraft.

Proc. ICLASS 335–9.

Yeo, D. (1961) Assessment of rotary atomisers fitted to a Cessna aircraft. *Agric. Aviat. Congr.* **3**: 131–5.

Young, B.W. (1986) The need for a greater understanding in the application of pesticides. *Outlook on Agriculture.* **15**: 80–7.

Young, B.W. (1991) A method for assessing the drift potential of hydraulic spray clouds and the effect of air assistance. *BPCP Monograph* **46**: 77–86.

Young, V.D., Winterfield, R.G., Deonier, C.E. and **Getzendaner, C.W.** (1965) Spray-distribution patterns from low level applications with a high-wing monoplane. *Agric. Aviat.* **7**: 18–24.

Zucker, A. and **Zamir, N.** (1964) Air carrier sprayers for cotton. *J. agric. Engng Res.* **9**: 188–93.

Index